A COMPANION
TO THE
LIFE SCIENCES

(Photo credit: Robert L. Wolfe, University of Minnesota)

A COMPANION
TO THE
LIFE SCIENCES

VOLUME I

Edited By

STACEY B. DAY, M.D., Ph.D., D.Sc.
Professor and Member
Sloan-Kettering Institute for Cancer Research, N.Y.
Head, Biosciences Communications and Education
Sloan-Kettering Institute, New York, N.Y.

VAN NOSTRAND REINHOLD COMPANY
NEW YORK   CINCINNATI   ATLANTA   DALLAS   SAN FRANCISCO
LONDON   TORONTO   MELBOURNE

Van Nostrand Reinhold Company Regional Offices:
New York   Cincinnati   Atlanta   Dallas   San Francisco

Van Nostrand Reinhold Company International Offices:
London   Toronto   Melbourne

Manufactured in the United States of America

Published by Van Nostrand Reinhold Company
135 West 50th Street, New York, N.Y. 10020

Published simultaneously in Canada by Van Nostrand Reinhold Ltd.

15 14 13 12 11 10 9 8 7 6 5 4 3 2 1

Library of Congress Cataloging in Publication Data

Main entry under title:

A Companion to the life sciences.

   Includes index.
   1. Life sciences.   I. Day, Stacey B.
[DNLM: 1. Biology.   2. Science.   3. Behavioral
sciences.   4. Medicine.   QH307.2 C737]
QH307.2.C65   574   78-8300
ISBN 0-442-22010-3

Introduction

STACEY B. DAY, M.D., Ph.D., D.Sc.
Sloan-Kettering Institute for Cancer Research
New York, New York

Rapid advances in the medical and life sciences, and the extraordinary increase in knowledge year by year, make it virtually impossible for even a well rounded scientist to keep abreast in every field. Such exponential growth has brought in its train need for new strategies and revised methodologic approaches to deal expeditiously and logically with analyses, syntheses, and integrations of newly accumulated knowledge information. Thus, while it may be argued that one whose field of interest is in *rehabilitation medicine* need little concern himself with advances in zoophysiology, one finds, not unremarkably, that much knowledge from zoophysiology, say at the level of the molecular biologic membrane, finds its way into rehabilitation medicine, via a better understanding of physiology of the muscles as they pass from a state of injury through repair to reutilization.

The integration and presentation of such diverse aspects of knowledge in dissimilar fields finds expression through such strategies of learning as presented in interdisciplinary projection of multidisciplinary studies developed within the embrace of communication of scientific information (biosciences communications). In that sense this present volume is both an experimental and experiential presentation in learning and transmission of knowledge information. Certainly a *Companion* is not a textbook. Nor is it an Annual Review whose goal is to cover in précis form as much as is possible of a field of scientific investigation. The conceptual size of this volume alone would prohibit that approach. Moreover one might question just how could an interdisciplinary volume master all current updated knowledge in any given number of fields? That would require for all essential practical purposes a synthesis of several annual volumes within the educational strategy outlined within this companion.

Bearing in mind inevitable limitations, the scope of the present volume describes efforts to present within limited descriptives selected broadly based contributions of current knowledge in the fields of the biosciences as well as in some of the sociological sciences (biosocial development) that may be of use to a wide variety of readers.

Within contemporary American medical reading, it would be fair to say, I think, that the *Companion* as a methodology for instruction, does not exist. Indeed, with the sole exception of the writing of the late professor Ian Aird's *Companion to Surgical Studies*, this form of approach to teaching seems to have been quite abandoned as a method of learning in the medical and scientific fields.

This volume owes much in its conceptual outlook to Aird's volume. It shares his philosophy, perhaps even more important these days, that while the universally accepted ideal for professional reading is the original journal article, paper, or monograph, contemporary pressures upon students, post doctoral graduates, clinicians and practitioners are such that within the constraints of time available to them, few in fact find this course possible. With the speed of incrementation of present knowledge moreover, both journals and textbooks generally, are virtually out of date almost as soon as they are published. This does not gainsay the importance of the value of these educational tools, but it must be pointed out, that to those who would wish knowledge at the forefront of present day advances, there is need for a more refined and more constantly updated system for presenting information from the constantly increasing knowledge data pool. It is no longer satisfactory enough to publish a textbook on the thesis of earlier years that it could be revised once every decade, presuming that within those years knowledge had generally moved slowly in direction from its original base. This is not so now. Indeed it is probably reasonably true to say that within ten years one medical generation is completely out of touch with its succeeding generation unless constant revision and updating of knowledge has taken place.

A *Companion* may serve as a friend in this updating of scholarship. It presents within brief scope significant broadly based, (that is general rather than specialized) advances in knowledge information, comprehensively written, dealing with subject matter that is presently at the research frontier of knowledge in medicine and science. This information, in my view, should be of wide interest to all those engaged in academic studies as well as to practitioners and those specialists whose disciplines or fields of knowledge fall outside the interdigitations of interdisciplinary learning.

With respect to the present *Companion*, it should be pointed out, that each essay or study section is presented by an author or scholar who is generally of proven worth; one who has, in most cases been personally associated with the field of learning; and most of whom have advanced knowledge in that same field or in associated disciplines.

Aird, in his original *Companion*, pointed out, as is quite generally known, that when one turns to a multi-authored book, there are inherent disadvantages. Not all presumed experts or specialists manifest the same facility or ability to present their knowledge information clearly, concisely, or without ambiguity. In this sense, as in any edited volume, essay dissertations tend to vary, no matter what an editor may wish or say. Style, content, method of presentation, cultural and literary expression, all may be anticipated to vary within the intercultural experience and academic background and scholarship of the pedagogic group who have assumed responsibility for the organization and presentation of material. There are dangers, of course, and inevitable hazards in constraining scientific colleagues to a ration of space obligating condensation of information. Variations in style may lead to perceived qualitative difficulties—over simplification, ambiguity, or unnecessary sophistication which may be irrelevant or mar content of knowledge for which the piece was primarily commissioned. Nor does multiple authorship make for easy reading at sequential sessions, nor do rapid changes in content of subject matter have appeal to more than a few devoted to source book reading as a leisure contentment. Yet it must be said that a *Companion is not designed for sequential reading.* It is a volume, as a friend, available within time and place, for selective consultation and updated advice. Like a good friend, a reliable colleague, a teacher at hand, a *Companion* is available to "help out" in academic situations that may arise suddenly, or take the post doctoral student, clinician or practitioner unaware, and in situations where he may not have readily available the original source references.

It should be stressed again that a *Companion* is *not* a manual or textbook. A *Companion*, in my view, ought to provide ready access to generalized but important *au courant* contemporary knowledge information in the biosciences, which knowledge may serve as the staging phase for subsequent detailed or specialised study investigation.

Aird, with particular felicity and keen insight, taught that a widely read doctor is a grace and benefit to his patients. I agree. A widely read doctor is a more reliable doctor. The clinician who has at his finger tips interdisciplinary learning and knowledge understanding is, in my view, a superior physician. I am of the belief that a *generalist* is in no way intellectually the inferior of a specialist, if he can command, on a broad base, the ability to synthesize, integrate and release or disseminate knowledge understanding from a diversity of fields of learning. A *Companion* is a deontological peer and friend of the specialist, and as important a knowledge consultant as any other information source committed to a strategy for the gradual extension of scientific knowledge on a wide front as it proceeds into the future through new learning and research scholarship.

A *Companion* offers the assurance of a widely read friend who has been persuaded to collect, collate, synthesize and interpret fundamentals of knowledge for use of the academic and practicing community.

THE PRESENT VOLUME

Subject inclusions in these several volumes of the *Companion* (of which this is the first general interdisciplinary source book) include:

1. *Biosocial Development*—the sociological perspectives of the biological sciences.
2. *Behavioral Sciences.*
3. *Biosciences Communications*, including information sciences, communication of scientific information and methodologies for human learning.
4. *Cell Biology*, including genetics, molecular biology, molecular pathology, biochemistry and biophysics.
5. *Developmental Biology* including ecobiology and study of the biosphere.
6. *Microbiology, Pathology and Immunobiology.*
7. *Nutritional Biology and Food Science.*
8. *Neurosciences.*
9. *Pharmacologic Biology, Drug Abuse, Addiction, Toxicology.*
10. *Clinical Sciences.*

In this first volume, general emphasis is referred upon:

(*i*) Broadly based although selected biological principles in contemporary research topics.
(*ii*) Aspects of innovations in several of the fields of information technology, human learning and biosciences communications (communication of scientific information).
(*iii*) *Biosocial Development*—contributions of the sociological and behavioral sciences (psychology, cultural anthropology) to evolution and the biologic development of man.

Because of the somewhat innovative approach of this book points (*ii*) and (*iii*) bear comment. The integrative philosophy as discussed above, important in the biologic sciences is, in my judgement, no less important for the sociologic sciences as they relate to biology and human evolution. To be quite frank, I see disputation between ideologues debating questions such as Nature versus Nurture as being only partially useful in assessing Man's so-called "place" on this our temporal planet. It seems to me rather unnecessary to urge conflict of genes versus dreams. It is better to achieve a synthesis that accommodates both the notion of classical Darwinian evolutionary biology as well as understanding and appreciation of Man's *sociological* evolution over the same time scale years of change. What I am saying, in effect, is that biological evolution is but one side of a coin, the reverse of which is transcribed in terms of *sociological evolution.* Sociological evolution and concepts of biosocial development add understanding to physiological concepts of the nature of man. In the sense that such parameters relate to biological perspectives, i.e., physiological functions, cellular changes, endocrinologic, nutritional, environmental and sociocultural factors in the biosphere, biosocial development has to do with man. Such sociological and cultural attitudes as complement biologic imperatives contribute to *quality of life*, and education for health must include, in my view, education in the social sciences. Health must be linked with development of society as much as it is linked with the evolutionary development of the anatomic body. If anything, disease and illness must, properly, only be seen against the interpersonal, intercultural, and socionational culture and psychology of the patient. Education for health far outweighs, in my opinion, education against disease. *Biosocial Development*—the sociological evolution of the person, the system and the state, (whatever may be the area of concern—economics, ethics, morality, legislation of health codes, or conditions of life)—make the evolutionary progression of man a sum addition of biologic imperatives as well as sociologic circumstances. Together they make a holistic sum fortified by preselection and intermarriage, immigration, and survival of the accumulated gene pool on the one hand, and by sociological circumstances of nurture, environment and biosphere on the other. Continual interrelationship between these variables is a guarantee of improvement for the species.

In conclusion, while the effort in this *Companion* has been to relate items of knowledge from within diverse fields of human endeavor, by selecting authors of wide research background to act as referees and sources for scholarship, I would hope that such writings as have been chosen, will prove to be useful and of value in understanding of their fields of knowledge information. If reasonable integration of pooled knowledge information can be synthesized and released again for the wider benefit and better understanding of those who use these *Companions*, the editors will be greatly satisfied.

Contents

B. Rural Health Perspectives

III BIOLOGICAL AND CLINICAL SCIENCES

A COMPANION
TO THE
LIFE SCIENCES

I
MEDICAL INFORMATICS AND BIOSCIENCES COMMUNICATIONS

A Model of Human Communication

K. S. SITARAM
Department of Communication
Utah State University
Logan, Utah

Several models of the communication process have been offered by specialists. Although no one of them is completely acceptable, each model has contributed to an understanding of a part of the process. One drawback in most models is the lack of explanation of what happens in the internal world of the persons that are communicating. Human interaction is being studied as it occurs in the external world, but what happens in the minds of those who interact? The purpose here is to offer a model which tries to explain not only the process that occurs outside the individual's mind but also inside. In the light of new theory, a new definition of communication will be offered and the old *S-R* formula will be examined.

The definition. It has been proposed by earlier speech specialists that communication is merely an *S-R* process or sender-receiver and stimulus-response situation where a person sends a stimulus in the form of a message and another person receives and responds to it. Many scholars now agree that communication is not the process of transference of meaning from the mind of a communicator to the mind of his/her audience. It is not possible to transfer meaning.[1] It is not a process whereby a person sends messages and the other just receives them as if he is a machine to receive and store. Communication occurs when a person has some experience that he wants to share with someone else. The experience has meaning to that person and it seems to him to be worthy of sharing. Therefore we can define communication as *a process whereby a person shares some of his/her experience with another person.* In order to share, he uses the medium of symbols he/she has learned or developed in his/her own culture. The person also understands another's experience with reference to his/her own experience and symbol system.

Thus, the process of communication is not only receiving and internalizing of symbols of another person but also sending symbols with the expectation that the other person receives and internalizes them.

A different formula. In communication, response to the received message is

3

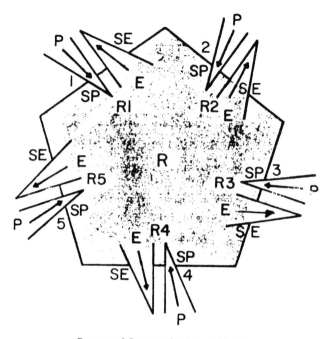

Process of Communication: A Model.

P = Perception   SP = Selective Perception   E = Expression
SE = Selective Expression   R = Retention   1–5 = The 5 Senses

more imporant than sending the stimulus. Because communication is expectation-oriented, the communicator expects his audience to react in a particluar way. Even when a person is asked to "pass salt and pepper," he/she is expected to pass them. Therefore, *expectation* is an important variable involved in the process of communication. However, most communicators do not realize that the actual response is not entirely the result of their stimuli. Response is the result of interaction of at least three factors. Stimulus is one factor. There are two others: culture and genetic characteristics. Cultural values, for example, play important roles in the way a message is received and stored. Genetic characteristics of a person play an important role in that person's communication. The way messages are received and interpreted between Black and White persons may be different from the way they would be exchanged between persons of the same color.

It is evident that response to a message is a function of stimulus, culture, and genetic characteristics. Therefore, the following formula represents this concept:

$$R = f(S + C + G)$$

where R = response, f = functions, S = stimulus, C = culture and G = genetic characteristics.

It is not easy to tell how much each of the three factors on the right hand side of the equation affects the response. Perhaps it depends upon the situation and the persons that are interacting. In order to measure the impact of each of the factors, we might apply certain techniques of measurement used by biostatisticians. They believe that biological and other observable characteristics or phenotype (P) of a person are the results of interaction between genetic characteristic or genotype (G) and environmental variable (E). This concept is represented by the formula:

$$P = f(G, E)$$

Although they have not perfected measuring the degree to which either genotype or environmental variable interacts to produce phenotype, they have at least made a beginning. Since studies of communication have not tried to measure how much each of the variable S, C, and G affect R, it is possible that biostatistical methods could be useful.

Response is the final stage in the process of communication. Before reaching that final stage, the individual passes through three stages. The first two stages occur within the individual or the internal world. It is the cherished property of each person. There he feels comfortable and secure, for no one else can see what is going on in there. That is where he stores information of all his conscious and subconscious experiences. Any information he stores must initially come from outside. Many Eastern philosophers do not share this idea. They believe that many bits of information that a person stores in his subconscious come from his previous life. Information that transmigrates may be revived in the subconscious state when one is awake or asleep.

Information may also be generated, they say, by meditation. There are others who believe information that is received is validated by intuition. Books have been written and schools have been developed to discuss theories of transmigration, meditation, intuition and understanding.[2] Because there are not sufficient scientific data to prove or disprove those theories, it is not feasible to discuss Eastern thoughts in this article. However, it is interesting to note that Easterners did try to explain communication as a process, several centuries before Christ.

The individual does not live in the internal world only. External world is as important as the internal world. That is where the objects of his perception exist. There he shares his own experience with others by acts of expression. The P-R-E model represents this new concept of the process of communication. It includes three stages: Perception, Retention and Expression.

Perception. It is the starting point of communication. All human communication activities start with the individual's contact with an external object. The ancient Hindus and Buddhists defined perception as the input of information that occurs as a result of mind-object contact. They believed that first of all mind of the perceiver comes into contact with another object via the five senses.[3]

Each person perceives the world around him based on his own beliefs. Information that supports or does not upset the balance of his own belief system is perceived favorably. Those situations that tend to make him question his own beliefs are not perceived or, if perceived, will be forgotten in the course of time. What is perceived casually is not always selected for serious perception. Information received from outside is sorted out, or decoded, and selected to be passed on to the next stage in the process. The selection process has been explained by many scholars under the heading of Selective Perception.

Retention. It is the process of selecting and storing the perceived information. A person's values, beliefs, expectations, and communicative skills all have their own slots in the storehouse. Communicative activity of a person is shaped mostly by richness or poverty of that storehouse. Samuel Johnson and Daniel Webster were able to develop dictionaries of the English language since they had a rich storehouse of information within themselves. However, the contention here is that even Retention is generally culturebound. Any information that supports the perceiver's culture is readily stored. Once a situation is selectively perceived, it becomes a part of the person's storage system. An example could be information related to snakes. Westerners do not like to look at long television programs on snakes. Even if they watch them, they try to forget the information. Snakes are vicious and ugly looking reptiles. They are also put down by the Bible. Somewhere in the back of their minds they have an idea that snakes are undesirable.

Expression. This is a sort of reverse perception. It is the process of sending out stored and encoded information. Based on a person's retention, his expressive abilities are shaped. Expression is also selective, perhaps more so than it is in the act of perception. We do not say everything we would like to say to a person. Many thoughts might come in our minds, but we say only those things we think are proper to say. Perhaps nonverbal expression is not always selective. A person might express his/her true intention nonverbally and not even be aware of it. But verbal expression is selective. Sometimes we cannot say everything since our symbol system (language) may not be adequate to help us put our thoughts in symbols. Foreign students in American universities may have experienced this problem. They do not know enough English to say everything that is in their minds. Yogis too have a similar problem. While two yogis can understand each other intuitively, they cannot share their experiences with lay persons since the existing languages are not adequate to help them communicate. This shows how important a symbol system is for selective expression.

Researchers have not paid much attention to the selective aspect of expression. The way we use symbols decide to some extent how our audience perceive us. We do not have sufficient research data to make definitive statements on the correlation between selective expression and successful communication. But in our classes in speech, radio, television, and journalism we do teach techniques of expression such as public speaking, radio announcing, television performance, and news reporting. These techniques are certainly selective and we teach them since we all have experienced the joy of saying the "right things" and seeing the expected results. The opposite of selectivity is spontaneity. We are spontaneous with those we trust. We do not conceal much of our thoughts from them, but the trusted ones are not many in each person's life. Most of the time we select what we have to say. In professional life we all are selective in expression. If a presidential candidate is not selective in what he says, he is in trouble, as when Gerald Ford made a spontaneous statement on Eastern European countries during the television debates. Communication specialists should be interested in all aspects of expression, rather than the variables which are only indirectly related to communication.

The five senses. They play important roles in developing our symbol system. Each of the five senses perceives a part of the situation. When two lovers are alone in a park, their noses can smell the fragrance of flowers, ears can hear songs of birds, eyes can see the trees, their skins can enjoy the cool breeze, and their tongues can taste fruits that are ripe on the trees. While each sense perceives the situation in its own way, it retains the information in the same way. When the lovers go to another park several years later, one can say, "Darling, something smells just like that flower in that park we went to several years ago." Or he might ask her, "Do you remember that breeze?" Information is received and retained in five forms: sound, smell, taste, touch and image. We use the same five methods to interact with others.

Information thus retained becomes an integral part of the existing storage. Each bit of information is associated with appropriate bits of other information. We do not associate the odor of ammonia with a rose. The fragrance of a rose is different. Although we have been studying verbal and sometimes visual communication, we have not paid attention to other modes of communication. We do communicate by means of touch, smell, and taste.

The P-R-E model. This model is applicable to intra as well as interpersonal communication. What happens within the "pentagon" (see Fig.) would be intrapersonal communication. It includes selective perception, retention, and encoding. When the message is perceived by another person it would be interpersonal communication. The process of communication is complete when the message is received and perceived by another person. Whether or not it is retained and the desired effect occurs depends upon several factors. One of them is the value system of the person. Values are parts of his retentive system and are shaped by his culture. As mentioned earlier, the *Response* depends upon values and other factors. This model also is applicable to intercultural communication.

Let us apply the *P-R-E* model to a specific situation. Consider, again, two lovers A and B interacting with each other. They are using all five means to communicate: sound, sight, touch, smell, and taste. Lover A expresses his thoughts in five ways and lover B perceives in five ways through the five senses. Whatever A selects to express is perceived by B and selected for retention or forgotten shortly after. B also repeats the process of selective expression. It is important to note that each does not know what is going on in the internal world of the other. If one of them happens to be expert in concealing real thoughts by not letting his/her nonverbal expressions betray him/her, the other person can never know what he/she really thinks. How many millions of disasters would occur if each lover could peep into the internal world of the other!

This model represents the process of communication which includes not only the act of making others understand us but also permits understanding others even before we try to interact. This model implies a never-ending process of human communication. The act of receiving, retaining and expressing is a constant process and goes on until the individual dies. Even after his/her death, the person will have left behind some information to remind others that he/she did exist.

References

1. Berlo, David K. *The Process of Communication*, New York: Holt, Rinehart, Winston 1960.
2. Sharma, Chandradhar *A Critical Survey of Indian Philosophy*, London: Rider and Company, 1960.
3. Saratchandra, E. R. *Buddhist Psychology of Perception*, Colombo: The Ceylon University Press, 1959.

Human Information Processing

D. R. LANGER
The Canadian Medical Association
Ottawa, Canada

There is evidence that record keeping has been practiced for about 14,000 years, when man engraved marks on an eagle bone to count off the days between new moons. As tribes became communities and communities were organized into larger states, record keeping techniques grew. Their purpose, historically, was for such things as the levying of taxes. When the census was revived around the middle of the 18th century, it was found that the public did not object to a head count, but was unwilling to reveal any information that was more personal.

The complexity of society today has required more sophisticated means of manipulation and storage of information and the computer has become the most recent solution to the problem. It is practically impossible to avoid becoming the subject of a record—as a worker, a student, a patient, a taxpayer, a guest in a hotel and so on. Most of the activities that generate records about individuals are presumably desired forms of participation in our society. There are three main types of records about people—administrative, intelligence and statistical.

Administrative records are often the result of a transaction, such as a marriage, obtaining a license, etc. Such records are usually treated as proprietary information by a private firm, but those held by the government are normally accessible to the public.

Intelligence records are exemplified by security clearance and consumer credit reports. Much of it is testimony of informants and observations of the investigators. Rarely made public, this is the area that concerns people most since it is generally impossible for anyone to determine the content (and even existence) of their own record to put a stop to it or to rectify any false or damaging information.

Statistical records are those made up of data usually gathered through a survey of some kind. Nearly always, the identity of the individual is separated from the record.

PRIVACY AND CONFIDENTIALITY

The computer is a nonmoral collection of responsive electronic equipment. It may or may not be discreet; it can and will act as a servant or a master; it does not question the motives of its users; upon receipt of a program or an instruction, it will tell you everything you wish to know.

When what you wish to know is the answer to multiplication problems pages long, or the financial status of your company for the first quarter of the year, very little objection is raised to the use of a computer to handle these jobs. The computer's capacity for timely retrieval and analysis of complex bodies of data can be of invaluable assistance to hard-pressed decision makers. Its ability to handle masses of individual transactions in minutes and hours rather than in weeks or months, as was formerly the case, makes possible programs of service to people that would have been unthinkable in the manual record keeping era. Medicare, for example, would be impossible to administer without computers to take over many routine clerical functions. However, when the information contained in a databank is of a personal nature, when it may be subjective or even worse, distorted, a moral issue is raised.

The tendency to develop coordinated computer systems has led to a concentration of data in central files or databanks. As far as the individual is concerned, this collection of personal data has brought about a conflict of interest, for instance, with regard to medical data. On the one hand, he is interested in the availability of those personal data which serve medical purposes because he hopes that this will improve or speed up therapeutic success. On the other hand, he must fear that the data might be exposed to unauthorized access or controlled transmission. Since in many areas personal data are being collected without consideration of the individual interest and without the consent of the individual, every citizen is entitled to data security and effective steps toward its realization.

DATA SECURITY

There are a number of ways to limit access to databanks. There can be controlled access to the computer room, locking up of data, distribution of data only to authorized persons etc. These might be called *organizational measures* and are most effective with batch operations, but difficult with time-sharing operations with terminals. In this case, *hardware measures* can be enlisted by locking terminals, requiring identification of readers at the terminal or identification of the data station required. But, *software measures* can still be necessary, which will give each user an identification and admit him to work only if a changeable password is named correctly. With software measures, certain data can be retrieved only by certain people, and furthermore can be modified, added to, deleted, etc. only be certain others. For example, a nurse might be allowed to read the medication file of a patient, but not to change it. Attempted violations of data security can also be detected by computer monitoring of users and the type and frequency of use of confidential data.

Difficulty arises when personal data are able to leave the controlled environment such as on magnetic tape or by telecommunication. A degree of security can be offered here if data are coded by substitution processes and cryptographic methods so that printouts are illegible to the unauthorized user.

It is intersting to note that the issue of privacy with regard to medical records blossomed with the use of computers. The old story that "whoever wears a white coat has access to a medical record" appears to indicate that manual records have been accessible for a long time.

FUTURE DIRECTIONS

There is little doubt that record keeping will be computerized to deal with our increasingly complex society. Rapid advances in technology certainly are making the computer more available. The question "to computerize or not to computerize" will no longer be the issue; the content of the records and safeguards to be used are now the focus. If doctors are to be expected to make use of the enormous amount of information available to them and apply it to the symptoms presented in each case, the computer will have to be utilized and we, as patients, will have to allow wider availability to personal information—a trade-off which will have to be considered to promote better health.

References

1. Bohm, K. Protection and confidentiality of medical data II: Simple methods for meeting the users' needs. *Conference Proceedings, Medinfo 74*, Vol. 1, pp. 193-196. Amsterdam: North-Holland Publishing Company, 1974.
2. Fischer, Th., and Helmbock, J. M. Data privacy and data security in Kiel KIS. *Conference Proceedings, Medinfo 74*, Vol. 1, pp. 197-199. Amsterdam: North-Holland Publishing Company, 1974.
3. U. S. Department of Health, Education and Welfare Secretary's Advisory Committee on Automated Personal Data Systems. *Records, Computers and the Rights of Citizens.* Massachuesetts Institute of Technology, 1973.
4. Westin, A. F., and Baker, M. A. *Databanks in a Free Socity.* New York: Quadrangle Books, 1972.

Information Retrieval in Science

ANN LEACH
The Canadian Medical Association
Ottawa, Canada

A BRIEF ANALYSIS OF INFORMATION SYSTEMS

It is unfortunate that social sciences have failed to keep pace with the physical ones. We know how to split the atom, but not how to forestall a nuclear holocaust; we have the means of controlling human fertility but we cannot persuade most people to use them; we have computers capable of performing split-second calculations, but can't seem to dispel the hostility which electronic gadgetry arouses among its potential users.

Information systems, from the classical point of view of an information scientist, may broadly be structured into two levels: information storage and retrieval systems (ISR), and control management information systems (MIS).

Both types of systems are well-covered in terms of published materials. Surveys of bibliography on information storage and retrieval systems may list well over 5,000 items and a recent survey of bibliography on management information systems consists of no less than 2,000 items.

In the past, information scientists, systems and management specialists were deluded by computers, by the intricacy of programming and by the refinement of interesting but not too practical mathematical and analytical methods. Therefore analyses of information flows within individual parts of systems resulted in total misunderstanding of the drastic changes under way. More important, the major goal of any information system, a satisfied user, has almost been omitted from the research, analysis, design, implementation, retraining and adjusting processes. For example, in the design of "total" hospital information systems (HIS), all aspects of hospital administration were included except the patient and his needs.

Does this mean that viable information systems (both ISR and MIS) are a mirage which will never be achieved? Bishoff in his work *Die Informationlawine* points out that there is an urgent need for effective information storage and retrieval systems due to the fact that information in some areas of human knowledge doubles at the following rates:

General Information	every 10 years
Information on chemistry	every 8 years
Information on electronics	every 5 years
Information on space	every 3 years

The well-known "publish or perish" syndrome—a very important factor in the lives of all academics and researchers as a means for promotion and fame, has contributed significantly to the abundance of information by the publication of many articles, papers and books dealing with topics which have no relevance to the real life information.

Information scientists are aware of the emergence of the information explosion (or pollution) since Luhn's permutated indexing technique developed in the early 1960's. They tried to mechanize and later to automate libraries by means of better housekeeping which included cataloging, indexing and autoindexing, filing, searching, ordering, and information dissemination. Many systems for university libraries and other knowledge-based institutions have been launched; all but few seem to fall far below initial expectations. Only very recently (and after a decade of efforts), MEDLARS, a typical batch processing medical library system, offered an online version of bibliographical search called MEDLINE. The general trend of library scientists has been to cope with mainly the methodological and procedural aspects of the library rather than the needs of the users.

MANAGEMENT OF INPUT

An ideal transformation of information into the computer would be a direct input from the material prepared for publication by means of a computer-aided typesetting or, from the tapes of government computers. The computer-to-computer input technique is still in the stage of new developments and therefore is, unfortunately, not yet available. The next best method of input to follow is that which most scientists, technical writers, and researchers have built— their private, personalized information storage and retrieval files aiding memory by notes, manual files, cards and so forth. When studying articles and monographs, or obtaining any information in the sphere of a specialist's interest, the source of the information, title of work, index (keyword) and the content is registered in a more personalized way than the usual bibliographic annotation. An experienced user of information also tries to note the "flavor" of the content, namely the richness of the author's language, new ideas presented, contribution to knowledge or personal enrichment. Each user, however, may have different styles of building up the necessary knowledge, as well as different approaches to this process of information storage. Therefore some sort of common classification is necessary.

BUILDING AN INFORMATION BASE

The following somewhat paradoxical findings point out many of the difficulties encountered when attempting to construct an information base:

- Despite tremendous efforts in terms of money and thoughts, there are very few successful computerized information systems available to the user;
- In the rapidly increasing volume of printed materials, books, monographs, textbooks, journals, papers, etc., it appears that the content of knowledge is decreasing because of lack of adequate quality measures;
- High expertise have been placed in computers, mathematical, linguistic, and information applications, all but few are of very little use in terms of real world projects;
- It seems that the major reason for the past failures of information retrieval systems stems from the misunderstanding of the end-users' needs in the current systems era. Until present, all efforts were dedicated to the technology in the broadest sense, i.e., computers, communication, programming, and terminals;
- Despite all the failures, there are a few successful applications which maintain the necessary level of optimism and promise for the future;
- Both researchers and practitioners have to accept the fact that traditional

printed documents are, and will be for a long time to come, the prime source of information.

In short, to avoid the past failures in many computerized information systems, any viable approach to information processing, namely both input and retrieval, must be human oriented. It should follow the path of the learning processes, methods of research, methods of annotations, techniques of writing and similar intellectual activities. Such research projects represent the advent of interactive, online computer-centered information sharing; a technological advance that is expected to change, and hopefully revolutionize mental attitudes of researchers, medical practitioners, health care politicians, administrators and students of systems and health care sciences.

Online computer-based information systems have only recently achieved sufficient technological maturity. It is the role of the end-user to implement new forms of creative information processing superior to the traditional batch data processing.

CONCLUSION

Through the use of existing software packages and commercially available data bases, it is possible for many modest operations to engage in searching files via computers. Also, there are more search activities that must support themselves, either within their own organization or by marketing their services outside. The theoretical in searching is giving way to the practical, as time, the users, and relentless economics winnow out the techniques that really work and the approaches that information consumers really want to take and pay for. In the course of the "popularization," we are arriving at an information community much more sophisticated than ever before in the ways of computer searching.

Regardless of the miracles of technology it is people who really make the difference between success and failure. We all pay lip service to this concept yet data processing literature barely acknowledges the existence of human problems generated by computers. Unless we start paying the same attention to people as we do to machines, the brave new world of super-efficient computer systems will remain an illusion.

References

1. Brandejs, J. F. *The C.M.A. Information Base—A Beginning of Operational Systems in Canada*, National Computer Conference, 1975.

2. Cadieux, J. Computer systems management—Human problems of computerization. *CA Magazine* 108: 66–68 (1976).
3. Cuadra, Carlos A. *Annual Review of Information Science and Technology*, American Society for Information Science, Volume 7, 1972.
4. Emmett, A. Futures: The computer meets the doctor. *Science Digest* 78: 85–87 (1975).

Data Base Systems in Life Sciences

OKAN GUREL

IBM
Westchester, N.Y.

Data Base is a term introduced in application of the computer technology to applied fields where extensive information is collected, processed, analyzed, and retrieved. Information may consist of both quantitative (numbers) and qualitative (words) data. If the size of data makes manual handling tedious the data may be entered into computers in coded forms. The efficiency with which these data are analyzed and retrieved is the optimality criterion for any data stored in a computer. This criterion determines the type of data organization. In many applications the same data are shared by more than one user thus the concept of data base becomes necessary.

For a specific application the design of the data base must answer the needs of various users. In most cases, the data base can be general enough to contain:

- the necessary data for different usage (integrated data base),
- the design to provide the flexibility for an end user,
- the security to assure the integrity of the data base as well as protection of privacy.

In the case of a multi-user environment, a terminal oriented *interactive* mode is the most suitable approach to data base systems. As it is the case with almost all practical applications, the end user of a data base prefers to have an easy access to the data base, without being forced to learn specialized computing languages. Conversational *high level languages* are introduced and developed for this purpose. The purpose of such a language is to allow the end user to converse with the computer almost in a daily speaking language.

Data base systems for life scientific applications follow these general lines:

- the same data base may be used for clinical, diagnostic, treatment, rehabilitation, research, and administrative purposes, thus, a total health care delivery can be achieved,
- the end users are not only many in number (terminal-oriented interactive computing) but also they are application oriented with different background and interest, thus a high level conversational language is needed.

Due to a wide variety of possibilities of using a life science related data base and large number of variables as well as unknowns, medical data bases may be extremely complex requiring more flexibility than the currently available data base concepts offer.

Currently some health care data bases are implemented in hospitals for administrative purposes.

Data Base—Identification of Needs

G. E. Alan Dever
M. R. Lavoie
Office of Health Services Research and Statistics
Division of Physical Health   ·
Department of Human Resources
Atlanta, Georgia

The allocation of health resources to meet designated areas of need is dependent upon a defined data set. The purpose of this paper, therefore, is to define an holistic framework on which to build a minimum data set. The data set, however, should also be prescribed to reflect the predominant measurement of the goals and policies of the responsible health agency. In addition, the data set should reliably supply information to the planning, budgeting, and evaluation processes and display cohesive characteristics for management analysis.

To define areas of need, the holistic approach to health status should be observed. That is, the measurement of need (and the data set) should reflect four main investigative areas, as defined by Georgia's New Health Outlook: environment, human biology, life style, and the health care system. In addition, benchmark analyses, employing traditional methods and variables, should be undertaken. Consequently, to identify areas of need, data must reflect the following broad categories:

1. Mortality
2. Morbidity
3. Disability
4. Fertility
5. Growth and maturation
6. Psychosocial well-being
7. Immunity

There is also a need for specific data sets, such as those relating to community health status. Specifically, these data should reflect the status of the community's health as a basis to provide a community health diagnosis. Appropriate measures should include:

1. Mortality rates
2. Disability rates (workday loss)
3. Incidence and prevalence of specific diseases
4. Morbidity rates

The above measures have been the traditional indicators of disease processes and continue to be important in establishing benchmarks for future evaluations. Some nontraditional data sets, however, need to be developed in order to address Georgia's New Holistic Health Outlook. Such data sets include:

1. *Health Systems Data*
   The data set should measure activity in the existing health system, such as:
   A. Health Service Utilization Data
      (1) Hospital admissions
      (2) Discharges
      (3) Ambulatory care visits
      (4) Patient-origin data
      (5) Residential

In addition, these data should reflect accessibility in terms of geographic, economic, and cultural factors, such as:

   (6) Distance to facilities in terms of:
      a. Time
      b. Mileage
      c. Social factors
      d. Cost

B. Facility and Management Data (availability)
  (1) Resources: Institutional
    a. Hospital beds (primary-care context)
    b. Nursing homes
    c. Health access stations (primary care)
    d. County health departments
  (2) Resources: Human
    a. Physicians
    b. Nurses
    c. Dentists
    d. Other essential personnel
C. Fiscal Data
  (1) Diagnosis and care costs
  (2) Indirect costs of lost wages and productivity
  (3) Indirect costs attributed to investment in individuals; i.e., premature illness or death
  (4) Government subsidies: Medicare, Medicaid, et al.

2. *Life Style Data*

Life style or health behavior data probably will need to be collected by a survey instrument (although not always) to delineate:
A. Reasons for seeking health care
B. Barriers to health care
C. Spatial restrictions
D. Life style components:
  (1) Drugs
  (2) Nutrition
  (3) Carelessness
  (4) Risk-taking
E. Marriage
F. Divorce

3. *Human Biology Data*

These data are necessary to identify target groups at risk of a disease or for providing health information via informational and awareness programs.
A. Demographic characteristics
  (1) Age (maturation, aging, internal systems)
  (2) Sex
  (3) Race
  (4) Ethnic group
B. Genetic risks
C. Intelligence (Comprehension)

4. *Environmental Data*

The spatial interaction of disease etiology is measured by the environmental variables which may be associated with disease patterns. Environment may be defined as the external, natural environment, or the internal "manmade" environment, such as homes, work places, etc.

A. Air, water, and soil conditions
B. General climatic conditions
C. Prevalence of rats and other pests presenting disease risks
D. General environmental quality
E. Data on the quality and type of housing to reflect:
    (1) Overcrowding
    (2) Inadequate plumbing facilities
    (3) Social restrictions due to inadequate planning
    (4) Unmet psychological needs or satisfactions.

The economics of data collection may pose a real problem if the process is not considered in a systematic context. Haphazard or disjointed collection efforts may result in an overburden of costs for data processing. The information processing state of the art usually dictates the cost of data processing, but adequate planning may reduce these costs to manageable levels. One answer is to utilize established district or regional geographic areas.

In brief, data collection and preprocessing may be accomplished at a district or regional level. Edited data could be retained for local analysis and, in addition, the "required core data" may be relayed to state central processing by a communication network for input to the budgeting, planning, and evaluation processes and other required state and federal reports. The cost of this system could be shared at all levels of fiscal management (county, district, regional, and state). The communication network could also be utilized to transmit completed reports or aggregated data to each district or region from the state level.

The determination of areas of need, allocation of funds, and the evaluation of health programs require a unified, reliable data system to provide the necessary information for the measurement of changing disease patterns. A timely and responsive communication network must also be established to provide information to widely spaced geographic areas. Without such a system, the assessment of health needs will remain segmented and inadequate as a basis to change disease patterns.

Dynamo: Continuous Modeling System

J. F. Brandejs, Ph.D.
The Canadian Medical Association
Ottawa, Canada

INTRODUCTION

North American society is rapidly approaching the stage in which human service delivery systems will prevail over industrial production. This society is called by some researchers, the post-industrial society. Providing services to people is a dynamic process taking place in a changing environment. With the emergence of computers during W.W.II, many new scientific techniques were developed to represent real or imaginary situations in order to understand the operational behavior of the post-war industrial society. This representation is called modeling.

The most widespread use of modeling is in a branch of science called operations research (OR) which is the quantitative study of operations in action. Work in operations research has induced the development of many theoretical models of various human activities. For example, queuing theory and, to some extent, linear programming have been found useful in the systems and bioengineering problems in hospitals, factories and government.

The whole subject of mathematical programming, with its subdivisions of linear programming and dynamic programming, was developed to solve systems problems of allocation, scheduling and routing of resources such as material, money, men, and information.

Models of systems, in general, are subjective, based on individual assumptions and are concerned with overall results. Because human variability and perception strongly influence the outcome of an individual operation within the modeled system, the theory of probability provides the basis of many of the mathematical models. For example, when the system is such that its condition at any time is expressible in terms of a denumerable set of states, then the dynamics of the operation can be simulated by advanced OR techniques.

Notwithstanding the advances in mathematical techniques, the major pitfall of OR is the inability to portray or model a situation or process in a complex dynamic environment. The models produced by OR are mainly static representations of how things look and not how they work. In addition, the models developed within the OR field require quantitative input and make many simplifying assumptions in order to arrive at some sort of feasible solution. Therefore, despite a vast quantity of published documentation, the modelling of life and social systems does not seem to be attainable through OR.

In real-life situations, decision making is usually made on value judgements, experience, and intuition rather than on hard facts. The human learning process is often based on trial and error which could be expressed as a positive or negative feedback loop. Human actions within our social, industrial, post-industrial, and world systems are governed by feedback.

UNDERSTANDING OF SYSTEMS

A "system" means a grouping of parts that operate together for a common purpose. An automobile is a system of components that work together to provide transportation. An autopilot and an airplane form a system for flying at a specified altitude. A warehouse and loading platform is a system for delivering goods into trucks. Management is a system of people for allocating resources and regulating the activity of a business. A family is a system for living and raising children. The human body is the most advanced system.

At the Massachusetts Institute of Technology (M.I.T., Cambridge) over the last 40 years, there has developed an approach to understanding the dynamics of systems. The foundation was laid in the 1930's when Vannevar Bush built his differential analyzer to solve the equations of certain simple engineering problems. In that same period, Norbert Wiener developed his concepts of feedback systems that were later given the name "cybernetics." In the 1940's Gordon S. Brown created the Servomechanisms Laboratory in which the theory of feedback systems was expanded, recorded, taught and radiated. In the 1950's Jay W. Forrester was director of the Digital Computer Laboratory and Division 6 of the Lincoln Laboratory where digital computers were first used for systems simulators; since 1956 he and a group of associates at the M.I.T. Alfred P. Sloan School of Management have extended these developments to cope with the greater complexity of social systems.

In primitive society, the existing systems were those arising in nature and their characteristics were accepted as divinely given and as being beyond man's comprehension or control. Man simply adjusted himself to the natural systems around him and to the family and tribal social systems which were created by gradual evolution rather than by design. Man adapted to systems without feeling compelled to understand them.

As industrial societies emerged, systems began to dominate life as they manifested themselves in economic cycles, political turmoil, recurring financial panics, fluctuating unemployment, and unstable prices. But these social systems suddenly became so complex and their behavior so confusing that no general theory seemed possible. A search for orderly structure, for cause and effect relationships and for a theory to explain system behavior gave way at times to a belief in random, irrational causes.

A structure (or theory) is essential if we are to effectively interrelate and interpret our observations in any field of knowledge. Without an integrating structure, information remains a hodge-podge of fragments. Without an organizing structure, knowledge is a mere collection of observations, practices, and conflicting incidents.

The concepts of "feedback" systems seem to be emerging as the long-sought basis for structuring our observations of social systems. For several years, the modeling of the feedback-loop structure of social systems had been known as "industrial dynamics" but later changed to "system dynamics." Applications of system dynamics have been made to corporate policy, to dynamics of diabetes as a medical system, to dynamics of dental and medical manpower, to social forces affecting drug addiction in a community, to the dynamics of commodity markets and to the behavior of research and development organizations. System dynamics groups are active in the United States, Canada, Japan and several West and East European countries.

DYNAMO SYSTEMS

The development of a new programming language, DYNAMO, was undertaken concurrent with the development of systems theory at M.I.T. The history of DYNAMO begins with a program called SIMPLE (Simulation of Industrial Management Problems with Lots of Equations), written by Richard K. Bennett in the spring of 1958 for the IBM 704 Computer. SIMPLE contained most of the basic features of DYNAMO, including the plotting routine, but the model specifications had to be stated in a rather primitive form and very few checks were performed on these specifications. These shortcomings were corrected by DYNAMO (DYNamic MOdels). In 1962, Jay W. Forrester suggested a relaxation of the requirements on initial values (N) which was implemented. When a time-sharing system became generally available, DYNAMO was modified to operate under it. This facility made it possible to create, debug, and run a model in a matter of hours, the output returning directly back to the console.

In 1965, it was decided to rewrite DYNAMO. Although the input language gave the appearance of actual equations, DYNAMO I was basically a macro expansion program. Simple algorithms for algebraic translation were now well understood and could be utilized in DYNAMO to relax the restrictions on equation formulation. Furthermore, higher-level languages (e.g., FORTRAN) had advanced to the point where they could be used as source languages to simplify the chore of rewriting the compiler. Finally, the third generation of hardware was rapidly replacing the equipment for which DYNAMO was written and some sort of rewrite would be required before long.

Models representing systems are developed one stage at a time through the accepted format of dynamic model construction as follows:

A. Verbal description of the system based on the mental model
B. Conversion of the verbal description into causal-loop algorithms
C. Preparations of DYNAMO-oriented flowcharts
D. Writing of DYNAMO equations from the flowcharts
E. Processing of the model on a computer equipped with DYNAMO language (compiler)
F. Analysis of the computer runs by asking the following questions:
  1. What behavior does the model show?
  2. How has behavior changed from previous computer runs?
  3. Why does the model exhibit the behavior?
  4. How can we alter the behavior?

It is interesting to note that users of DYNAMO modeling systems very often are satisfied with the conversion of their mental model into causal-loop diagrams which describe the feedback relationships among the systems variables. Application of causal-loop diagrams as a modeling tool is called "systems intervention."

Causal-loop diagrams play two important roles in system dynamics studies. First, during model development, they serve as preliminary sketches of causal hypotheses. Second, causal-loop diagrams can simplify illustration of a model. In both capacities, causal-loop diagrams allow the analyst to better communicate the structural assumptions underlying his model.

The causal-loop diagramming process begins with identification of the relationship between individual pairs of variables. When a change in one variable produces a change in the same direction in a second variable the relationship is defined as positive. When the change in the second variable runs in the opposite direction, the relationship is defined as negative. The variables are linked together by an arrow to form the feedback loops of the system. The polarity of a loop is determined by assuming all else remains constant and tracing the results of an arbitrary change around the loop: (1) reinforcement of the change indicates a positive feedback loop; (2) opposition to the change indicates a negative feedback loop.

The next step in developing a DYNAMO model includes the construction of one or more flow charts showing which variables affect which others. Perhaps the first diagram in a model's development is a crude one that shows how the key variables affect one another. DYNAMO language uses a very simple vocabulary for the basic variables such as levels, auxiliaries, and rates, which are indicated in the flowchart by three various shapes. Rates which originate outside the system or flow out of the system are indicated as coming from a "source" or terminating in a "sink" respectively. Generally the user may invent his own symbology.

Once the model begins to take form a more informative diagram is useful. One-to-one correspondence between flowchart symbols and program equations is beneficial. The writing of DYNAMO programs (equations as well as insertion of tables, constants and notes) is based on flowcharts and requires special programming skills obtained through formal training. Similar skills are required for the compilation (processing) of the programs because of a wide variety of error statements. The analysis of the results is based on a printed output which is usually produced as a DYNAMO graph. For the user, it is very important to understand the scale of the graph. The scale can be modified by the computer during the repeated runs.

CONCLUSION

We are on the threshold of a new era in life sciences—a better understanding of the nature of our social systems. For the next years we can expect rapid advancement in understanding the complex dynamics of our social systems, but only with effort. Advancement will require research, the development of new teaching methods and materials and the creation of appropriate educational programs.

Progress in developing a new approach to social systems will be slow. There are many crosscurrents in real life that will cause confusion and delay. A new professional field may also be emerging—the profession of social dynamics.* This new profession will deal competently with the uncertainty of mental models and our inability to anticipate the consequences of interactions between the parts of a system. This uncertainty may be alleviated by models such as DYNAMO.

References

1. Forrester, Jay W. *Principles of Systems.* Cambridge, Mass.: Wright-Allen Press, Inc., 1973.
2. Goodman, Michael R. *Study Notes in System Dynamics.* Cambridge, Mass.: Wright-Allen Press, Inc., 1974.
3. Levin, Gilbert and Roberts, E. B., et al., The Dynamics of Human Service Delivery. Ballinger Publishing Company, Cambridge, Mass., 1976.
4. Meadows, Dennis L. and Donnella H., eds. *Toward Global Equilibrium: Collected Papers.* Cambridge, Mass.: Wright-Allen Press, Inc., 1973.
5. Pugh, Alexander L., III. *DYNAMO User's Manual Fifth Edition.* Cambridge, Mass.: The MIT Press, 1976.
6. Roberts, Nancy H., Dynamic feedback systems diagram kit. Paper, M.I.T., Cambridge, Mass., 1975.

*Or as called by Professor Stacey Day—*Biosocial Development.*

Programming Languages

A. GELLMAN

The Canadian Medical Association
Ottawa, Canada

The programming language is the communicative medium by which a problem must be presented to the computer for solution. Available languages range from the low level machine language written entirely in binary codes to the sophisticated and powerful high level languages whose single statements incorporate the functions of many lower level language statements. Of the higher level languages one can choose from those developed for use in very specialized fields to the more general and all-purpose language with many applications.

DEVELOPMENT OF LANGUAGES

Since, in their final usage, all languages, simple or complex, are broken down into a basic machine language it seems most reasonable to begin by looking at these. The most common machine language is binary, in which all instructions and numbers are represented as strings of binary digits, called "bits." The first programs were written entirely in this form and it was quickly realized that this was a very cumbersome technique. The first improvement came when programmers were allowed the use of certain mnemonic codes for representation of instructions to the computer. Further development was made allowing the writing of numbers in their decimal form and the addressing of memory locations by mnemonics instead of numbers. Such developments as these led to the creation of the first of the higher level languages, developed to supply a solution to certain basic needs of the programmer.

The programmer's needs were for a language that was natural—meaning that it approached English in its form and notation—and that it be more readily understandable than the machine languages. Furthermore there was a demand for pre-written library functions and routines that would eliminate the necessity of rewriting such programs as "square root" and trigonometric functions that were needed in many mathematical processes. It was found that too much time was spent in explaining the increasingly complex tasks required of the computer to the programmer and then having him program the desired function. Yet, the actual programming was too complicated for the person with the problem to learn in a short time period. Thus, the desire was for English-like languages that would be easy to learn and still be efficient.

By 1957, several higher level languages had been developed such as A-2, A-3, PRINT and BACAIC. Although these languages did allow several machine language statements for each of the program statements, they had very rigid formats and often the notation for mathematical expressions was far from natural; better systems were needed. Of the four languages to be discussed in this section, most were developed between about 1960 and 1965, each often taking several years to be fully developed.

While there is no rigid definition of a high level programming language it does have certain defining characteristics that separate it from the machine or assembly language. One such characteristic is that the programmer need know nothing of the actual processes of the computer to perform his task adequately. The language is machine independent. However, the programmer who wishes to use the computer to its maximum of efficiency will find it necessary to know something of the inner mechanisms. Since the higher level language is relatively machine independent this implies that one is able to run the same program on two separate computers with a minimum of rewriting. Another characteristic of the high level language is that its statements each encompass many machine language statements. Further, it is easier to learn than the machine language in that it is far more natural in its approach to problem solving and in its actual statement notation. The languages discussed here are FORTRAN, BASIC, PL/1 and COBOL: all are very successful and popular high level languages.

FORTRAN

One of the first of the major high level languages to be developed was FORTRAN, IBM's FORmula TRANslating system, first discussed in 1954. It was developed by 1957 for the IBM 704 computer and was intended for use in mathematical and numeric problems having provision for the handling of many variables and the computation of formulae. The original FORTRAN, a relatively machine dependent language, was slowly developed into FORTRAN IV which first appeared in 1962. It is natural in its algebraic notation which is its major attraction as a computational language and is easy to learn.

FORTRAN's technical characteristics include a character set consisting of a full alphabet, the ten digits and ten other symbols. Variable names consist of a letter followed by up to four alphanumeric characters and up to two subscripts with special provisions that make all variables beginning with the letters I, J, K, L, M or N integers. It has a library of mathematical functions and provision for the calling of subroutines within programs. Because almost all aspects of FORTRAN have been duplicated and in many cases improved upon in recent years, its major importance lies in the fact that it is easy to learn and that it was the first language developed that was usable on the machines available at the time.

BASIC

BASIC (Beginners All-purpose Symbolic Instruction Code) is very similar to FORTRAN and is also a very simple language for the learner. It was developed at Dartmouth College in 1965 by John Kemeny and Thomas Kurtz as an online programming language intended specifically for students, but has been found useful in various business and industrial applications. It is very similar in many ways to FORTRAN with one of the distinctions being that it allows only two-character variable names (a letter followed by a digit), has no integer variables and, apart from supplying a library of functions, it provides for user-defined functions. BASIC consists of two types of online commands: those that call for some function of the computer, such as RUN or LIST a program, and the program statements involving calculation or input-output functions within the program while it is being run. Using a relatively small range of statements the programmer is given extensive computing capabilities. These statements are of several types: arithmetic, logic, input and output, loop and function and subroutines. All statements in BASIC programs are preceded by a statement number and the program runs in ascending order of the numbers. Following each number is an English word defining the type of statement and then the numbers or variables involved in the statements. BASIC's main importance lies in its simplicity yet wide applicability and the fact that it is an online system involving direct interaction between the user and the computer from a terminal keyboard. This allows the programmer much faster and easier editing and debugging capabilities than are available on any batch system and thus increases its capabilities as a learning system.

PL/1

PL/1 (Programming Language 1) is the most modern of the languages discussed here. Many people found FORTRAN limited in some of its alphanumeric data handling capabilities and work was done within IBM to remedy this. At first, attempts were made towards the extension of the FORTRAN IV system itself, but this was found to be impractical and thus, after a committee study of several other languages, including COBOL and ALGOL, the PL/1 language was produced in 1964. A little later Allen-Babcock Computing Inc. developed RUSH, an online language based on PL/1 whose statements are, in fact, a subset of PL/1. PL/1 itself was found to be very successful and it was thought by some that it would fulfill its original objective of replacing FORTRAN, COBOL, and the other major languages.

The technical characteristics of PL/1 include two character sets; one of 60 characters and another of 48 characters in which some of the symbols in the larger set are replaced by letters. Its complexity and wide range of uses arises from a very large variety of statements available to the programmer and a vast library of preprogrammed functions. It attempts to bring together all the desirable features of the other major languages such as good data handling capabilities, manipulation of matrices and arrays, and good computational facilities. As a result it offers the experienced programmer extremely powerful tools but is at the same time very complicated for the beginner. It is possible, however, to write programs using only a portion of the capabilities available. Another shortcoming of PL/1 is that while its programs tend to be relatively short in length, the complexity of the language itself requires a very large compiler for their processing. Nevertheless, PL/1 is growing in popularity and is probably the best of the attempts to provide the capabilities of many specialized languages within one language.

COBOL

Of the languages so far discussed, none approaches natural English as closely as does COBOL (COmmon Business Oriented Language). COBOL was created as the result of the work of a committee of representatives of a large group of users and manufacturers and first appeared in 1960. As is indicated by its name, COBOL was developed for use in businesses and has good data handling capabilities such as documentation and table creation. A COBOL program consists of four distinct sections: the identification division, the environment division, the data division and the procedure division. The identification division merely holds the program name and comments. The environment division assigns the hardware that is used in execution. The procedure division defines the procedures that are to be performed on the files and data described in the data division of the program. COBOL is the first English-like language to be able to handle files and yet still remain relatively machine independent. Moreover, its approximate likeness to English makes it relatively easy to learn and easy to read.

FUTURE DEVELOPMENT

With the continual advances being made in computer technology it is felt that the actual programming languages used on modern machines are outdated and the progress in developing languages has fallen behind the machines themselves. Future improvements will probably lie in three different areas: (1) the programming languages, (2) the programming theory and technology, and (3) the hardware itself.

Whether actual language developments of the near future will be of any major value is questionable. Development can lie either in the area of specialized languages or in the creation of a universal language. A universal language is not something that can be expected in the immediate future because the theory and the understanding of languages is not available and, furthermore, such a language would require a certain amount of regimentation and new and possibly very large compilers. When one considers the number of users at present, all with different systems, it seems unlikely that any new language of the near future should gain such a foothold as to become "universal".

In the region of specialized languages one can find hundreds of systems available, all developed to handle specific needs. Systems such as the MAC-360 have appeared with the ability to write equations in two dimensions as they should be with true exponents and subscripts rather than writing two dimensional equations in linear form. Unfortunately, the area of specialized languages is too vast for one to expect major improvements in this area.

Potential for improvement does appear to lie in the area of programming techniques and theory. On the theory side, more efficient methods of using the available languages can be found; in the area of programming techniques the trend must move toward online programming through terminals and time-sharing systems.

As can be seen in Fig. 1 the online system is far less cumbersome than the off-line and is thus time-saving. Past justification of the off-line method has been its cost, but with present day increases in man-hour costs and decreases in hardware costs, the online systems are becoming more cost-efficient and attractive.

In the region of hardware, while improvements in the computer are continually being made, improvements for the programmers lie in the development of faster and more efficient compilers. This involves not only developments within the compiler but also in language theory and understanding.

Long-range developments include the approach to natural language that will open up the use of the computer to all. The question then arises concerning whether or not the computer should be available to all, in which case a natural language would be needed; or, whether it should be left in the hands of a small group of highly skilled programmers, in which case it is necessary to have a high level, highly efficient language. This will be determined by social trends of the future. Research is even being done into verbal communication with the computer but this is a very long range project and is inapplicable at present. Thus, we feel that developments for the near future lie in improvements in programming theory and in the online systems and in improvements in compilers and actual language theory and understanding.

Programming Techniques

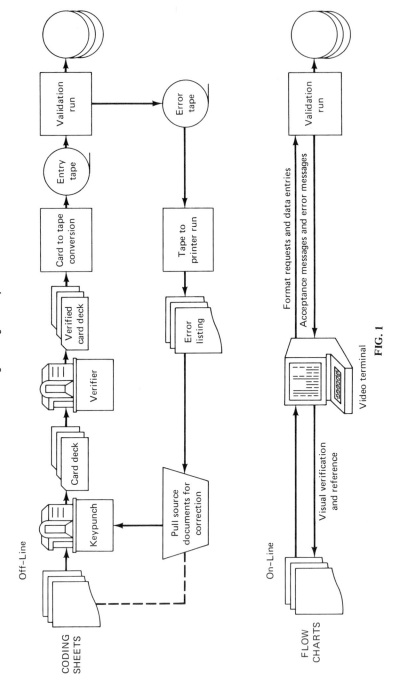

FIG. 1

References

1. Pollack, S. V. and Sterling, T. D. Procedure-oriented languages. In *Encyclopaedia of Computer Science*, A. Ralston and C. L. Meeks, eds. pp. 1112–1156, New York: Petrocelli-Charter, 1976.
2. Sammet, J. E. *Programming Languages: History and Fundamentals.* New Jersey: Prentice-Hall, Inc., 1969.
3. Schur, L. D. *Time-Shared Computer Languages.* Reading, Mass.: Addison-Wesley Publishing Co., 1973.

Digital Computers:
Babbage's Difference Engine to PET

A. GELLMAN
The Canadian Medical Association
Ottawa, Canada

Techniques and methods of data processing have developed incredibly in the few millenia of historic times, from the clay tablets of ancient times to the computers of recent decades. There has always been a need for data processing, both of numbers for problem solving and of information records of businessmen. Early methods of mathematical problem solving included a Roman technique of finger multiplying and the use of such manual machines as the abacus which can be employed, in adept hands, to perform all four basic functions. The keeping of records can be dated to prehistoric times with the scratching of information on rock. Mention of business records is made in the code of Hammurabi, the ruler of ancient Babylon in the 23rd century B.C. The Egyptians, Greeks and Romans also had quite advanced accounting systems involving extensive inventories, receipts for delivery of goods or payment of debts and business ledgers. The Romans had budgets for the army and the emperor and levied taxes accordingly. In England, the Exchequer was created in the 12th century to keep records of taxation. But beyond these early accounting methods there were no real advances in data processing techniques until the advent of practical machines in the 19th century.

One of the first multiplication aids was a set of rods called Napier's Bones, each having numbers written on them in such a way that by correct manipulation of these rods, long multiplication problems were reduced to the addition of these numbers. The first mechanical adding machine was built in 1642 by Blaise Pascal, its internal workings based on a set of rotating wheels. In 1694 Gottfried Leibniz built a machine capable of performing all four arithmetic functions but it was somewhat unreliable. The first successful calculator was not built until 1829 by Charles Thomas. Calculating and accounting machines were developed into the 20th century, some with the ability to print and others with electric motors, but each was very dependent on the presence of an operator at every step of the calculation and was in no way automatic.

The idea of preprogramming a machine through the use of punched cards had its beginnings in 1801 in a machine built by Joseph Jacquard. By following a set of instructions punched into cards the machine was able to weave elaborate designs into cloth. The idea was adopted by Dr. Herman Hollerith who used punched cards to mechanize the United States census of 1890 and in so doing accelerated the processing of the collected data by several times. Later as punched card machines grew in popularity and capability, Dr. Hollerith started the Tabulating Machine Company which, after several mergers, evolved into the International Business Machines Corporation (IBM).

One of the first major steps toward the actual development of the computer was the Difference Engine designed by Charles P. Babbage for the creation of mathematical tables. A model of this machine had been built by 1822 and Babbage received grants to subsidize the building of the actual machine from the British Government. However, the Difference Engine was left uncompleted as Babbage turned to the development of a more general purpose machine, the Analytical Engine. The Analytical Engine was to have an arithmetic unit, a storage unit and a control unit and had it been developed it would have been the first digital computer. Babbage worked until his death on the Analytical Engine, but was only able to build parts of it. He was however, one of the first to see all the problems that the building of such a machine entailed.

The next noteworthy steps in automatic data processing after Babbage's death were not until the 1940's when, after much work, the computer truly came into being. In cooperation with IBM, Prof. H. Aiken built the Harvard Mark I in 1944, the first digital computer. It was an electromechanical device with input and output on punched cards. The Mark I did its calculations in decimal form and was used mostly in the field of scientific calculations. It was the first in a line leading to the Mark IV but the line was surpassed by the much faster ENIAC.

ENIAC (Electronic Numerical Integrator and Calculator) was the first of the entirely electronic computers. The machine itself was built in 1945 and was a massive affair containing 18,000 vacuum tubes and weighing 30 tons. It operated in binary and was capable of performing 300 multiplications/second where most electromechanical devices could only do one. Although it was originally programmed by the interconnection of various panels by wires, it was later converted to punched card programming. It was first used for the solution of ballistics problems, problems involving atomic energy, and other scientific computation, and was in use until 1956 when it was housed in the Smithsonian Institute.

The EDVAC (Electronic Discrete Variable Automatic Computer) was the first computer that did not require programs to be externally wired by hand. It was smaller than those preceding it having only 4,000 tubes. Its major importance was that it was the first computer to be designed on the principles of Dr. J. von Neumann which are used in many of today's computer designs. Although it had been built by 1949, the EDVAC did not become fully operational until 1952 and even then required much maintenance (during that year over twice the total number of tubes had to be replaced). It was used at the Aberdeen Testing grounds until 1962.

The first commercially available computer was the UNIVAC 1 (Universal Automatic Calculator) which was produced by the Sperry Rand Corporation in 1951. It was the first general purpose computer for data processing with the capability of storing and manipulating both digits and characters. Input and output used magnetic tape and data was input through a keyboard or punched cards. Between 1951 and 1958, 46 UNIVAC computers were delivered. All have since been phased out.

The surge in computer use was dramatic and a 1965 survey showed that there were more than 23,000 computer installations across the United States. All of the tube-using computers have been surpassed by the advent of computers using solid state technology. This makes the modern computer far more powerful, compact and reliable than its ancestors. Moreover, the technology that did this has brought down the price of the computer so drastically that it is now available to even the relatively small business.

The computer, although it is not independent of human direction, once given a problem and a set of instructions, can solve that problem without the intervention of humans. Once entered, the set of instructions is stored by the computer and can be called upon repeatedly to solve problems. While the program itself is a specialized operation code and in running a certain program the computer becomes for a while specialized, it is really a very general purpose tool for it is capable of following an infinite variety of programs to solve almost any problem. Its usefulness lies not only in the wide range of its capabilities but in the speed at which its electronic components allow it to function. As solid state technology advances, the speed of the computer increases and the size decreases.

Although there are two types of computers—digital and analog—the majority of computers are the digital type because of its ease of data handling, which is required in the business and administrative world. Analog computers are used in engineering problems and differ from the digital computers in that the representation of numbers is in the form of voltages of varying magnitudes rather than binary pulse strings.

Although all computer systems differ according to the needs of the user, certain generalizations can be made in defining portions of the system. The input section is the means of communication from user to computer and can be in the form of a card reader, terminal, or tape reader. Output section involves communication in the opposite direction and can be through terminal, line printer, or various other devices. Also external to the computer is the file section which is in the form of discs, or magnetic tapes. Within the computer itself is the main memory, the control section, and the arithmetic and logic unit. The main memory holds the information for the control unit which directs the flow of data to the output units or the arithmetic unit which can process it and work with it. Depending on the users' needs, each section will have certain apparatus, e.g., while one user may need a plotter for output, another may need a certain type of VDT and some may require more disc memory space than others.

The development of computers can be divided into generations where each successive generation is made more powerful and compact by advanced technology. The first generation is considered to have lasted from 1937-1958, the second from 1958-1964, the third from 1964-1975 while the fourth is just in its opening stages. These generations are not randomly picked time periods but signify advances in the technology that goes into the designing of the computers.

The main characteristic of the first generation computers was the use of vacuum tubes which made them very large. Memory was in the form of mercury delay lines and electrostatic storage systems using the Williams tube. Large scale memory storage was on magnetic drums. The UNIVAC was the first of the commercially successful first generation computers but the most successful was the IBM 605. A major development during the first generation was that of core memory which was far more practical than the available systems, and which was immediately installed on the IBM 704 and 705 then quickly adopted by other manufacturers. The next step forward was the use of solid state components which announced the second generation.

Transistors, although developed in 1948, could not be efficiently implemented for computer use until about 1958. Their advantages over vacuum tubes were many but their use was delayed by mass production problems. The transistor was smaller than any tube, required less power because it had no heating filament, could operate faster, and was more reliable than the vacuum tube. Thus, the second generation computers were both more powerful and yet more compact than those of the first. All had core memories and the auxiliary memory was supplied by discs, magnetic drum or magnetic tape. The second generation lasted only a short while and ended with the development of the integrated circuit.

The third generation began in 1964 and was characterized by the use of integrated circuit technology and advances in system designs. The ability to place many solid state components on one small semiconductor chip as opposed to connecting them individually with wires on a circuit board greatly reduced the size of computers. Probably the most important computers of this generation were the IBM 360 line which was introduced in 1964 and the 370 line introduced in 1970. One of the significant developments that went into the 370 line was the use of the cheaper and faster MOS (Metal Oxide Semiconductor) memories rather than the core memories. Another important product of integrated circuit technology is the minicomputer such as Digital Equipment Corporation PDP-8 and PDP-11 computers. These incorporate much power into a minimum of space. Still more compact is the microcomputer sold as programmable pocket calculators by such firms as Hewlett-Packard and Texas Instruments. Here the use of large scale integration has produced powerful capabilities on just one or two semiconductor chips that can be made cheaply so that the final product can be sold at remarkably low prices.

The fourth generation is at present in its early stages and is represented by such computers as the IBM 3033 and the Amdahl 470 V/6. Future trends through the fourth generation are uncertain but there are some far reaching possibilities. New types of memory storage are being developed using lasers that will probably replace the conventional systems and production of cheaper computers. Greater miniaturization brought about by large scale integration may take the computer into the home and make its use available on a personal basis. Microcomputer calculators have been on the market for several years but are mostly used for scientific and mathematical applications. However, such machines as the PET (Personal Electronic Transaction) manufactured by Commodore Business Machines and available at little over the cost of the most advanced calculators have far greater ranges of capabilities. The PET is desktop sized and yet offers memory space, has a video display and will even play chess. Whether the future lies in such independent units or terminals connected to a much larger computer by telephone or even a combination of both is unpredictable, but certainly advances will make the use of computers available to an increasingly greater range of people.

References

1. Morris, G. J. Digital computers: General principles. In Encylopaedia of Computer Science, pp. 463–466. (A Ralston & C. L. Meeks, eds.) New York: Petrocelli/Charter, 1976.
2. Randell, B. Digital computers: Origins. *Ibid*, pp. 487–490.
3. Rosen, S. Digital computers: History. *Ibid*, pp. 475–487.

Computer Networks

S. CONSTANTINOU

The Canadian Medical Association
Ottawa, Canada

The term *computer networks* has been used to describe:

(a) Geographically remote terminals and Remote Job Entry (RJE) connected directly to a central computer.

(b) Geographically remote smaller computers used for minor tasks which store information on magnetic tapes. The magnetic tapes are then sent by post or messenger to a central computer for further processing.

(c) A central computer directly connected to smaller machines capable of performing specialized functions. The smaller machines provide the central computer with services such as storage and communications.

(d) Independent computer systems connected with each other in a way that enables direct communication as well as the sharing of hardware, programs or data.

Even though computer networks should not be confused with information networks the latter can be an integral part of computer networks.

The greatly increased reliability of computers, the availability of low-priced minicomputers, and major advances in the communications technology are some of the most important reasons which have led to the proliferation of computer networks in recent years. Experts feel that many characteristics of future networks are currently feasible. Even though inexpensive intelligence, all digital carrier facilities, efficient protocols, satellite and distributed networking techniques can be utilized at present, it will take some time before they can be coordinated in an efficient manner. Hence their impact will not be fully felt for some time.

Technological advances are making the addition of intelligence to terminals less expensive. As a result many central computing functions can now be done at local sites. It is expected that distributed networks will keep expanding at the expense of centralized networks. In many cases the concept of online networks is gaining over batch processing and is reducing the prevalence of punched cards processing. Online information display has certain advantages: (1) it allows the user to select the relevant data and to eliminate on the spot information that is too general or not needed for a specific task, and (2) it allows faster decision making and, in a fast changing world, this is an important asset.

The majority of the present networks utilize complex languages that do a lot of thinking for operators and users alike. An online interactive system should allow these people to use their intelligence while at the same time taking advantage of the immense computer power available to them. Even though knowledgeable programmers are indispensable within the broad computer network, they should not make themselves the key to the operations of the system by building artificial intelligence at the detriment of basic human logic and intelligence.

In applications in the medical field, coding should be avoided. How users respond and think should be an integral part of the software and hardware configuration of the network.

The move from centralized to distributed networks will increase the danger to the confidentiality of the information transmitted within any computer network. In the same manner that microwave telephone lines can be easy prey to the spies of foreign powers, corporations, or criminal groups, so could computer communications.

It is felt that the potential advantages of computer networks are as follows: load and resource sharing, better transfer of data, lower degree of obsolescence, improved access to large data bases, upgraded processing and data distribution, as well as economies of scale. However, before computer networks can become commercially viable, a number of problems that exist in all computer systems must be solved. These problems become accentuated in the case of computer networks and include the following: security, confidentiality, and accountability; reliability and cost effectiveness of communication links; standardization of data formats, communication, and user protocols, etc.; adequate disaster back-up; and a more advanced understanding of the concepts of data and process distribution.

A number of computer scientists believe that there will be substantial growth in computer networks by 1985 but they will not predominate data processing methods by then. They expect remote access facilities to expand at a much faster rate.

The mind boggling advances in telecommunications and computer technology are currently used interdependently to promote better and cheaper computer networks ostensibly for the sake of better management, economic growth and development. Yet most types of telecommunications are open to abuse and sophisticated safeguards need to be designed and built into the distributed networks as they are developed, not after the network is in operation.

The enormous power provided to humans by these computer advances raises some very basic questions as to what *ought and ought not* be done by computers, which transcend what *can* be done by computers. A few experts believe that projects aimed at replacing human functions that entail interpersonal respect, emotion, and understanding by a computer system should not be pursued. The same applies to computer applications that may clearly have irreversible and not totally predictable side effects.

References

1. Dolotta, T. A., et al. *Data Processing in 1980-85: A Study of Potential Limitations to Progress.* John Wiley & Sons, Inc., 1976.
2. Ralston, A. and Meeks, C. L. (eds). *Encyclopaedia of Computer Science.* New York: Petrocelli/Charter, 1976.
3. Rarp, H. R. and Lapidus, G. *Executive Guide to Data Communications.* 1976.
4. Weizenbaum, J. *Computer Power and Human Reason.* W. H. Freeman and Company, 1976.

Computers in Medicine

S. Constantinou
The Canadian Medical Association
Ottawa, Canada

INTRODUCTION

The applications of computers in medicine are viewed by many experts as potentially vast and far reaching. There are enough examples of specialized applications to support this view. Yet computer technology has made an impact on the health care delivery field which is minimal in comparison to its effect on most other major industries. This paper can only provide an overview of some of the major applications, the reasons why medicine has proven a hard field for applications, and some insights into what can be expected in terms of future developments.

THE POTENTIAL AND THE PROBLEMS

The potential benefits of computer applications are, of course, determined by the nature and objectives of the health care delivery system. These objectives are to provide for the health of the consumer in a most effective, cost efficient way through hospital, ambulatory and home care.

Computers can assist in better meeting this objective in a number of ways:

business applications in hospitals and physicians' offices; biomedical engineering; automated medical records; multiphasic health testing; decision support systems; laboratory and pharmacy systems; hospital information systems, and computer-aided ·instruction in medicine.

Just a few illustrations should suffice to further clarify the advantages computerization will offer. A well planned Hospital Information System can improve and provide savings in the following areas:

- accounting (billing, general ledger, balance sheet, etc.)
- admission scheduling
- patient identification and label printing
- diagnostic testing, medication, dietary ordering and dispensing
- central supply ordering, maintenance and control of inventories
- appointment scheduling (in-patient and out-patient)
- health statistics and benefit profiles
- internal manpower planning.

A multiphasic health testing can save lives by detecting dangerous diseases in their early stages. It can also save money because preventive care is usually less costly. Computerized medical records could facilitate the treatment of patients and if properly utilized free physician time for direct patient care. So could the computerization of appointments, billing and accounting in physicians offices.

Yet, progress in all these areas is very fragmented and at best extremely slow. The major obstacles to progress in computer applications in medicine are as follows:

1. Physician and patient resistance to automated techniques of health care.
2. The structural characteristics of the health care industry which are not, as a rule, conducive to computer applications.
3. The inability of computer scientists to fully appreciate the above unique characteristics and/or properly account for them when designing health care information systems.
4. Financial and managerial problems. Computers are becoming less expensive and more powerful with time, but their medical applications still represent a considerable investment for those contemplating them. Even if their cost is not prohibitive in absolute terms, as is usually the case, their successful implementation requires good managerial skills on the part of hospital or clinic administrators who decide to use them—an intelligent machine cannot do the job itself, it needs intelligent managers who will take full advantage of its capacities and use them in innovative ways.

The health care industry is a quasi-public one. It is, hence, not subject to, nor operated under, the same competitive conditions that prevail in the private sector industries. Even though hospitals are accountable, political considerations rather than economic determine their budgets and they are in no danger of going out of business if they run into deficits. As a result the motivation to maintain cost efficiency in hospitals is not nearly as high as that exhibited in the private sector. It follows then that computer applications which produce better management of resources and better balance sheets are more sought after by the private sector.

Hospital administrators have a difficult task to perform. They are operating in a system with incentives for excessive utilization (otherwise their budgets will be cut). They have to meet ever present demands for the best available medical equipment both from patients and from physicians. Accusations of costly duplication of expensive and not fully utilized equipment are commonplace among students of health care costs. Many physicians on the other hand have not fully realized the potential of computer applications in medicine. They tend to put more emphasis on the negative aspects of computer applications such as the dangers to the confidentiality of sensitive medical information and the likelihood that computers may interfere with their relationship with the patients. All these are legitimate concerns, but they can be minimized if a computer application is carefully implemented.

SOME EXAMPLES OF COMPUTER APPLICATIONS IN MEDICINE

Many examples of computer applications in medicine can be found. Yet it should be kept in mind that for every successful application one can easily find two unsuccessful ones. This fact is due to the problems mentioned above.

One can start by mentioning a few rather well known applications such as computer aided electrocardiographs (ECG), computer aided tomography (CAT), radiology and radiotherapy systems, computer based unit dose systems, online admission lists, computer aided monitoring of patients, etc. One challenging and promising application currently being tested in the United States is the MYCIN system. The system's objective is to assist physicians in selecting the optimum therapy for patients with bacterial infections. It utilizes computer techniques from the field of artificial intelligence (AI) and incorporates the expert medical knowledge available in its decision making subsystem. The system is fully conversational, easy to use, can accept new information in a flexible manner and has the capacity to justify the decisions it reaches! Because of its flexibility and good modeling it has the potential to expand the knowledge in the field and to become an excellent educational tool. According to the experts, MYCIN has an excellent chance of improving drug selection in the treatment of infectious diseases.

The Indian Health Service, Health Information System (IHS-HIS) is an example of how the delivery of health services can be improved and made more effective by the use of computer technology. This system is at the core of the primary source of health care for about 12,000 Papago Indians in Southern Arizona. The system accepts input about patients from IHS physicians and paramedics, from public health nurses and from outreach health workers selected and supervised by the tribal government, including mental health technicians and nutrition aides, to list only some of the professions involved. All data are then merged into a single, integrated data base which is available to providers of health care to the Papago reservation with all due respect to security and confidentiality.

Current reports indicate that a growing number and variety of computer based systems are appearing in the North American marketplace. These systems have been designed to run both on mini and big computers, in batch and real-time environments. They usually start with financial applications but are capable of expanding modularly into patient care by using online terminals strategically located within the hospital.

It is generally felt that a mini computer is the most advantageous for hospital applications. It offers not only economies of scale but the capacity to add, at a relatively low cost, another processor as backup. One of the latest such minis has been installed at Mary's Help Hospital in Daly City, California. The system enables the admissions department to enter patient information through a keyboard. It provides a real-time census of the beds and the business office has access to the patient file to process patient charges. From terminals at nurses' stations, medication can be ordered from the pharmacy, a schedule of medication to be given patients can be produced for each ward and the nurse can indicate that the medication was, indeed, given. The pharmacist is also able to get information of drug allergies, etc. from his own terminal.

A similar system exists at the El Camino hospital in California and a number of other hospitals in the U.S. The philosophy underlying these systems is essentially similar. They are online systems that perform the admit/transfer/discharge functions and link the pharmacy, labs, and housekeeping department. They also enable the dietary department to know of patient movements and of orders for tests that affect meal planning. Numerous hospitals in Canada and the U.S. have found the basic business applications of computers a good investment but it appears at present that clinical applications are a much more difficult extension of the computer technology.

How the changing structure of health care can affect the use of computers in the hospital field can be demonstrated by a few examples in the U.S. and Canada. The introduction of Medicare legislation in the U.S. forced almost all hospitals to utilize the computer for financial processing, patient billing, and accounting. It is now felt that the proliferation of PSRO's will make it imperative for U.S. hospitals to computerize their clinical operations as well.

In Canada, National Health Insurance has been in operation since 1958 for hospitals and 1970 for medical care and the recently demonstrated willingness of provincial governments to accept computerized billing has led to the expansion of computers by clinics and group practices. A number of groups in Canada are currently in the process of installing inhouse computers for appointment, billing, accounting, and automated medical record purposes. The success of these projects may be the signal for a much wider acceptance and use of the computer in physicians' offices as well as hospitals.

References

1. Brandejs, J. F. *Health Informatics, Canadian Experience*. Amsterdam: North-Holland Publishing Company, 1976.
2. Shortliffe, E. H. *Computer-Based Medical Consultations: MYCIN*. American Elsevier Company, Inc., 1976.
3. Van Egmond, J., de Vries Robbé, P. F. and Levy, A. H. (eds). *Information Systems for Patient Care, Review, Analysis and Evaluation*. Amsterdam. North-Holland Publishing Company, 1976.
4. HCIB, *Health Computer Applications in Canada, Catalogue and Descriptions.* **Vol. III.** Ottawa: Health Computer Information Bureau, June 1976.

Artificial Intelligence

D. R. LANGER
The Canadian Medical Association
Ottawa, Canada

A broad definition of artificial intelligence (AI) might be: "the intelligence of any machine that performs a task that a century ago would have been considered a uniquely human intellectual ability." Usually:

(a) the machine is a digital computer or is controlled by a digital computer, and

(b) the task involves symbolic reasoning or "thinking" rather than arithmetic calculations or information storage and retrieval.

The beginnings of the field of AI, a branch of computer science, are attributed to A. M. Turing who, in 1950, suggested that a machine is intelligent if an individual is unable to decide whether he is interacting with a computer or with another human using a teletype.

There are approximately eight application areas that encompass most of the work of AI: game-playing; math, science and engineering aids; automatic theorem proving; automatic programming; robots; machine vision; natural language systems; and, information processing psychology.

Today, AI research focuses on the problem of describing information so that it can be used by a computer to perform tasks which require intelligence in humans. It is assumed that to perform such tasks, the computer needs information similar to that used by humans in that task. Often human reasoning is "intuitive" and based not on fact but on past experience, environment, context, etc. This makes it difficult or impossible to instruct a computer in the "logic" or even provide it with the necessary information.

Physicians, for example, seem to use an ill-defined mechanism for reaching decisions despite a lack of formal knowledge regarding the interrelationships of all the variables that they are considering. This mechanism is often adequate, in well-trained or experienced individuals, to lead to sound conclusions on the basis of a limited set of observations. This unformalized expertise is often referred to as the "artistic" component in the science of medicine. Models of inexact reasoning have been worked out and can be applied to programming of computers so that decisions can be made on the basis of evidence which is not entirely mathematical.

Computer-assisted medical decision making is an area of medical computing which fascinates numerous researchers, partly because modern medicine has become so complex that no individual can incorporate all medical knowledge into his decision making powers. Complex systems have been developed, such as MYCIN, which incorporates diagnostic, prognostic, treatment planning and educational programs. The name of the project reflects the central concern of the program, namely the selection of an appropriate therapeutic regimen for a patient with a bacterial infection. There is evidence that physicians often do not choose antimicrobial therapy wisely due to time required for culture reports, etc., and often prescribe in a way which is significantly different from that which would be recommended by infectious disease experts with the same limited clinical clues. Since professional resources are often overburdened in today's hospitals, a computer-based system that could serve effectively in a consultation role to the nonexpert would help bridge the gap between general practitioners and, in the case of MYCIN, experts in infectious disease therapy.

There are some problems which must be overcome before professionals will use computers. At the moment, users such as physicians must take the initiative in asking for an interactive session with the computer. It has been demonstrated that physicians seldom choose to use computers for tasks they feel they can do themselves.[3] It might therefore be more successful to have programs which would act as monitors and generate warnings when appropriate. Automated language understanding will also have to progress so that unrestricted discourse can take place. Development is very costly at the moment for the limited range of applications of AI, and the memory now required for storage of programs is considerable.

In addition to solving the above-mentioned problems, some long-term objectives of AI research are: to automate the processes that make the decision on a particular representation after the specification of the task has been given, possibly in natural language; to write highly efficient programs of "universal" applicability. At present as the level of complexity rises, with increased applicability, efficiency in solving individual problems drops.

It is often observed that it will be years before machines can perform problem-solving tasks at a level approximating that of humans. However, at the present stage, if researchers select real-world goals within the current limitations of the AI field, a fund of experience can be built upon to obtain systems which are effective, feasible in terms of cost, and which contribute to the betterment of mankind.

References

1. Mylopoulos, J. and Perreault, C. Man-Machine communication: The artificial intelligence approach. *Canadian J. Info. Sci.*, 2(1) (1977).
2. Shortliffe, E. H. *Computer-Based Medical Consultations:* MYCIN. New York: American Elsevier Publishing Co., Inc., 1976.
3. Startsman, T. S. and Robinson, R. E. The attitudes of medical and paramedical personnel toward computers. *Computers in Biomedical Research*, 5: 218-227 (1972).
4. Turing, A. M. Computing machinery and intelligence. In *Computers and Thought* (E. E. Feigenbaum and J. Feldman, eds.) pp. 11-38, San Francisco: McGraw-Hill Book Company, Inc., 1963.

Telemedicine

RASHID BASHSHUR

Professor
Department of Medical Care Organization
School of Public Health
University of Michigan
Ann Arbor, Michigan

Telecommunications have been employed in the delivery of health services for a long time. Patients are accustomed to calling their physicians or clinics by telephone to make an appointment, to refill a prescription, or to receive advice in an emergency or for a minor health problem. Indeed, due to the volume of telephone usage, some providers have contemplated billing patients for "telephone encounters." Similarly, the radio has been utilized in various parts of the world to link geographically remote areas with central locations, to provide advice, make a remote diagnosis, or dispatch aid. Nonetheless, the use of two-way, interactive television (IATV) is new, as is the addition of telemetry and computers to its capability.

Despite variations in its usage, the prominent concept of telemedicine today is that it is an organized system for the delivery of health services via interactive television and telemetry. Its irreducible requisites include geographic separation between patient and physician (as in telediagnosis) and between one physician and another (as in teleconsultation); heavy reliance on telecommunications technology to bridge time and spatial barriers; development of organizational structures that are capable of efficiently utilizing the technology (as in multi-site delivery settings); and the use of mid-level health practitioners in the delivery of care.

Variations in the speed of transmission and display of picture frames provides several technologic options that can be utilized for communications: *realtime* as in interactive television; less than realtime as in slow scan TV; and *static* as in the use of video tape. Audio-visual transmissions provide the facility to overcome barriers of time and distance with a full set of informational signs. Thus, video capabilities provide information about a psychiatric interview which would have been lost if conducted over the telephone. Medical data, such as radiographic images, can be transmitted to one or more sites simultaneously and without the degradation in quality associated with more conventional analog-to-digital-to-analog conversions. Moreover, the relay system permits signal enhancement and, hence, additional information to be gained. Microscope attachments, electronic stethoscopes, zoom lenses, and instant replay may increase the information obtained over that of the conventional face-to-face encounter between physician and patient.

Telemedicine systems have been installed in such diverse locations as nursing homes, prison health clinics, mobile field centers, Indian reservations, urban neighborhood health centers, airport emergency rooms, hospital surgical suites, classrooms, primary care physicians' offices, first-aid centers in remote Alaskan villages, on board ships, and in outer space.

An equally great diversity of types of persons have participated in the use of telemedicine, including primary and specialty care physicians, engineers, physician assistants, nurse practitioners, family nurse associates, nurses, social workers, administrators, emergency medical personnel and patients.

The origins of telemedicine lie in both the manned space program of N.A.S.A. and in the work of earthbound, but equally visionary, individuals in medical care. Research, development, and demonstration of the capabilities of hardware for remote biomedical telemetry were major achievements of the space program. In anticipation of longer space flights and of the manned orbiting laboratory/shuttle, N.A.S.A. has continued to provide technological achievements in the development of complete medical delivery systems for remote diagnosis and treatment.

The issues of the long-term impact of telemedicine on the delivery of medical care have been identified as follows:

> ... the distribution of medical manpower in the United States, its relationship to the proliferation of medical specialization and fragmentation of medical responsibility; the clinical role of the non-M.D. provider that will not jeopardize the quality of care rendered to the patient; the effect on current barriers to the receipt of care for those who are at locational disadvantage in relation to health facilities; and finally, telemedicine as a mechanism for collegial interaction among health professionals, serving as an instrument for quality control in administering health services.[1]

While there were not substantive data to determine telemedicine's merits at this time, the theoretical and conceptual problems related to these issues were discussed through the defining parameters of telemedicine: (1) geographic separation; (2) telecommunications equipment; (3) staffing arrangements; and (4) organizational structure. Evaluative data must be collected under normal or optimal operation before a decision can be reached about the merits of telemedicine. Without such measures, a temporary arrangement to help solve current problems in medical care could become a permanent institution introducing new problems into the health care system.

Some potentially negative consequences of increased use of health care communications systems have been identified. These include: (1) increased alienation and isolation; (2) patient feelings of inferior care; (3) loss of privacy and confidentiality; (4) a health system inundated with well patients because of eased access; and (5) actual neglect of continuing education by providers because of the ease of specialist consults.[5]

The primary economic consideration of a telemedicine system is cost effectiveness. While proponents of telemedicine have suggested that the system will organize medical services so that the gains will outweigh the substantial initial capital outlay for equipment, there is the additional cost of adding new components, such as the non-M.D. provider, to the medical care system. Whether or not the difference will be compensated for by savings in physician time remains to be seen. Telemedicine has been defined as a production process that yields medical care outputs. The introduction of new technology will change both the production process and, as a result, the set of outputs.[2]

Considerable emphasis has been directed toward analyzing the most cost-effective combinations of non-M.D. providers and technology. Categorization of non-M.D. providers by degree of training and ability indicates varying requirements of physician/technology support. Another study has suggested that relatively low-cost training of personnel might substitute for more costly advances in the sophistication of equipment.[3]

The coalescent relationship of non-M.D. providers and telecommunications is a primary feature of nearly all telemedicine applications. Increasing attention has been directed toward analyzing the most effective manpower-technology combinations, especially as this applies to utilization of non-M.D. providers. Protocols which determine the specific functions and activities of the non-M.D. providers have become an increasingly important attribute of this type of health care delivery.

Communications aspects of telemedicine have been discussed by Park.[4] He pointed out that because the medium of interactive television is different than face-to-face interaction, methods of communication are different. Hence, specific behavior patterns or codes of behavior must be developed to facilitate successful communication in the interactive television medium. The physical properties of a television system that influence communication include: the process of scanning the image; the two-dimensionality of the picture; frame size; gray scale; color values; switching; lighting and sound. The psychological properties include tension produced by the frame, notion perceptions, and interpretation in terms of editing film to illustrate or explain. He defines several cultural properties of an interactive process which must be redefined for communication over television. An example is the problem of culturally acceptable distances—proxemics—and how these are affected by the television medium. More experience is necessary to develop appropriate codes of behavior for effective communication in telemedicine.

A significant new direction in telemedicine is the effort to make it pay for itself. Thus far, telemedicine projects have had to be heavily, if not totally, supported in order to survive. New trends in telemedicine point to a system that could derive some revenue, making this mode of delivery viable for the future. It has been suggested that any subsystem of telecommunications such as telemedicine cannot be economically viable on its own; rather a systems approach must be taken, emphasizing the various uses of telecommunications, especially in rural areas.[6]

References

1. Bashshur, Rashid L. Telemedicine and medical care. In Bashshur, Rashid L., Armstrong, Patricia A., and Youssef, Zakhour I. (eds.), *Telemedicine: Explorations in the Use of Telecommunications in Health Care.* Springfield, Ill.: Charles C. Thomas, 1975.
2. Berki, Sylvester E. Telemedicine: Some economic implications. *Ibid.*
3. Dickson, Edward M., in association with Raymond Bowers. *The Video Telephone: Impact of a New Era in Telecommunications.* Praegar Publishers, 1973.
4. Park, Ben. *An Introduction to Telemedicine.* New York City: Alternate Media Center at the School of the Arts, NYU, 1974.
5. Rockoff, Maxine. The social implications of health care communication systems in Institute of Electrical and Electronics Engineers, *Transactions on Communications* Vol. COM-23, No. 10 (October 1975): 1085–1088.
6. U.S. Congress, Office of Technology Assessment. *The Feasibility and Value of Broadband Communications in Rural Areas: A Preliminary Evaluation.* April 1976.

TEL-COMMUNICOLOGY: Outreach Services for Rural Americans with Communicative Disorders

GWENYTH R. VAUGHN, Ph.D.
Birmingham, Alabama

INTRODUCTION

More than twenty million Americans have hearing and speech disorders.[1] Many of the communicative disorders are related to critical shortages in rural health care delivery programs. In 1976, President Ford stated that 1400 counties and regions, mostly in rural areas, suffered from a lack of health services.[2]   At the same time, Presidential Candidate Carter pointed out that almost a third of the American population lived in rural areas and was served by less than 20% of the professional health personnel.[3]   Rural residents with hearing and speech disorders are still underserved or unserved because they live in areas remote from audiology and speech pathology centers.

Accessibility

Travel barriers limit accessibility to services and are inherent in traditional institution-centered health care programs. They are reflected in the expenditures of time, effort, and money by persons with communicative disorders or by members of their families who have to accompany them, thus losing a day at work.

Travel barriers affect the very young, the old, and the ailing. Restrictive physical conditions and severe weather often make long trips exhausting and inadvisable.

Inpatient care is generally not provided for persons with communicative disorders, unless the disorder is related to a medical or surgical problem.

Availability

Not only are services inaccessible to rural populations, they are also unavailable. Although the number of professionals has increased, many facilities do not have funds to employ them. If the vast numbers of persons with speech, language, and hearing disorders were referred to the speech pathologists and audiologists presently employed by rehabilitation centers, hospitals, or public schools, the facilities would be overwhelmed. New approaches to the problem of health care delivery for persons with communicative disorders need to be considered, if quality care for rural populations is to become a reality.

TEL-COMMUNICOLOGY: INNOVATIVE OUTREACH SERVICES

The lack of availability and accessibility of health services for rural populations is a critical issue and must be addressed not only by private facilities but also by public agencies.

The Veterans Administration, as the largest health care system in the Western World, focuses upon developing ways to provide improved, expanded, and extended health care. The veteran population, including survivors and dependents, numbers 99.6 million. Great numbers of veterans reside in rural areas remote from treatment facilities with audiology and speech pathology centers. Veterans are an aging population; older persons tend to have increased hearing, speech, and language problems.

TEL-COMMUNICOLOGY

TEL-COMMUNICOLOGY is an innovative health care delivery system that utilizes the telephone to provide supportive audiology and speech pathology services to veterans with hearing, speech, and language disorders. The intent of the TEL-COMMUNICOLOGY programs is to enhance, not replace, traditional health care delivery systems.

Special procedures, materials, and adaptive devices are being developed that make it possible to treat many types of communicative disorders. The system is patient-centered rather than institution-centered.

Pilot studies show that veterans receiving the *outreach* programs of TEL-COMMUNICOLOGY improve in their communicative skills in nearly every instance. Formal evaluation of TEL-COMMUNICOLOGY as an effective health care delivery system is in progress.

HUMAN COMMUNICATION AND ITS DISORDERS

Communication enables and enhances most human activities. Functional communication contributes greatly to the achievement of educational potential, optimal job placement, and effective personal/social adjustment. Communicative disorders appear as isolated problems or in conjunction with other disabilities; they result from disease, injury, aging, or psychosocial problems. Some communicative disorders signal the need for medical, surgical, psychological, or educational intervention. A voice disorder, for example, may indicate the presence of laryngeal cancer.

Hearing Impairments

Hearing loss is sometimes referred to as the great masquerader. Loss of hearing in the young may be confused with mental retardation, aphasia, or psychogenic problems. Reports on the high incidence of middle ear disease and accompanying hearing disorders among children on Indian reservations and in Alaska point up the need for increased rural health care.[4,5]

Some rural youth develop hearing defects as the result of noisy farm machinery. Others may have losses caused by music played at excessive levels, by working around jet airplanes, or from exposure to gun fire.

Associates of adults who were formerly normally hearing but who have had a gradual reduction in auditory functions may think the latter have lost their business acuity and social skills, or that they have become deliberately inattentive. In old age, severely hard of hearing adults are sometimes thought by others to be stubborn, "ornery," or suffering from psychiatric problems.

Speech Impairments

Speech impairments are many. They include articulation defects, stuttering and other rhythm disorders, and voice disorders. Cleft palate, cerebral palsied, and mentally retarded individuals often have communicative disorders that are composed of mixed articulation, voice, language, and hearing problems.

Alaryngeal Speech Programs. There are 5,000 new alaryngeal persons in the United States each year.[6] Appi oximately 40,000 alaryngeal persons live in the United States.[7] The impact of inadequate services for laryngectomees is often reflected in their inability to communicate with their families and friends and to achieve their vocational potential. Local speech pathology services for alaryngeal persons living in rural areas are nearly nonexistent.

Language Impairments

Aphasia and other similar conditions are often associated with strokes, head injuries, and prenatal and perinatal oxygen lack. Persons suffering from these types of disorders may understand what is said to them but may not be able to express themselves satisfactorily. They may have impaired ability to read, write, or use numbers. The high level of frustration of these persons and those around them usually increases even more, if the afflicted persons are also unable to understand what is being said to them.

Aphasia Programs. One of the communicative disorders in which long-term care is crucial is aphasia. In 1965, the Joint Council Subcommittee on Cerebrovascular Disease of NINDS and the National Heart Institute indicated that the Nation's population included at least 2,000,000 individuals who had survived strokes. No reliable data were available concerning how many of these individuals were aphasic, but it appeared that 20% of the stroke population had aphasic sequelae. If these numbers were added to the total for persons whose language disorders were due to tumor or to traumatic accidents, an estimate of 600,000 aphasic American adults would not be unrealistic. When this number is combined with the 1,500,000 children estimated to have a neurological involvement, the total of central communicative disorders would be well over 2,100,000.[8]

Specific Telephonic Systems

TEL-C. TEL-C is being used very effectively with veterans who have various types of speech, language, and hearing disorders. For sessions with the majority of veterans, only a telephone line and regular desk sets are needed. The system is installed and maintained by the local telephone company.

TEL-AUD. TEL-AUD is a procedure utilized in TEL-COMMUNICOLOGY that allows the audiologist to call persons who have been fitted with hearing aids in order to check on how they are doing, to find out if the aids are functioning satisfactorily, and to identify specific problems they may be encountering. Thirty-eight veterans out of 179 who were called in a recent survey at the Birmingham VA Hospital were found to be having some difficulty with their aids and were recalled for further counseling and hearing aid check-ups.

In view of the cost involved in buying and fitting hearing aids, it is critical that proper, professional skills be available to the user in order to assure optimal results. TEL-AUD offers that kind of delivery system for the user, regardless of distance from the treatment facility; it also provides opportunities to counsel family members concerning some of the problems encountered by hearing impaired persons.

STARR. A system for testing audiological responses remotely (STARR) has also been developed by the TEL-COMMUNICOLOGY Project. The Touch Tone® pad is used to control a specially designed audiometer that is placed next to the person being evaluated. This portable audiometer can be located in rural nursing homes or places central to children and adults needing services. The test can be administered by professionals from their offices in remote facilities. If the persons being tested need further evaluation, they are called into the facilities for traditional evaluation procedures.

TEL-PLUS. In order to treat children and adults with aphasia and other language problems, a device known as TEL-PLUS (telephonic programmed learning utilization system) was developed at the Birmingham Veterans Administration Hospital. TEL-PLUS turns the telephone into a teaching machine. When the person at one end pushes a button on his Touch Tone telephone, a light appears in the clinician's console. This indicates the person's answer to a question that was posed by the clinician.

FUTURE APPLICATIONS

Telephonic *outreach* delivery systems should be implemented in rural educational programs.

Preschool Programs. Children who are born with complex communicative disorders need early access to differential diagnosis as well as to continuity of treatment. Screening programs are essential if children with hearing deficits are to be identified. Programs of amplification, language intervention, and speech development must be initiated early in the children's lives. Language needs of hearing impaired, retarded, or aphasic children must be monitored. Parents of atypical children need guidance if they are to provide optimal language environments. Through TEL-COMMUNICOLOGY, many of the counseling services for parents and monitoring of young children in rural areas can be achieved on an *outreach* basis.

School Programs. Special skills are needed to meet the needs of school-age children with communicative disorders. In the past, most deaf, blind, and mentally retarded children were sent to state schools. Some parents moved to urban centers where day schools for atypical children were located.

The recent trend toward mainstreaming of atypical children has created a demand for services that are seldom available in rural areas. Few school systems are in a position to offer the variety of atypical programs needed by special students.

The possibility of providing telephonic education services should be considered, particularly in rural areas where trained personnel is usually not available nor can be funded adequately. TEL-COMMUNICOLOGY is an innovative approach to meeting the standards set forth in the new legislation aimed at providing comprehensive services to all handicapped children.

Adult Programs. Middle aged and older adults often have communicative disorders that require frequent and intense clinician/client relationships, long-term treatment or maintenance, and periodic check-ups.

The cost of delivering traditional health care is beyond the reach of most clients, their insurance companies, and their government agencies. TEL-COMMUNICOLOGY offers *outreach* programs that can supplement and enhance ambulatory care, hospital based home care, and independent living programs.

SUMMARY

Rights to Services

Atypical rural populations of children and adults have rights to services. If these rights are to be respected, innovative health care delivery systems, such as TEL-COMMUNICOLOGY, must be designed and evaluated. The traditional institution-centered services must become person-centered; specialists must be made available and accessible to children and adults who need special educational programs and treatment for communicative disorders.

The TEL-COMMUNICOLOGY Project of the Audiology-Speech Pathology Service at the Birmingham VA Hospital was funded by the Veterans Administration Exchange of Medical Information Program. Four years of experience demonstrated the effectiveness of telephonic systems for supportive treatment and interim evaluation of veterans with hearing, speech, and language disorders. Accessibility and availability of quality services characterized this innovative health care delivery system. Utilization of telephonic systems reduced expenditures of time, effort, and money, and eliminated travel barriers.

Future applications of TEL-COMMUNICOLOGY will include the delivery of services to rural areas, so that children and adults with hearing, speech, language, and educational problems will have quality programs, and thus, their rights to services will become a reality.

References

1. Data obtained in a telephone conversation with a representative of the American Speech and Hearing Association in May 1977.
2. President Ford's views on rural health. *Rural Health Communications.* University, Alabama: Clearinghouse for Rural Health Services, October, 1976.
3. Jimmy Carter on rural health, a personal essay by the Democratic Presidential Candidate. *Rural Health Communications.* University, Alabama: Clearinghouse for Rural Health Services, October, 1976.
4. Comprehensive planning: A clearinghouse. *ASHA* **11:** 557 (October, 1969).
5. Northern, J., ed. Ear disease and hearing loss in American Indians. *Hearing Instruments* **26** (April, 1975).
6. Information obtained in a telephone call to the Services and Rehabilitation Department of the American Cancer Society, August, 1977.
7. *The IAL News* **22,** 1 (June, 1977).
8. National Advisory Council of Neurological Diseases and Stroke: *Report of the Subcommittee on Human Communication and Its Disorders.* Bethesda, Md.: National Institutes of Health, 1969.
9. Personal data. Gwenyth R. Vaughn, Ph.D., Chief, Audiology-Speech Pathology Service, Veterans Administration Hospital, Birmingham, Alabama; Director, TEL-COMMUNICOLOGY Project; and Professor, Schools of Medicine and Dentistry, University of Alabama, Birmingham.

II
BIOSOCIAL DEVELOPMENT

A. Person and Patient
B. Rural Health Perspectives

A. Person and Patient

Abortion

Jules S. Terry, M.D.
Atlanta, Georgia

Abortion became a viable form of pregnancy control with the Supreme Court decision in 1970 which provided legal sanction for this procedure. A later Supreme Court decision declared abortion to be a matter to be resolved by the patient and her physician during the first trimester of pregnancy. Since that time, the number of abortions has increased steadily, so that in 1975 abortions represented one-fourth the number of live births in the state of Georgia.

While abortion should be available as one of the means of controlling fertility, it actually represents a failure in the utilization of conventional contraceptive measures. Abortion is preventative in that a living child does not result from this pregnancy. However, since abortion, or any surgical procedure, is not without risk of morbidity and mortality, basic effort should be directed toward the prevention of conception. In addition to the risks involved in abortion, the cost of the procedure itself far exceeds that for conventional oral contraceptives or the several mechanical barriers to conception.

The birth rate has been declining in all groups except in the under age 19 population. In that group, there has been both an absolute and relative increase in the number of conceptions. This is reflected in the increased births to teenagers and the increase in abortions in this same group. Much rhetoric and literature has been devoted to the needs of the teenager. However, these concerns for the teenager have produced very few service delivery systems designed for this particularly vulnerable group. Moreover, it is insufficient to provide special prenatal services to teenagers. The thrust should be to the young adolescent before he or she becomes sexually active.

To reach the adolescent, the approach must be directed to the education of the schools, the community and the providers of service. To provide a suitable background to permit individuals to change their behavior and values, youngsters and adolescents should be provided with basic facts on the reproductive process and its consequences to the adolescent. The school curriculum must not be limited to the basic physiology of pregnancy, but must detail the psychological, social, and economic results to the teenage parent. In addition, young people are in search of a suitable role model for parenting. Since this model is often not available in the home situation, skills in parenting, interpersonal relationships, and basic communications must be an integral part of every school curriculum. Similar learning experiences should be available in the community for those who are already parents.

A specific effort to change the attitudes of the providers of health services, is an essential component of creating the climate conducive to the acceptance of those changes. Too often, health care providers tend to be judgmental in their relationship with adolescents, particularly in the areas of sexuality and pregnancy.

Undoubtedly, some patients will always require abortion services, and these should be readily and inexpensively available. As long as the presently known contraceptive measures are not 100% effective all of the time, abortion should be available as a backup measure. However, our efforts should be directed to the present and future reproductive groups, so that they may have a clear understanding of the complexities of reproduction, and can therefore make objective decisions regarding sexuality, contraception, and abortion.

Accident Prevention

THOMAS W. MCKINLEY, BSA, MPH
CHARLES B. MOSHER, M.D.
Atlanta, Georgia

A preventive approach to the disease entity of accidents in the 1980's will entail changes in both lifestyle and environment, tailored specifically to reflect the *biology* of the population at risk.

An estimated 50,000,000 accidents annually in this nation result in 114,975 deaths (the fifth leading cause) and an unstudied, but presumably massive number of disabilities. Although motor vehicle accidents take the greatest toll, this paper will be limited to the home environment.

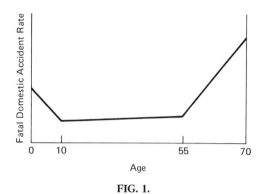

FIG. 1.

THE BIOLOGY

As illustrated in Fig. 1, fatal domestic accidents occur more frequently below age ten years and above age 55 years.[1]

Additionally, those afflicted with certain chronic illnesses are at higher risk of suffering the acute insult of an accident than the normal population.

THE LIFESTYLE

Accidents in the home may reflect attitudes and way of life as much as the environment. Epidemiological data on the relationship among attitude, lifestyle and accidents are currently inadequate. Since changes in habits have the greatest potential for decreasing morbidity and mortality, it is proposed that community health workers undertake studies of accidents within their communities by developing a surveillance and reporting system to gather not only basic identifying information relative to person, place, and time, but some lifestyle and attitudinal information as well.

The design and implementation of an effective epidemiological model for identifying circumstances, attitudes, and lifestyles associated with domestic accident hazards will present a creative challenge. Where evidence indicates that attitudes or lifestyle play a role in accident causation, a meaningful approach to future prevention might include some self-analysis of risk factors among recent accident victims when motivation for prevention is high.

Education and motivational efforts should be aimed at key groups who can, in turn, reach and motivate high-risk target populations identified through data analysis. Such groups include supervisors of small children (parents, teachers,

[1] *Theory and Practice of Public Health*, W. Hobson.

older siblings) and the elderly. Where hazardous behavior is found to be widespread, legislation plus education aimed at behavior modification would seem to be the best approach. Marketing techniques could be utilized in schools, at health facilities, on public transportation, and, of course, through that lowest common denominator of education, the family T.V.

THE ENVIRONMENT

Environmental manipulation offers the most direct approach to accident prevention in the home. It is anticipated that through the 1980's fewer American families will build and own homes due to economics. The environment of the growing majority will be condominiums or apartments. These could be constructed according to accident hazards for certain age groups:

(a) *Recommended for childless adults*—high accident risk; maximally decorative and functional with regard only to ease of access.

(b) *Recommended for young families*—low risk to children; minimal ornamentation, protected fireplace and heating devices, locks on all cabinets within three feet of the floor, no electric outlets within three feet of the floor, stove top recessed back from counter edge for difficult access, no sharp edges.

(c) *Low risk to elderly*—heavily padded wall-to-wall carpeting, railings throughout, no steps, no cabinets above five feet from the floor, automatic turn-off device on heating appliances, automatic maximal temperature mechanism in hot water system, telephone within reach of bathtub.

Mobile or prefabricated homes will also increase in the 1980's due to economics. Fire safety of these units will be a major concern.

Prevention of domestic accidents in the 1980's will present a challenge. Human biology, lifestyle, and environment must be considered.

Motor Vehicle Accidents and Emergency Medical Services

CHARLES B. MOSHER, M.D.
Atlanta, Georgia

The entire medical care system is built on the primordial human fear of death. People are willing to invest lots of money and absolute trust in physicians, hospitals and other providers of curative medical care. Never is that fear more acute than in the disease called trauma.

Yet, to keep a perspective, the gory motor vehicle accident deaths are less frequent than death from heart disease, cancer or stroke. Spinal cord damage from multiple sclerosis and cancer is 20 times and 100 times more frequent (respectively) than motor vehicle related spinal cord injuries. So why consider motor vehicle trauma a major community health problem? It is because:

1. most victims are *young*, making trauma a disease that lowers the average American life expectancy.
2. most victims are *men*, thus affecting the whole family's security.
3. all accidents are *preventable*.
4. trauma affects *quality of life*.

Following the military evacuation and treatment system developed through the Vietnam era, and the Maryland Trauma Center model, optimal EMS (Emergency Medical Service) system response to trauma is:

a. rapid access to the EMS (medical care) system, via 911, C.B. radio or dedicated phone number.
b. basic life support (BLS) provided to the victims by trained citizens or police—the first on the scene.
c. ambulance personnel (Energency Medical Technicians) stabilize the victim using BLS or advanced life support (ALS) techniques. I.V. (intravenous) therapy and spinal immobilization are provided, and *the patient has now entered the medical care system.*
d. the patient arrives at an *appropriate* facility, previously designated as having personnel and equipment capable of managing the patient's most severe injury.

e. skilled triage. Physicians and nurses specializing in emergency care diagnose and stabilize the trauma victim's problems using ALS techniques. Blood is administered, chest tubes inserted, central venous I.V. lines placed.
f. definitive care is rendered. Surgery as appropriate with subsequent intensive care is given.
g. rehabilitation.

RESPONSIBILITIES AND ORGANIZATIONS

Because of the tremendous demands in training and financial support for such an EMS system, total community committment to the goal is essential. The three major barriers to implementing an optimal EMS system will be:

1. *economic* turf guarding. "Private" providers of medical care don't want patients who can't pay.
2. *ego* turf guarding. Some physicians and hospital administrators refuse to recognize (at least publicly) their inability to handle certain types of cases.
3. *Public Health/Private Medicine* turf guarding.

Overcoming these three barriers will require skills in interpersonal relationships in addition to a broad knowledge of the technology and economics of EMS systems. Community resources must be identified with both the human and the technological aspects in mind.

ECONOMICS

There are two economic models available for providing a competent EMS system to a community.

(a) *Patient-Subsidized Model:* The philosophy is that everyone is entitled to the same, ultimate level of EMS care regardless of ability to pay or geographic location. EMS services are regionalized, with critically injured patients stabilized at the scene, then transported to the appropriate hospital. The immense costs involved are met by federal subsidy of the patient (National Health Insurance).

The advantage is optimal care for all. The disadvantage is economic insolvency of the design since costs are now extreme and would grow in quantum leaps as technology advances the "optimal" level of care.

(b) *Stratified Health Care System Model:* The philosophy is that Emergency Medical Care will be available to all regardless of ability to pay, but the level of care available will be determined by the local economy. EMS services are provided in three degrees of sophistication: Basic, Intermediate and Advanced. The choice of which level of EMS care is desired would be made by each person when he chose where to live, just as each of us chooses the level of police and fire protection and school system we wish when choosing where to live. One would elect his lifestyle (smog and ALS vs. clean air and BLS).

The advantage is economic solvency by placing no burden on the federal tax dollar for EMS health care. The disadvantage is a slightly higher mortality and disability rate for rural areas.

EMS is a community-based medical care delivery system addressing one of the diseases of greatest concern to Americans. Each community must plan and implement *its* EMS system in response to the needs and resources of the human beings which are that community.

Alcoholism

GEORGE P. DOMINICK, M. DIV.
Atlanta, Georgia

Alcoholism is considered today to be the fourth largest public health problem in the United States affecting approximately eight million people. It has been only since 1923 with the establishment of the Yale School of Alcohol Studies, and now the Rutgers School, that this disability has been and continues to be scientifically researched and reported on.

E. M. Jellinek in his classic book *The Disease Concept of Alcoholism*, reintroduced the illness concept and the debate, pro and con, has literally been carried on in each issue of the *Studies Journal of Alcohol* for years.

The American Medical Association recognizes alcoholism as an illness and not as a symptom of an illness. Though alcohol is a necessary element in the cause of the illness, alcohol is not considered today to be a sufficient cause of the illness. There is no indication at present that there is a single cause, but there are indications arising in the literature that combinations of biological factors in different degrees, combined with internal and external stress, set the stage for a life style to be established in which a person skillfully learns the habit of self-medication. This habit practiced over a long period of time with alcohol which has the capacity to build tolerance, taken in sufficient quantity, will develop the illness—Alcoholism.

So, today the *Criteria for the Diagnosis of Alcoholism*, published in 1972, by the Medical Committee of the National Council on Alcoholism, clearly spells out the major and minor criteria involved in making this diagnosis. When the abuse of alcohol progresses to the advanced condition of alcoholism, persons are severely affected in their total functioning.

The emotional dimension of alcoholism is seen in the long established, skillfully learned habit of self-medication. A life style is developed by which the individual has one answer for every condition he or she encounters—that is, alcohol—a drink. If it's hot, it cools you off, if it's cold, it warms you up; if you are "down", it lifts you up; if you are high, it relaxes you and calms you down. The core of one's feeling life centers around the presence or absence of the availability of alcohol.

The social dimension of alcoholism is recognized by the change in how one feels and thinks about self. When alcoholism is developed, significant other people and as well as the victim see it as a "drinking problem"—rather than a person who has problems and feelings as do other people. There is a false identity as a "drinking problem", that replaces the former felt personhood. The person is referred to as "just a drunk". This conveys a distorted view of self.

The spiritual dimension of alcoholism is seen in a change in the values by which a person lives. That which is worth living for is that which can be taken or experienced now—immediately. Goals to be obtained by working a little longer continue to decrease in importance. "I want, what I want, when I want it" becomes the measuring stick for achievement. This essentially is a breakdown or erosion of a value system and is part of the spiritual change which takes place.

FUNDAMENTAL PRINCIPLES OF TREATMENT

This holistic concept means that treatment must be geared to intervene at several levels, over long periods of time, with specific treatment modalities.

The principles of such intervention currently carried out in many different ways are as follows:

1. Alcoholism treatment requires physical and emotional stabilization, social and spiritual reorientation.
2. Alcoholism treatment requires the use of many disciplines working together. Treatment situations should bring together psychiatrists, medical internists, psychologists, nurses, social workers, clergymen, alcoholism counselors, vocational rehabilitation counselors, judges, and probation officers, etc. Multidisciplinary intergration works in Georgia in a model in which all staff and patients of a given center are divided into a number of teams. The staff members have added roles as treatment coordinators

working as enablers and managers of the patient's entire treatment plan. Decision making authority for what happens to the patient is focused in the team members.

3. Alcoholism treatment requires the recognition that the illness is chronic, yet has acute phases. It is present even when the illness is in remission.

4. Alcoholism treatment produces results through changing habits of self-medication.

5. Alcoholism treatment needs to involve significant persons around the alcoholic.

6. Alcoholism treatment requires abstinence from alcohol, recent research notwithstanding.

7. Alcoholism treatment requires the appropriate balance of love, justice, and compassion (Kellerman, J. L., 1975).

8. Alcoholism treatment is most effective where peer identity and program identity are accomplished. Close identification with others seeking the same goals through a program that all are working on together have been singularly most successful.

An assessment of where we are in our expanding understanding of the illness and in new treatment approaches is always valuable. First, research in the biochemical field is indicating that alcohol abuse alters the metabolism patterns of enzymes and that there may be inherited differences in metabolism patterns. Second, where once we spoke of alcoholism as a singularly linear progression with slips and recoveries, we might more accurately speak of several different kinds of alcoholism. Third, success in treatment is greatly increased through early identification, early treatment and follow up. Fourth, much remains to be learned about treatment of certain age, sociocultural and ethnic groups of alcoholics, and it is indeed in the thorough study of these groups that we may learn answers to other problems in this field. Much remains to be studied and learned as well from those who have, and remain, successfully recovered by their having arrested the illness.

Anxiety

FRANK M. BERGER, M.D., D.SC.
New York City, N.Y.

Anxiety can be defined as an unpleasant feeling of apprehension or dread which occurs in the absence of a recognizable threat, or is brought about by a threat that is, by reasonable standards, quite out of proportion to the emotion it evokes. Anxiety must be differentiated from fear, which is an emotional response to a clearly recognizable and definite danger that threatens the individual.

Freud considered anxiety as the central problem of the neurotic state. He believed that anxiety is evoked when a person perceives his or her unconscious wishes and impulses as the source of danger. Anxiety has also been considered as the original reaction to helplessness that is subsequently reproduced in situations of danger. The behaviorist school of psychology considers anxiety a learned, conditioned response to symbols representing painful experiences. In the past, anxiety was considered a motivational force because individuals tend to avoid situations causing anxiety. Modern research concerning the nature of anxiety based on factorial analysis of specific, measurable manifestations, found that there is a single, general reaction pattern of behavioral responses characteristic of the anxiety state. Among these are a lack of confidence, a sense of guilt, an unwillingness to venture, a dependency, a readiness to become fatigued, irritable and discouraged, suspicion of others and general tenseness. There is only one kind of anxiety which is qualitatively the same in all subjects suffering from it. The result of these studies also indicates that anxiety is not a specific drive or motivational force, but is a symptom of a disease, such as neurosis, depression, or schizophrenia.

In anxiety states the interneuronal circuits in the brain are in a state of hyperexcitability. The interneurons and the circuits they form are interposed between the sensory input receptors and the motor output terminals of the central nervous system. Drugs that decrease conductivity in the interneuronal circuits are of value in the control of anxiety and tension states. Drugs cannot be expected to resolve conflicts or ease the problems of living that may be the basis of the psychoneurotic state. By temporarily alleviating the disruptive influence of anxiety, the drug may make the subject more accessible to psychotherapy and enable him to face his problems.

Interdisciplinary Approach to Child Abuse

MICHAEL KOCH, M.D.
Saint Paul, Minnesota

Since early reports about the "battered child syndrome" many Americans have been concerned about the problems of abused and severely neglected children who, for various reasons, are receiving less than adequate care and protection. The most severe form of child abuse is seen in the "battered child syndrome". Concern for children suffering from inappropriate rearing practices has brought us beyond this condition to different and less lethal forms of abuse including "failure to thrive" due to inadequate care and nutrition, children with repeated minor injuries inflicted by adults, and cases of sexual abuse.[1] Currently, all 50 states have enacted child abuse reporting laws which require physicians and frequently, nurses, teachers, social workers and law enforcement personnel to report suspected child abuse.

The number of annual child abuse case reports in the United States has increased from about 7000 in 1967 to over 200,000 in 1974.[2] When prevalence is considered, involving cases already being followed in treatment or agency supervision, the estimates are much higher and indicate the major burden this places on departments of social service in the area states. An accumulating literature on different aspects of child abuse suggests changes in professional behavior and agency policies regarding the management of abusive families.[3,4] It is generally accepted that comprehensive, coordinated services are needed for evaluation and treatment of child abuse and neglect. It is not possible to effectively deal with multidisciplinary problems with only a single discipline service unit, regardless of whether the service unit be in medicine, social services, juvenile courts, or the educational system. As a result, the interdisciplinary child abuse team approach has evolved in many childrens hospitals and child protection agencies.

Psychological, social and organic factors play a role in etiology;[5,6] consequently, treatment by professionals from a variety of disciplines is necessary to prevent future abuse. Barnes and Starbuck[7,8] have discussed the advantages of a collaborative team approach. Starbuck found that confirmed reabuse in team-handled cases recurred less frequently than in nonteamed cases. Financial costs of the collaborative team approach is a little greater than without a team and this approach has also resulted in an improved climate of cooperation and communication among the medical and social work professionals in the community.

Help for abused and neglected children and their families may be required in several levels or phases. One is related to the acute problem itself, for example a burn or fracture, whiplash injury to shaking an infant, of sexual molestation. Medical/surgical treatment, usually in a hospital setting, involves relatively standard procedures. However, the interpersonal relationship between patient and staff established at the time of entry to hospital is critical to the family's receptiveness to medical treatment and to ongoing services. A necessity for coordination is obvious and a pediatrician or family practitioner usually assumes this role.

Another phase of aid is that related to early recognition or pre-abuse intervention. Again, the health professional represents a crucial focal point in the core program. Obstetricians, pediatricians, and nursery personnel can detect abnormalities in the early mother-child attachment process. Other physicians and professionals in the community, including social workers, teachers, and visiting nurses may also be concerned about potential abuse in an adult, parent to be, or even a babysitter.

Coordination of professional assistance is also important in the third dimension of this problem, that aspect of management which is concerned with protecting children from repeated physical consequences of family crisis. Child abuse usually represents a family problem requiring family and frequently, community action, for satisfactory long-range solution. Remedial efforts frequently involve more than one professional group, and physicians can be most helpful if they are able to work with other disciplines. They may be expected to document the injury or neglect which has occurred, indicate possible or probable cause of injury, and describe the psychological profile of the child, parent or suspected abuser. Other difficult questions are: Will abuse recur? Can the family be helped? If so, by whom? Usually the physician alone is not qualified to provide these answers and he or she best functions in collaboration with other professionals. Despite the apparent advantage of this approach, one of the major obstacles to effective intervention is that, while a multidiscipline approach is needed, professionals typically find it difficult to work together across agency and discipline lines. Each profession has its own language, goals, and set of role expectations; these can become barriers to the understanding and trust necessary for a good working relationship. Helfer discusses barriers to physician involvement in child abuse and neglect.[9] Physicians have not been given appropriate training about child abuse and are frequently isolated from other health disciplines. Work in the area of a child abuse may be very time consuming and with minimal personal rewards. Physicians may be also reluctant to become involved in the adversary legal process of the court room.

Because of the importance of both medical and social factors in child abuse and neglect it is important for physicians to become involved in the problem. They can refer individual cases and cooperate with community services in the evaluation of cases referred by others. Their cooperation in the child abuse team process is necessary for evaluation and remediation of the abuse or neglect situation.

References

1. Kempe, C. *et. al.* The battered child syndrome. *JAMA* **181**:17–24, (July, 1972).
2. Newberger, Eli H. and Daniel, Jessica H. Knowledge and epidemiology of child abuse: A critical review of concepts. *Pediatric Annals*, **5** (3): 15–21 (March, 1976).
3. Kempe, C. H. and Helfer, R. E. (eds). Helping the Battered Child and His Family. Philadelphia: J. B. Lippincott, 1972.
4. Gil, D. *Violence Against Children, Physical Child Abuse in the United States.* Cambridge, Mass., 1970.
5. Gelles, R. Child abuse as psychopathology: A sociological critique and reformulation. *Am. J. of Orthopsychiatry* **43**(4):611–621 (July 1973).
6. Lystad, M. H. Violence at home: A review of the literature. *Am. J. Orthopsychiatry* **45**(3) 328–341 (April, 1975).
7. Barnes, G. B., Chabon, R. S., and Hertzberg, L. J. *Team Treatment for Abusive Families. Social Casework.* pps. 600–611, December, 1974.
8. Starbuck, G. W. Collaborative Team to Non-Accidental Injury and Neglected Children. Department of Pediatrics, University of Hawaii School of Medicine. Unpublished paper.
9. Helfer, R. Why most physicians don't get involved in child abuse cases and what to do about it. *Children Today* **4**:28 (1975).

Child Health Assessment

MARGARET PARK, B.S.
FRANCES R. HANKS, M.S.
Atlanta, Georgia

The development of a comprehensive child health assessment program has broad implications for community health because it will provide a "health profile" of the child population and, through appropriate management, can result in allocating resources to high risk areas as a means of changing disease patterns. Since many of today's major health problems begin at an early age and influence adulthood, programs that begin early in life which promote changes in lifestyle, address the hazards of the environment, and modify the health care systems will have the greatest impact.

The complete child health appraisal includes a health and developmental history and a physical, nutritional and developmental assessment with appropriate health education and counseling, laboratory tests and immunizations.

The health and developmental history should include general identifying information, history of family illness and personal illness, and maternal and neonatal history, as well as other information appropriate for age.

The physical appraisal should include assessment of anthropometric measures (height, weight, and head circumference) and an unclothed head to toe physical examination, including a dental inspection.

A standardized developmental screening tool, such as the Denver Developmental Screening Test, should be incorporated to indicate deviations in psychological, neurological, emotional, or speech and language development.

The nutritional assessment is an essential component deriving from an interpretation of data from the physical examination, laboratory data, and dietary evaluation. The level of nutrition cannot be assessed with dietary intake alone. Anemia, failure to thrive, hypertension, obesity and dental caries are indicators of poor nutritional habits.

Routine laboratory tests appropriate for age and as indicated by history and physical assessment are a necessary part of a comprehensive program.

Health education and counseling is needed to support good health practices and modify health patterns which adversely affect the health of the child.

It is implicit that some components would change with age and that the assessment would be performed periodically to maintain surveillance and identify developing problems.

The child health assessment can be provided by professional groups other than the physician. A primary care team would include nurse practioner, generalized nurse, nutritionist, health educator, and physical therapist. Georgia has designed and implemented this model for providing services to the medically underserved areas of the state through public health primary care centers. Essential in a model using nurses as the primary provider of child assessment is the medical backup to monitor the quality and skills of service delivery. The plan must also include the concept of continuing care and a referral system for detectable abnormalities and other services not available from the primary care center.

One goal of any assessment program is to identify problems which can be corrected or the effect diminished or maintained. Therefore, it is imperative that any assessment program have a data-base which will produce a "health profile" and will identify the incidence of abnormalities by age, sex, ethnic origin, and geography. It is this identifying information which provides the tools for planning and resource allocation (reallocation) to first diagnose and treat present problems and secondly, assist health workers to mount a program of health education and prevention.

Evaluation of such a program should be focused in two major areas: outcome evaluation (to determine the effectiveness of the program in relation to childhood disease patterns) and process evaluation (to determine the integrity of the assessment process, its value and its cost effectiveness).

The net result of both types of evaluation and further fiscal evaluation would give input into cost effectiveness of the program. The greatest value of a well-designed child health assessment program, however, is the increased human potential which results in a higher quality of life.

In order to preserve the outcome of healthier children, communities must recognize and promote an accessible health delivery system which offers health education and prevention, quality primary care, and diagnostic, referral, and treatment services.

Community Health in the Eighties

JAMES W. ALLEY, M.D.
THOMAS F. GIBSON, M.P.H.
Atlanta, Georgia

It has been said that true caring is one beggar telling another beggar where to find bread. There is enough bread in our affluent country today for all, but it is poorly distributed, and controlled by a comparative few.

The diseases of today are far more epidemic than those of yesterday. Our medical technology is unsurpassed, but the infant death rate in this country is 17th among the developed countries of the world. In Georgia, with a population equal to that of Denmark, the homicide rate is over 20 per 100,000, while in Denmark it is less than one. Our number one killer, responsible for one-third of all deaths last year, and the major killer of all males over 35, is heart disease. We boast of more professionally trained personnel than any developed country in the world, but quality health care is still unavailable, unaffordable, or inaccessible to vast numbers of our people.

How is it that this country with 6% of the world's population, 45% of the world's wealth, and which utilizes nearly three-quarters of the world's natural resources cannot better direct its knowledge, skills, and technology to solving these epidemics? This unenviable record did not always exist. At the turn of this century, nutritional deficiencies, parasitic infestations, and childhood infectious diseases made up the disease patterns. We were a rural society with a strong family focus. Most families were large, because a rural society needed workers. The high fertility rate was necessary because the childhood diseases took their toll, and the cycle was repeated.

The development of county health departments in this country was a phenomenon developed specifically to combat the existing disease pattern of the day. It was imminently successful. The environment was cleared up, and pit privies and water chlorination marked the beginning of the end for parasitic infestations. Immunizations were developed against the major killing childhood infectious diseases. Nutrition education and resulting diet changes for mothers and babies conquered the major nutritional diseases of the day. And so a major disease pattern has been all but conquered.

As all of this was happening, our life spans were increasing, and our society was becoming urbanized. The extended family structure was changing, and people fell prey to an entirely new complex of diseases, comprising a totally new pattern.

Heart disease, cancer, stroke, accidents, sexually transmitted diseases, maternal and infant deaths, alcoholism and drug abuse, nutritional abuses, and dental diseases are now the major killing and crippling diseases making up today's disease pattern. And we know that the epidemiological model which was so successful in overcoming the disease pattern of yesterday is no longer valid. There is no immunization against heart attacks, cancer, accidents, and maternal and infant deaths. The nutritional abuses of today in no way resemble the nutritional deficiencies of yesterday. Active rural lifestyles have been largely exhanged for sedentary urban living. Excesses of alcohol, drugs, tobacco, and food now characterize many of our affluent lives, and low income still inflicts its deprivations on too many of our peoples' lives.

Political decisions in the form of national legislation have shaped our present health care system, with the emphasis increasingly on curative medicine, (and a corresponding decrease on the principles of prevention), and the responsibility of the individual for his own health. Billions continue to be spent on research and technology, on facilities, on health planning, and on medicare and medicaid, and the disease patterns remain relatively unchanged. Quality health care is still unavailable, unaffordable, and inaccessible to unacceptably high numbers of Americans.

Human resources are too precious to be wasted. None of the diseases making up today's complex of killers and cripplers have specific etiologies which lend themselves to the approach of the classic epidemiologist. It must be attacked using our available human and other resources. Today the physician, nurse,

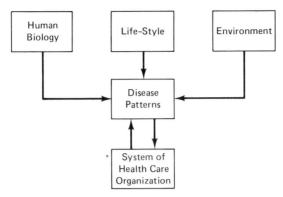

FIG. 1. A model showing the relationship of Georgia's New Health Outlook to the disease patterns of the 1980s. (Adapted from G. B. Hill and J. M. Romedes, "Health Statistics in Canada," paper presented at the Society for Epidermiological Research, Toronto, Ontario, June 18, 1976, p. 6.)

educator, social worker, administrator, and statistician must share their knowledge, and each support the efforts of the other. We must develop interdependence, and this can be best accomplished through an organizational structure which brings these skills together at the level of service delivery and promotes coordination of services to the individual. This permits a better opportunity to study today's disease patterns through the eyes of several professionals and to reallocate resources where necessary.

Georgia's New Health Outlook attempts to evaluate the disease patterns of today by putting all of the foregoing together, and to give appropriate emphasis to four basic components: lifestyles, biology, the environment, and the present health care system. It is not a finished model, but it recognizes that the old model is no longer satisfactory; it says that caring is important, and that a big part of our personal and professional responsibilities is in the more equitable distribution of the bread.

Community Mental Health

JAMES DODD, Ph.D.
Atlanta, Georgia

Four potent forces have emerged and coalesced within the last two decades to change dramatically the complexion of mental health service delivery. The first of these, and undoubtedly the most significant, is the development of chemotherapy. The second is the entrance of the Civil Rights Movement into the mental health arena. As early as 1960, Dr. Martin Birnbaum in the American Bar Association Journal articulated the legal right of a "mentally ill inmate" in a public institution to adequate medical treatment for his mental illness. In 1972, in "Wyatt vs. Stickney," the federal courts established the landmark principle of the right to treatment in the least restrictive environment. The third force is the mounting research that demonstrates that traditional psychotheraputic approaches to treating the mentally ill are only "modestly effective" with the neurotic population and ineffective, even detrimental, with the severely disturbed or chronic schizophrenic population. The fourth is the development of federal community mental health legislation, first passed in 1963, which established categorical federal funding for the creation of community-based comprehensive mental health service systems.

The meshing of these forces propelled mental health services delivery towards community-oriented care systems and away from large state mental institutions. "De-institutionalization," "community treatment alternatives," "out-patient services," "crisis intervention," "alternative living," etc., became the new language to replace the old language of hospital-oriented treatment. States rushed to open the "back-doors" of mental institutions and release patients to community treatment centers. Concomitantly, institutions were pressed to guard carefully their "front doors" and keep people out. State hospitals were soon faced with empty beds, onslaughts against their budgets, and lost prestige. No more "back wards," no more "snake pits," no more "cuckoo's nests," became the dream.

Such movements notwithstanding, some old problems remain and some new problems have emerged in mental health care. Community mental health centers and state mental institutions are too often placed in adversarial positions, each fighting for their existence and each defending their virtue as health care systems. Few mental health professionals have the expertise to operate effective and efficient comprehensive community treatment programs. Thousands of emotionally unstable people have been released to inadequate or nonexistent community programs. With little or no services available to them, they have been unwanted, feared, exploited, uncared for, and thrust into marginal deplorable living conditions. Back wards have often been transferred from an institutional to a community locale. Stable, sensible, and adequate funding sources often have not been developed for community programs. The same is true of accountability systems. Because of these problems, mental health professionals have been forced to look beyond the rhetoric of the last few years that has vilified institutional care and idealized community care to the broader and more basic issues of what constitutes effective, efficient, and adequate mental health care delivery.

Georgia is an example of a state in which the managers of its mental health system over the past five years have attempted to deal with the real and complex issues and problems briefly discussed above. There has emerged an approach to mental health care delivery which seeks to avoid the bind of "community" vs. "institution" and attempts to link community and institutional programs into comprehensive, coordinated, and balanced systems. Ten MH-MR institutions, ten health districts, and 34 comprehensive community mental health area centers have been organized into eight regional planning and management systems or "consortia." Their tasks are to assess the needs of their region; plan for comprehensive and balanced services to meet those needs; coordinate the management of the mental health care system so that there is continuity of care; and, advocate for adequate resources. The eight consortia work towards a basic service goal: to help individuals served by the system to live as self-sufficiently in society as possible. The guiding principles of planning and management to be followed in delivering this comprehensive system of care are:

1. There should be a continuum of service functions geared to client treatment needs from stabilization to growth.
2. There should be a continuum of treatment facilities from inpatient to alternative community living so that clients can be served in protective, supportive, or natural environments as need dictates.
3. The least intervention possible to meet client need should be the treatment of choice.
4. The least restrictive environment possible to meet client need should be the treatment locale of choice.
5. Resources should be continually assessed and strategically deployed to the components of the system in greatest stress.
6. Generalist and specialist staff roles should be utilized.
7. All disability and age groups should be served.
8. Hospital and community managers of the system should plan and manage in consort so as to insure continuity of care throughout the system.

As mental health systems throughout the country struggle, as Georgia is doing, with the issues and problems of mental health care delivery systems that are efficient, effective, and adequate, a new language to explain the care system is emerging. "Balanced service delivery," "crisis stabilization," "sustenance," "growth," "prevention," "protective, supportive, natural environment," "competency training," "consumer of service," "citizen involvement," etc., are components of the new terminology of mental health service, replacing the old terminology of disability and age groups and facilities that have inaccurately been used to describe services.

Emerging patterns of mental health care and related issues and problems determine the tasks ahead for mental health professionals. The major tasks appear to be:

1. To continue to develop methods to help mentally disturbed people function as independently as possible in as natural an environment as possible.
2. To advocate for sane systems of funding that will support comprehensive mental health service systems and allow for the maximum flexibility in deploying resources.
3. To develop information systems that will help mental health system's managers manage more effectively and efficiently.
4. To develop accountability procedures for mental health systems based on desired outcomes relating to client needs and not on arbitrary and meaningless standards imposed by those far removed from the service delivery system.

Hopefully, the stamina, intelligence, dedication, and determination to accomplish these complex and important tasks will be present in the mental health systems of this country.

Drug Abuse

BETTY JAMES
Atlanta, Georgia

The most serious problems resulting from drug abuse in American society are waste of individual abilities, energy, time, and money. The indirect as well as direct glorification of mind-altering drugs by the press, advertising media, police and parents, the ready availability of many of these drugs, the hedonistic tendencies in our society, and the widespread alienation of young and old have brought us to our present predicament. Millions of Americans are unable to find meaning or purpose in life, to be happy, or to relate to other human beings without using drugs.

All drugs are not equally dangerous, destructive or costly. The magnitude of the problem is related to the frequency and quantity of use, and reduction is dependent upon limiting availability of drugs, providing treatment and rehabilitation to compulsive users, and developing effective prevention techniques.

We have accumulated a vast pharmacopoeia ranging from deadly poisons to life-saving, bacteria-destroying antibiotics. We have mind-influencing drugs, such as anesthetics, pain relievers, tranquilizers and psychic energizers. We have alcohol, caffeine, nicotine and other depressants and stimulants. Drugs to regulate mood and emotion can alleviate mental illness.

Today, the most obvious "drug problem" stems from the sheer multiplicity of drugs. Thus, drugs that are innocuous or were constructive when employed on a small scale suddenly assume frightening aspects when employed on a mass scale. Legal drugs may cause the greatest physical problems, and discretion is indicated in their use. There is a need for tighter controls on pharmaceuticals, especially barbiturates and amphetamines.

For the many Americans who are deeply troubled by the phenomenon of drug abuse, the natural first question centers on the extent of the problem. Surprisingly, for all the attention given to the "drug dilemma" over the past several years, this question is not easy to answer with any precision. Statistics on the scope of drug abuse and addiction are estimates based on a large variety of sources, including reports from law enforcement agencies, surveys, questionnaires and hospital studies. Obviously, these sources vary in reliability. Whatever figures they produce must be regarded as approximations of reality, and must be used with caution.

However uncertain the totals, current statistics do establish two clear trends. The first is that abuse of drugs in the United States is increasing, and the second is that drug use is increasing in progressively younger age groups. Not long ago, problems of drug use and addiction were almost wholly associated with big cities and with poverty. Today, such problems can be found in all economic groups and areas all across the country.

Drug abuse is believed to be a symptom, an index of the confusion and uncertainties which affect increasing numbers of people. The chronic social change of contemporary society is a deep source of stress to Americans. It splits generations apart, for if the experience of the past cannot be applied to the tasks of the future, then the older generation has nothing "relevant" to teach the younger. Chronic, rapid change tends to create the cult of the present which takes many forms, including the quest for kicks, speed, sex, and stimulants. The greater the involvement with drugs, the less the involvement with people, and the greater the risk that communities will not be able to survive as healthy institutions.

From the earliest systematic investigations to the present, researchers have regularly noted the importance of social rather than pyschological factors in the etiology and continuance of illicit drug use. For a practice of drug use to continue, being illegal and at times strongly suppressed, a system of supply, secrecy and recruitment is required. All of these requisites are fulfilled through the operation of social organizational processes. These data clearly indicate that social situation, as defined in terms of types of groups, is a significant factor in conditioning drug experimentation.

In the past, total abstinence from drugs was regarded as the only criterion of success in treatment. Today, however, reduced drug use along with improved social functioning is regarded as a degree of success. The fact remains that no treatment method yet developed has solved, or promises to solve, all the complex problems involved in drug abuse. There is still confusion and controversy about the nature of drug dependence and how society should deal with it. It is clear, however, that a variety of methods and approaches must be available to help the various types of drug abusers that exist today.

Society is ambivalent about the efficacy of primary preventive programs. A major source of resistance is to the idea that real prevention must make major social system changes. A second force operating to resist efforts at primary prevention is our societal commitment of privacy and the personal freedom of choice.

The major methods of drug abuse treatment in use at the present time are:

Methadone maintenance has been in recent years the most widely used method for treating opiate-dependent persons. In addition to maintenance, methadone programs also provide outpatient detoxification.

Therapeutic communities are residential treatment programs which attempt to deal with the psychological causes of addiction by changing the addict's character and personality.

Drug-free outpatient treatment programs differ but may include some or all of the following services: group or individual psychotherapy, vocational and social counseling, family counseling, education, and community outreach. This modality may be more effective with youth who are experimenting with drugs than it is with hard-core addicts.

Emergency treatment, usually called crisis intervention, is sometimes required for acute adverse reactions resulting from non-opiate drug use.

Free clinics provide a variety of general medical and social services in addition to treatment of drug abuse emergencies.

Potential treatment methods—Some researchers are working with behavioral techniques, such as aversive therapy or negative conditioning. Others are using biofeedback techniques to attempt to train people to control internal states and body processes. Much attention is currently directed toward developing alternatives to drug abuse, which may include meaningful activity or pursuits in which young people can become involved instead of resorting to drugs.

The major themes of federal strategy for community-based drug abuse treatment are:

- continued support for heroin treatment programs,
- expanded treatment emphasis on polydrug abuse,
- outreach through the criminal justice system to bring into treatment those drug abusers who would not seek it of their own volition,
- new rehabilitation and vocational training efforts,
- nationwide school-based early intervention programs,
- examination in terms of stress, the damage that drug addiction can do to American families.

References

1. Camps, Francis E. Drugs of dependence: A social and pharmacological problem. *Int. J. Addictions*, 5(1): 131-153 (March, 1970).
2. Einstein, Stanley and Garitano, Warren. Treating the drug abuser: Problems, factors and alternatives. *Int. J. Addictions*, 7(2): 321-331 (1972).
3. Fort, Joel. Social problems of drug use and drug policies. *Int. J. Addictions*, 5(2): 321-333, (June, 1970).
4. Kessler, Marc and Albee, George. Primary prevention. *Ann. Rev. Psychology*, 243-256 (1975).
5. MacLeod, Anne. *Growing Up in America—a Background to Contemporary Drug Abuse*. National Institute on Drug Abuse, DHEW Publication No. (ADM) pp. 75-106.
6. National Clearinghouse for Drug Abuse Information: *Treatment of Drug Abuse:* An Overview, Report Series 34, No. 1, April, 1975.

Decentralization of Resources

THOMAS F. GIBSON, M.P.H.
RICHARD CUNNINGHAM, B.S.
Atlanta, Georgia

Most organizations today are moving toward a greater centralization of resources as one means of coping with a super-industralized society. With the development of almost instant communications, complex computer and data networks, and greater mobility, we tend to think that a centralized organization is more responsive to such rapid change.

In developing a health system in this country we find numerous programs which have a national or state focus, as opposed to a local one. The error in this kind of centralization of resources is that it does not allow for unit break downs, since many systems are interdependent in the distribution of an organizations' resources and can ill afford to operate independently. Each aspect of the organization has developed a function that is inclusive of only its particular mission and usually does not understand the mission of the other units within the organization. Usually this will result in a crisis, because there is conflict over who is setting the mission and purpose of the organization. It will result in an organization that is spending much of its energy in resolving conflict rather than implementation of its goals.

One alternative to this situation is to decentralize resources.

In reviewing "Georgia's New Health Outlook," we discover an epidemiological model that considers the health delivery system as an important part of solving the disease patterns of a super-industrialized society. In order to create a service delivery system that will be responsive to the problems of a local community, we must bring together three important components of decision-making—authority, responsibility and accountability.

The authority to make decisions is usually the controlling point in the distribution of resources. It is authority to be able to say that today we face a health problem relating to expectant mothers at high risk and that we should direct resources into preventing such a problem even if it means diverting resources from established programs. But in centralized operations we find that *priorities* cannot be changed because the authority does not exist at the program level involved. Decentralization means we are willing to give away to the local community this power in decision-making. We must be willing to develop information systems that will allow local communities to monitor changes in disease patterns in order to facilitate effective decisions. Budgetary controls should be at the local level to allow adjustments in manpower, expenditures of money, and the acquisition of facilities if we expect local communities to effectively address their changing disease patterns.

A key element in decentralization of resources is in allowing local communities the right of being responsible for their own health. Each community will not decide to address the same issues. A centralized organization will usually predetermine by allocation of resources what activities the total organization will be responsible for and thus stifle community responsibility. This type of centralized predetermined responsibility for health will not permit very much initiative, creativity and commitment on the part of those responsible for service delivery.

By refusing to vest authority and responsibility at the service delivery level, we are also refusing the local community the right to be accountable. It is a mark of organizational maturity to be willing to share with others the right to be accountable for decisions. It is too easy for the centralized organization to point the finger at some program unit down the line and say, "You made a mistake." It is even more common for this organization to say, in effect, to a successful program unit, "Your success is largely the result of our efficient, centralized operation."

A final important aspect of accountability is in allowing a community to review and profit from its decisions. A renewed interest and energy in shaping their own destiny may well be the result within the community.

We must review and modify any centralized health system which does not encourage citizen involvement and decentralization of authority, responsibility, and accountability. It is time to face the reality of centralized resources as Alvin Toffler did in *Eco-Spasm*, when he said, "It has been argued that the multinational corporation is itself a product of the dying industrial commitment to linear growth and that we may soon move toward new kinds of production based on greater decentralization, smaller production units, and low-energy technologies." We can no longer expect the resolution of health problems to come out of a centralized network of health systems. We must build local resources which are capable of responding to problems as they develop. Unless we are willing to take the risk of decentralizing resources we cannot expect to make much impact on the killers and cripplers of our super-industrialized society.

Dental Health

JEANNETT LEVERETT, R.D.H.
Atlanta, Georgia

We are in a new era of health care. This era should be characterized by a deep awareness on the part of many as to why there are differences in health status among people. For a long time this interest has been concentrated almost entirely on the system for delivery of health care. Regarding dental diseases, the greatest factor in controlling them is prevention. Among the factors involved with prevention are: (1) fluoride in water systems, in mouthrinses, and other topical applications, (2) preventive educational programs and (3) availability and utilization of the existing clinical services.

The use of fluoride in preventive dental care has shown that when the fluoride content of the water system is at the optimum level (0.7 ppm to 1.2 ppm, based on average daily temperature), there is a marked resistance to dental caries. Fluoride is a nutrient essential to the formation of sound teeth and bones, and can be made available to the tooth structure by two general means: systemically by way of the body's circulation during the developmental period of the tooth enamel, and topically applied directly to the enamel surface of an erupted tooth. Also, a fluoride rinse program is an added measure in the reduction of tooth decay. The significance of fluoride in the clinical practice of dentistry and dental hygiene is that it will lessen the incidence of dental caries and tooth extractions.

From our experiences and those of others, Georgia is looking at prevention of dental disease as it relates to people's attitudes toward their individual oral cavities. The ADA Council on Dental Health has stated that dental health education is widely advocated by dental professionals as an effective measure for the prevention and control of dental diseases. It is considered an essential aspect of primary prevention. Public information and dental health education, urging improved personal oral hygiene practice, dental visits (public or private), and lower sugar intake have been carried out over a long period of time. It is more effective to plan dental health educational programs concurrent with water fluoridation and topical application programs along with routine detection and treatment of the existing diseases. Secondly, there should be seminars and workshops for decision makers, consumer groups, and dental professionals regarding the effectiveness and practicality of available preventive procedures at the primary, secondary, and tertiary levels of prevention. Their major purposes should be: (1) provide accurate information about various dental diseases, and prevention and control measures available, and (2) stimulate group decision making, placing emphasis on getting the local news media, schools, and families involved.

Realizing the great need for dental care for people across the nation, the availability and utilization of dental services is crucial. Having dental services available does not always mean the services are being used or that the services have quality. The maldistribution of dentists and the under-utilization of the dental hygienist's skills have created a large gap in the availability of dental services. One of the solutions to this problem is enlarging the duties of the dental hygienist and other auxiliary groups. Also, laws related to the practice of dental hygienists need review and modification. A well trained dental hygienist does not need to work under the direct personal supervision of a dentist. Many duties could be performed under general supervision, within an existing protocol.

Many factors are involved regarding utilization of dental services, such as cost, transportation, socioeconomic status, and behavior. Additionally, we need improved statistics on the utilization of dental services. Only 40% of U. S. citizens see a dentist once a year for any reason. To improve this statistic, people must be encouraged to establish a personal value system. One method of accomplishing this is through local and state professional organizations sponsoring workshops and seminars directed to dental needs. The impact of such well-directed community efforts can have a far reaching effect on the social, political, and economic awareness of thousands.

Americans are moving forward in the field of dental health, becoming less complacent, less content with themselves. Their active involvement will bring about improvements in total health care, not just in dentistry. Much time has been spent promoting the various aspects of preventive dentistry. What we must remember is that without the combined efforts of the citizens and the public and private professionals, this era of improvement cannot exist. No proposal to answer every need has been found, but a realistic look is being taken at some of the accomplishments which can be reached.

Education for Health

RONNIE S. JENKINS, M.S.
Atlanta, Georgia

The concern of health education is with the prevention of illness and disability and promotion of positive health habits. Health Education is an integral part of quality care. It motivates an individual to translate health information into a personal responsibility—to keep oneself healthier by avoiding actions that are harmful and by forming beneficial habits.

People need knowledge which illustrates the relationship between health and human biology, lifestyle, the environment and the present health care system. It is through education in health that we can understand the interrelationship of these four components and how this information provides insight and direction so that we have sufficient knowledge of health maintenance and proper utilization of existing health services.

The disease patterns of the 70's show a distinct relationship to biology, environment, lifestyle and personal health care services. Education for health can have a role in improving the health status of those at risk. These causes of death or disability are coronary heart disease and stroke, chronic obstructive pulmonary disease, sexually transmitted disease, cancer, motor vehicle accidents, suicide, homicide, alcoholism, dental diseases, and nutritional abuses.

In designing a program of education for health relative to these diseases, it is necessary to identify risk factors and programs aimed at changing consumer lifestyles, improving the environment, and coordinating the efforts of the various providers of health services.

FACTORS THAT AFFECT MULTIPLE HEALTH PROBLEMS

An analysis of the current disease pattern shows that a number of factors usually contribute to the cause of the disease. A classic example is lung cancer in which cigarette smoking, environmental pollutants, and perhaps even individual genetic and metabolic factors are causally related. Frequently a number of diseases are influenced significantly by one activity, such as the contributory relationship between suicide, homicide, motor vehicle accidents, metabolic diseases, and a high degree of alcohol consumption. The latter is an example of a self imposed risk which can be positively modified if individuals elect to change their lifestyles.

Other risks, frequently resulting from the decisions of various societal groups may also have multiple impacts. Pollution of the physical environment may result from industrial group decisions, which can contribute significantly to cancer and to chronic obstructive pulmonary disease. "People pollution" may also result from individual and group decisions thereby contributing to emotional stress, motor vehicle accidents, sexually transmitted diseases, cardiovascular problems, homicide, suicide, and alcoholism. Additional external risk factors are the persistence of poverty, racial discrimination, slum housing, urban squalor, high crime rates, and lack of access to quality health services. Related to the latter factor also are the lack of interest and knowledge in disease prevention, and the apparent inability of many health agencies to coordinate these efforts in the provision of accessible, available, and affordable health care.

There is a need to conceptualize and implement programs which address the occupational, recreational, and consumer needs related to biology, the environment, and our present lifestyles. The compelling problems of poverty, education, social customs, and the accessibility, availability, and affordability of quality care profoundly influence many areas of personal health, such as premature births. The ignorance of many consumers, patients, and even some providers surrounding certain disease conditions, such as those associated with pregnancy, must be addressed. Social, cultural, and economic environments, unless modified, will continue to spawn diseases, and feed the present disease pattern. Organizational, personal, and societal resources must be better marshalled and directed to prevent further squandering. And finally, advocacy groups to support legislative changes and minority rights must be organized and given support and direction. These changes are basically educational in nature, and are required if quality health care is to be every person's right.

Individual behavior, then, becomes of overriding importance. Changes in lifestyle must be addressed if poor health status is to be changed and rising health care costs are to be contained. Health promotion, with its emphasis on the education and motivation of the individual, must be seen and addressed as a top priority.

There are many different agencies and groups which should play a role in the successful implementation of community-wide programs of health promotion. These may be grouped into the following broad categories: (1) providers; (2) schools; (3) media; (4) business and industry; (5) civic, social and religious organizations; (6) government; and (7) voluntary health organizations.

THE ROLE OF THE PROVIDER IN HEALTH PREVENTION

Hospitals, local health departments, HSAS, third party payers, professional organizations, community mental health programs, emergency medical service systems, physicians, dentists, nurses, pharmacists, health educators, nutritionists, and other health-related workers are primary sources of service and knowledge. A major role of all should be to provide leadership in planning and developing health promotion programs—in concert, when possible. Their attitude toward consumers should be one of receptivity which recognizes the consumer as a partner in the responsibility for achieving and maintaining good personal health.

The coordination of programs to reduce costs and avoid duplication should be addressed jointly by providers through such mechanisms as Joint Health Promotion Councils. Finally, providers have the responsibility to develop and participate in community outreach programs which carry appropriate services to areas of need.

THE ROLE OF THE SCHOOL IN HEALTH PROMOTION

For too long we have modified, added on, and patched up the health curricula in our schools. What is needed is a totally new educational outlook in school health, one that is based on affective learning, is conceptualized from new perspectives, and which incorporates content into a meaningful, continuous, and related sequence. To do this, three basic areas within school systems must be fundamentally redrafted: health curricula, the school environment, and school health services.

Colleges and universities need to revise and enrich their curricula in the area of teacher training. Many such curricula have little or no relationship to today's disease patterns, and too much emphasis is placed on physical education at the cost of comprehensive health education.

Schools also have a role in making available public forums and discussions, and credit and noncredit courses in health promotion areas. The school facilities should be made available for evening health promotion activities.

And through innovative approaches which relate to the student's own life experiences, continuous and meaningful courses in the prevention of disease, nutrition, hygiene, smoking, drugs, human sexuality, human relations, sexually transmitted diseases, and emotional problems should replace the all too frequent current pattern of isolated and unrelated talks or classes involving only a portion of the student body. The worth of the individual should be the cornerstone of curriculum-building, and the dignity and self-esteem of both students and faculty should be reflected in the school environment.

THE ROLE OF THE MEDIA IN HEALTH PROMOTION

Newspapers and periodicals, radio, commercial and educational television, and organizational public relations departments are uniquely situated to convey health promotion concepts and principles. Their messages can sensitize people in the area of personal health, and through localized information, provide a balance between local, state, and national programming. The media also has a role in determining appropriate product promotion. New approaches, perhaps through public school systems, should be developed for bringing educational television programming in acceptable forms to far greater segments of the public.

THE ROLE OF BUSINESS AND INDUSTRY IN HEALTH PROMOTION

Employee health programs should include appropriate screening and referral, and have formalized relationships with private and/or public providers. The employer should provide the mechanism for appropriate follow-up. Innovative health promotion programming in many areas of personal health should be made available to employees. Clear and supportive information regarding health insurance and benefits, and personal health services should be made available through the employing organization.

Business and industry has a responsibility to concern itself with *living wages*, with providing accident-proof work environments, and with establishing an overall atmosphere within the organization which minimizes emotional trauma, and, to the extent possible, eliminates boredom and promotes and rewards efficiency.

THE ROLE OF CIVIC, SOCIAL, AND RELIGIOUS ORGANIZATIONS IN HEALTH PREVENTION

These organizations provide a means of reaching specific target groups with particularized health promotion programs. Areas of personal health concern should be related to the basic purpose of the organization so that capitalizing upon an existing interest of the member may be accomplished.

Maintaining up-to-date library and resource material for their members and for the public in areas of personal health should be a program function of these organizations. Their physical facilities, if any, should be made available to the community, through their membership, for health promotion activities. And the value of good personal health habits by the organization membership to both the organization and the community should be a periodic programming theme of civic, social, and religious organizations.

THE ROLE OF GOVERNMENT IN HEALTH PROMOTION

Government agencies, whether health related or not, can play a critical role in health prevention involving both their employees and the public they serve. As employers, government agencies carry the same responsibilities as business and industry. As servants of the public, they have an even greater role in developing an agency attitude toward promotion of health which is communicated to the public it serves. No smoking areas, clean surroundings, and efficient and personalized services are leadership activities which the public should expect of government agencies. Such agencies have a responsibility to continuously review their program activities for any elements which are not supportive of good personal and public health promotion.

THE ROLE OF VOLUNTARY HEALTH ORGANIZATIONS IN HEALTH PROMOTION

These organizations generally provide health and health-related services to specific target groups and/or to the public at large. A major part of their effort is directed to public education, and as such, they are unique as an added resource in health promotion. Such organizations should expand their influence and potential through searching for cooperative programming areas as an obligation to the public which supports them, and minimize the present attitudes of some organizations which are characterized by isolated program activities and turf-guarding. Such organizations are uniquely American in concept, and have societal obligations which transcend their organizational identities.

Georgia is currently developing the health promotion concept as previously described. Individual differences in communities and geographic areas will be recognized and allowed for in programming efforts. Strategies are being developed to improve the health status of Georgians through modifying current lifestyles. Coordination mechanisms for health promotion within geographic areas are being studied with emphasis on local responsibility.

This new health outlook will require a change in perspective on the part of both consumers and providers, and these changes will probably be seriously challenged by some who would maintain the status quo. Such changes will not come easily or readily, but as consumers take on more and more of a decision-making role, and as providers act with increasing responsibility, the needed changes in our health care system will occur.

The Environment

THOMAS W. MCKINLEY, BSA, MPH
Atlanta, Georgia

That the environment plays a significant role in the distribution of health and disease has long been recognized. In his book entitled *Airs, Waters, and Places*, Hippocrates wrote that in order to predict the kinds of diseases which would occur in a community, a skilled physician would have

> ... due regard to the seasons of the year, and the diseases which they produce; to the states of the wind peculiar to each country and the qualities of its waters; who marks carefully the localities of towns, and of surrounding county, whether they are high or low, hot or cold, wet or dry; who, moreover, takes note of the diet and regimen of the inhabitants, and in a work, of all the causes that may produce disorder in the animal economy.

Hippocrates' teachings dominated western medicine until the latter part of the 19th century when the germ theory of disease began to gain prominence and the work of numerous investigators ushered in what has come to be known as the "golden age of microbiology." Speculation regarding disease causation gave way to scientific discovery. Diseases rapidly began to be traced to specific microorganisms rather than being assigned to weather, location, or season.

EARLY ENVIRONMENTAL HEALTH PROGRAMS

By the end of the 1920's, the route of transmission for most of the major enteric, parasitic, and vector-borne diseases had been discovered. Health departments began to be formed over the United States for the purpose of controlling diseases in the community. Engineers and sanitarians became the foundation or cornerstone of the public health effort. Pit privies and later septic tanks were constructed in rural areas to insure the sanitary disposal of human wastes. Individual wells were properly constructed, sealed, and protected from surface contamination. Municipal water supplies were filtered and chlorinated and municipal sewage was treated and chlorinated before being released to surface waters.

Swamps were drained to prevent the breeding of mosquito vectors. Systematic garbage removal to sanitary landfills was begun. Rodent harborages were cleaned, rats were trapped and buildings were rat-proofed.

In 1943, the U.S. Public Health Service published the first model "Milk

Ordinance and Code." This code which required pasteurization of milk was quickly adopted over the country and milkborne disease rapidly disappeared. Approximately four years later, in 1947, a manual entitled "Recommended Ordinance for Eating and Drinking Establishments" was published by the U.S. Public Health Service. This, too, was widely adopted across the country and great emphasis was given to cleaning up the nation's public eating and drinking establishments.

Morbidity and mortality figures attest to the almost precipitous decline in such diseases as malaria, typhoid, typhus, and dysentery. Initial reductions in communicable diseases in the early 1900's were probably due to economic improvements which allowed people to screen houses, buy shoes for children, and obtain better food for their families. However, there can be no question that the activities of sanitarians and public health engineers markedly hastened the decline. By the mid-1950's, communicable diseases were well under control and a new disease pattern was emerging.

ENVIRONMENTAL HEALTH PROGRAMS TODAY

Once the threat of major communicable disease outbreaks had been set aside, environmentalists (sanitarians and engineers) fell into a maintenance role. Food sanitation, tourist accommodations, sewage disposal and water supply programs continued as major programs.

In the early 1960's, environmentalists in the health field sought to fluoridate water supplies to prevent dental caries among children. They also raised concerns relative to the mounting problems of solid waste disposal, air pollution, and water pollution. This resulted in a minor increase in funding, but it was grossly inadequate to really deal with the problems. Finally, in the late 1960's, pressures from numerous public groups who joined the ecology movement were successful in bringing about the creation of an Environmental Protection Agency at the Federal level to address the mounting problem of environmental pollution. Air, water, sewage, and solid waste program responsibilities were assigned to this agency. Many states quickly followed suit by gathering these responsibilities from health and placing them into a state environmental protection agency.

The need for a consolidated, aggressive approach to preventing environmental pollution and restoring the environment through an agency, such as the Environmental Protection Agency, is unquestioned. The question is, what part health departments should play in the environment in order to change today's new disease pattern of heart attack, cancer, stroke, chronic obstructive pulmonary disease, obesity, accidents, homocides, suicides, alcoholism, addiction to drugs, and sexually transmitted diseases?

ENVIRONMENTAL HEALTH PROGRAMS IN THE FUTURE

Occupational health appears to be one of the most fertile areas for environmental health to commit resources in the future. Exposure of workers to accidents, dust, noise, and chemicals in the work place clearly contributes to today's disease pattern. Work related accidents for blue collar workers are about three times higher than for white collar workers, with construction workers having the highest incidence. The increased risk of cancer recently found among asbestos, polyvinyl chloride, kepone, and rubber workers raises very frightening questions as to what will be the end result of exposure to the more than 15,000 chemical agents now used by industry, any one of which could be carcinogenic. High noise levels in the work environment are productive of both stress and hearing loss. While the responsibility for occupation health has been assigned to the Occupational Health and Safety Administration (OSHA) at the Federal level, it is clear that this agency cannot manage the job alone. States will have to increase or divert resources to address the problem. This should be done in a state health setting.

Radiological health (inspection of X-ray machines and inspection and licensure of radiological materials users) is also an important environmental health activity aimed at reducing the risk of cancer. Persons who had X-ray treatment to reduce the size of the thymus gland 15 to 20 years ago when this treatment was standard practice are now known to have an increased risk of cancer of the thymus. Dentists who have had repeated exposure to X-ray over the years are also known to be at increased risk of cancer and questions have recently been raised about the propriety of mammography for women under 50 because of the possibility for increasing the risks of breast cancer as a result of the procedure.

The wisdom of continuing to invest significant public health resources in certain other programs such as the inspection and permitting of food service establishments and tourist accommodations must be weighed very carefully against the ability of these programs to change today's disease pattern.

Control of certain of today's diseases such as accidents, homocides, suicides, alcoholism, and sexually transmitted diseases will require a studied, calculated effort in the social environment. The social environment which is comprised of community organizations and the constructive and destructive forces which operate therein has generally been neglected by environmental health workers. Housing, land use planning and transportation are all vital parts of the social environment.

Poor housing has been associated with homicide, suicide, and alcoholism. Although it is difficult to determine whether the association is causal, it stands to reason that depression and a sense of hopelessness cannot help but be exacerbated by dismal surroundings and overcrowding. Urban renewal programs have often only added to the problems of those living in poor housing because they have been dislocated from surroundings which, even if dismal, were nevertheless familiar. House renovation programs which do not dislocate people have proven to be more rewarding in some settings. Environmental health needs to devote more resources to housing whether this is done directly or acting in the form of a conscience or catalyst to other agencies.

Land use planning appears to be the key to prevention of many housing and slum development problems. Environmentalists have long recognized that a poorly designed piece of kitchen equipment is not as functional as it should be, and when it gets soiled from food particles, may be difficult, if not impossible, to clean and maintain in a healthful state. However, if the equipment is well designed, not only will it be functional and easy to clean, it also will have features which make using it correctly a process that comes naturally. Mistakes in use will be difficult to make.

The same is true for a well planned community. It will have a balance of homes, industries, schools and recreational areas, will be easy to keep clean, and will make healthful living a natural process. Health workers may extol the virtues of exercise and even convince many that this is a healthful thing to do. However, where the makeup of the community does not provide suitable space or facilities for exercise the most determined resolve to carry out an exercise program is soon abandoned.

Transportation is a very basic and vital part of our social environment. Availability determines where people live, their access to jobs, shopping places, recreation and even health care. Transportation is also a major factor in environmental pollution, stress and accidental injury and death. In addressing today's disease pattern, there would appear to be a definite role for the environmental health worker in assessing transportation needs, studying accident patterns, and becoming more aware of the stress and pollution problems caused by transportation. While primary responsibility for these areas will not likely be and probably should not be assigned to health, the environmentalist can make recommendations relative to these areas and serve as a catalyst for bringing pressure to get things done.

SUMMARY AND CONCLUSIONS

Environmental health workers enjoyed phenomenal success in controlling infectious and vector-borne diseases through their efforts working in the physical and biological environments. These diseases are now under control and have been for several years. A new disease pattern has emerged comprised of social and chronic debilitating diseases. With certain exceptions, major areas of the physical environment which are related to this disease pattern have been consolidated into Environmental Protection or Control Agencies. Environmental health should look for a new role in the social environment in order to change today's disease pattern.

Evaluation in Community Health— An Overview

G. E. ALAN DEVER, Ph.D.
CHARLES M. PLUNKETT, Research Associate
Office of Health Services Research and Statistics
Division of Physical Health
Department of Human Resources
Atlanta, Georgia

Evaluation is essentially ascertaining the value or amount of success in attaining specific goals and objectives. The actual processes of establishing goals and objectives involve several facets, depending on the current level of operations. Using an holistic model—emphasizing the interrelated aspects of *environment, life style, human biology*, and the *system of health care delivery*—entails the design of a system capable of reaching goals through an analysis of problems for potential prevention and intervention measures. Thus, emphasis is placed on a comprehensive health policy, reflecting ultimate goals plus desired program objectives.

The method by which this may be accomplished is through data analysis and the political process. Data analysis examines existing conditions, trends, and forecasts to determine community needs. Essential to this is the establishment of baseline or benchmark measures. This problem may be approached in two ways:

1. Lacking prior data on the health status of the population being investigated, the first year of operation may be utilized to establish a baseline for the following years.

2. If timely and accurate community data are available, baseline or benchmark studies can establish present and past levels, making it possible to set realistic objectives to be attained at single and multi-year intervals.

At this point, two very critical questions must be asked. What kinds of goals and objectives should be selected for inclusion in the health plan? How do we determine what levels or attainment of goals and objectives would be realistic?

The answer to the first question is a direct consequence of the adopted, hopefully holistic, health policy. If this policy is broad and comprehensive—including elements of the environment, biology, life style, and system of health care organization—then the goals and objectives should reflect such a policy, together with the identification of nontraditional, as well as traditional, goals and objectives having a decided impact on changing disease patterns. A parallel approach to all this is, of course, to identify the disease patterns (specific program problems) for each community—an obvious need but not always met.

As to the second question, the level of attainment of goals and objectives may result from realistic assumptions or may be determined quantitatively. In addition to the need for baseline or benchmark measures which aid in the setting of goals and objectives, it is imperative that trends or standards be determined for the specified health status measures. For example, "to reduce infant mortality by 20% within the next year" may be a totally unrealistic objective when we have been able to reduce it by only 2% per year over the last ten years—and, further, when this objective is already below the national rate. Thus, a review of past trends and promulgated standards may reveal appropriate targets or standards for attainment of objectives. Another facet to be considered is what will happen if one does nothing? "Will infant mortality continue to decrease by 2% anyway?" Accordingly, if a new program is to be implemented, then possibly a 3% decrease might be expected.

Finally, objectives should be set knowing that a rate may change within certain probability limits. Readjustment of objectives is always feasible, based on unforeseen issues which may surface and have an impact on the program.

The political process in establishing goals and objectives recognizes the trade-off between virtually unlimited needs and desires of a community and the scarcity of available resources. The goals and objectives of any health improvement program must be placed in the framework of providing the greatest benefit for the least cost. It is likely that resources available to communities will decrease further in the '80s, making it imperative that efficient and cost effective programs be developed. Data analysis will provide the framework for decision-making; but, ultimately, the political process enters into the determination of appropriate allocation of resources based on a range of alternatives.

Without the establishment of goals and objectives that answer the questions "what," "how much," "who," "where," and "when," evaluation is impossible. Evaluation attempts to determine changes in health with reference to programmatic effects. The results obtained become input to the continuing process of planning, budgeting, and evaluation. Within the evaluation component of the cycle are three areas of concern: costs, activities, and outcomes. These three are commonly called fiscal, process, and outcome evaluation.

1. *Fiscal Evaluation* must focus on and determine cost accountability.
2. *Process Evaluation* must determine program activity in terms of the population receiving the program by age, sex, race, or other demographic variables, program organization, staffing and funding, or program location and timing. It is a measure of program efforts, or proposed activities, rather than program effects.
3. *Outcome Evaluation* must delineate program objectives in terms of effects to determine if a change in health status has occurred as a result of the program.

EVALUATION DESIGNS[1]

Each of these measures can apply to one or many of the evaluation designs. It is appropriate to evaluate programs from a process, fiscal, and outcome perspective, no matter what the evaluation design.

1. Before vs. After Program Comparison (Design 1, Fig. 1)
   Compares program results from the same geographical area, measured at two points in time: before the program begins and after implementation. This is a very common design, but it is difficult to separate the effects of program activities from other influences.
2. Time Trend Projection from Pre-Program Data vs. Actual Post-Program Data (Design 2, Fig. 1)
   Compares actual post-program data to estimated data from a number of time periods prior to the program. Changes are identified as the actual differences vs. what they are estimated to be by the projections.
3. Comparison with other Geographical Areas or Population Segments not served by the Program (Design 3, Fig. 1)
   Compares data from the geographical area or population group where the program is operating with data from another geographical area or population group where the program is not operating.
4. Controlled Experimentation (Design 4, Fig. 1)
   Compares preselected, similar groups, some of whom are served by the

[1] Harry P. Hatry, Richard E. Winnie, and Donald M. Fisk: *Practical Program Evaluation for State and Local Government Officials*, the Urban Institute, Washington, D.C., 1973.

Design 1

BEFORE vs AFTER
PROGRAM COMPARISON

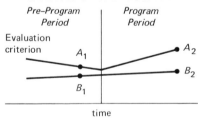

$A_2 - A_1$ = estimated program effect

Design 2

TIME TREND PROJECTION
OF PRE-PROGRAM vs
ACTUAL POST-PROGRAM DATA

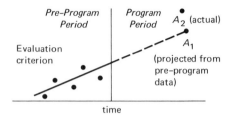

$A_2 - A_1$ = estimated program effect

Geographical area A has a program;
geographical area B, the comparison
area, does not.
$(A_2 - A_1) - (B_2 - B_1)$ = estimated program
effect (or rate of change might be used rather
than absolute amount of change)

Design 3

COMPARISON WITH OTHER
GEOGRAPHICAL AREAS OR
OTHER POPULATION SEGMENTS

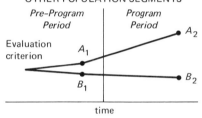

Design 4

CONTROLLED EXPERIMENTATION
COMPARISON OF PRE-ASSIGNED
SIMILAR GROUPS ONLY ONE OF WHICH
IS SERVED BY THE PROGRAM

Design 5

PLANNED vs ACTUAL PERFORMANCE

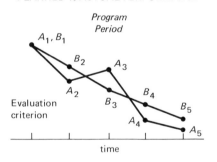

$A_2 - B_2$ = program effect

Planned performance (B)
Actual performance (A)
$(A/B) \times 100$ = percent of performance level
achieved or percent of
objectives targeted

Source: Modified from Hatry.

FIG. 1.

program and some of whom are not (or are served by alternate programs). The critical aspect is that individuals are randomly assigned to program and comparison groups before program implementation so that the groups are as similar as possible, except for the program treatment.

5. Comparison of Planned vs. Actual Performance (Design 5, Fig. 1)
   Compares preprogram, planned targets to the actual program performance.

As a health program management tool, evaluation will demonstrate its worth through reduced costs or increased benefits (fiscal evaluation), improved effec-tiveness (process evaluation), and changed health status (outcome evaluation). It can provide the impetus for a reallocation of resources to meet community health needs and a redirection of efforts along the lines of an holistic approach to improving the health of the community. Although it is seldom that one can be absolutely certain a change in health status has been brought about by the program, this is the goal of evaluation. We ultimately must determine that goals and objectives are appropriate, whether or not they are achieved, and the extent to which the program has changed the health status of the community. The degree to which changes directed at life style, environment, biology, and the health care system can ultimately change health status is reflected in the ability to measure such changes.

The Family—Critical Issues Facing Institutions of Medicine

EMILY H. MUDD, Ph.D.
Haverford, Pennsylvania

Relationships between men and women whether within or without the prevailing institutions of law, religion, or the traditions and customs of society, have in-evitable impact on the health of every individual and this, on the institutions of medicine. In this connection recent decades have witnessed a variety of chang-ing roles from those earlier accepted by families in Western cultures. These changes are closely interrelated with a variety of factors: increasing longevity, movements for women's rights, acceleration in the cost of living, larger welfare enrollment, and more fundamentally, with the newer freedom for women in their potential control over the reproductive functions of their own bodies. In this last area medical institutions in particular are closely involved.

Population in the United States in 1977 is nearly stable with the highest number of births in the teenage unmarried group. Specifically this means less demand for obstetrical services and more engagement with ethical problems in servicing large numbers of minors, with or without parental consent, in the promotion of health in the broadest sense.

Drastic modification in laws has made the early interruption of pregnancy a woman's right of choice—again precipitating physicians and institutions of medicine into new and often difficult decisions and patient relationships. In many of these newer demands and opportunities physicians have had little exposure and orientation in their earlier medical education.

All physicians see many cases when symptoms do not subside as they should under the usual methods of treatment, when convalescents do not convalesce, when recovery from accidents or surgical procedures is unusually protracted for no apparent reason. Symptoms also seem to arise for no obvious cause. Physicians are aware that somewhat emotional factors must be entering into the condition. Conflicts and tensions accompanying marriages and family life and man-woman interrelationships are common sources of emotional and physiological disturbances with which the physician is called upon to deal. Often his knowledge of psychology, sexual needs and behavior, and changing social customs does not enable him to deal effectively with these problems, in addition to the time limitations. The use of a variety of collateral helping resources must be recognized and considered: marriage and family therapy, workshops for education in sexuality, encounter groups, and self-appointed ways of living together without institutional blessing.

Many professional and lay men and women still believe in the importance of some stable or family structure for a range of functions, including childbearing, child rearing, and the psychological, physiological, and economic needs of men and women. For these reasons continued interdisciplinary research is recommended into functions performed by the family, and particularly the ethical and psychological implications of existing and alternative types of male-female relationships, and of family organization.

Building the Human Community— Developing and Maintaining Health Interpersonal Relationships

JAMES W. ALLEY, M.D.
THOMAS F. GIBSON, M.P.H.
LYDIA WALKER SEVASTEN, B.A.
Atlanta, Georgia

The purpose of this paper is to give a brief description of developments in the behavioral science field that have applicability to community health management and services in the 1980's.

The systematic study of human behavior in groups is a relatively new science. Along with psychology, sociology and anthropology, behavorial science has brought new insight and perspective to community health. Certain values and behaviors, previously the province of religious teachers and philosophers, have been examined by behaviorists and sociologists. Research has documented increased productivity where human needs have been recognized and met. Some conclusions have been reached that can be applied and relied upon to guide planning and delivery of health services. These are briefly discussed below.

Human beings must meet individual needs for basic survival and security before they can make a collective investment in a society; i.e., before they can share goods and resources, influence and leadership. Next, persons must feel a sense of belonging and community support before they can invest in status-seeking and achievement pursuits. When these needs are met they can grow toward the fulfillment of their creative potential. It follows that persons whose needs for survival, security, belonging and self-worth are being satisfied in a work setting are more productive, creative and enjoyable as co-workers.

Humans tend to behave in predictable ways in order to feel included, influential, and trust-worthy in any new group of persons, no matter what the purpose of the group. Humans tend to behave in predictable ways to protect themselves from physical and psychological vulnerability in relationships.

Persons who are included in decision-making by having the opportunity to give input have increased ownership of and investment in the goals of that group or society.

Persons respond to space and architectural design in predictable ways. All of the above behaviors are only slightly modified by cultural-educational influences.

Making use of these data, health professionals in the Division of Physical Health in Georgia have built specific programs in human relations and group process into in-service training.

The value inherent in this allocation of time and resources is the belief that humankind has a basic right to total health and that the psychic health of human service workers is as important as that of the client. This entails more than adequate salary for survival and security. Each person has the right to be valued as a person as well as a role or function in the work arena. Each has the right to adequate physical and psychic space for creative work. Each has the right to self-expression and to give input to decisions that will affect his/her health, environment, and lifestyle.

The assumption is made that healthy, satisfied personnel will deliver services more productively, more efficiently, more responsively, and more happily. The total health of the client will be affected positively by the humanistic style in which the services are offered.

Finally, since stress-related diseases are on the increase, it is essential that steps be taken to prevent or minimize stress-induced diseases within the organization whose mission it is to treat and prevent disease in the larger community.

Fortunately, the business of how persons relate to one another need no longer be left to "human nature"—that mythological rationale that for centuries has been used to excuse various social ills. It is now possible to systematically apply principles of behavioral science to relationships within people-systems, from the family to the largest organization, and to consciously and intentionally create environments that are trustful and affirmative of persons in these systems. It is possible to apply behavioral principles to efficient managment without sacrificing human rights and values. It is possible to train individuals in new behaviors that are constructive, rather than destructive, in relationships. It is possible to apply collaborative methods of problem-solving to facilitate change in people-systems. These methods, based on human values, do not deny human dignity and inhibit creativity but instead enhance them. Health care personnel can learn styles of relating to consumers that will gradually change their role from provider to enabler.

The operational value underlying human relations training is that every person is one of worth. Development of basic personal skills in relationships and groups is the primary focus of this experience. These skills include communication, consultation, goal setting, motivation meanings, values clarification, self-affirmation and creative use of conflict. Team-building involves the above plus techniques for problem-solving, contract-building collaboration and resource sharing.

One example of the application of H. R. training to community health can be observed in the Division of Physical Health, Department of Human Resources,

State of Georgia. In 1973, the new Director and his management team began a program of in-service training in human relations for their key staff people. Their goal was to improve service delivery by attending to the human needs of the professionals, staff, and technicians who deliver services. They recognized the importance of responding to stress factors caused by the environment and began to sensitize themselves and key people in the organization to the holistic approach to health.

In the three years' since the inception of the training program, close to 500 personnel, at all levels of management, have gone through basic human-relations or team-building workshops. Some have had specialized and advanced training, such as values clarification, race relations, creative conflict management and women in government. In addition, an outside consultant is available on a part-time basis for process observation in planning groups, third-party crisis intervention, organizational development expertise, and individual consultation. A retrieval system now being put into effect will offer useful data to assess long-range efficacy of such training. In the meantime, the Division leadership relies on subjective evaluations written at the end of each event. Thus far, the following observations can be made:

1. Immediate response of the participants to the training opportunities has been overwhelmingly positive.
2. Such training has facilitated team development in management and staff groups that can benefit from shared leadership, decision-making and resource-sharing.
3. Some individuals have expressed personal benefits and a new sense of supportive community and productivity on the job.
4. Some small but measurable gains have been made in affirmative action. Though percentages are small, more women and blacks have been hired at higher levels of management.
5. Response and interest has come from other state health programs, indicating the attractiveness and potential of the program.

Good management, however, requires more than focus on producing good feelings and satisfying the needs of personnel. The Division team uses Management by Objectives for goal setting, planning and evaluation. They expect firm and decisive management behavior from each other and from other supervisors. It is sometimes necessary in order to carry out the task, because of pressures from above or from federal politics, to make arbitrary decisions with less than desirable time or input from a wide base. The ultimate goal of humanistic management is to achieve a balance between efficiency and productivity, on the one hand, and good communication and maximum involvement of persons on the other hand.

At this point it is important to mention some problems in implementing programs of this nature and to raise a few cautious notes. Two related obstacles are the general lack of knowledge and understanding of applied behavioral science and training methodology. For this reason, support funding from higher levels of government is slow in coming. Patience and persistence is needed to include these persons in the planning for such training.

It is not in the interest of good management of people to open a system to more personal risks on the part of personnel without the concurrent commitment to responsible follow through of such training. It is irresponsible and counter-productive to apply such training in a coercive and dogmatic manner. Care must be taken as to the methods employed for implementing human relations training so as not to be hypocritical and contradictory. It is important to choose consultants and practicioners with care as there are some unethical and untrained people operating in the field of humanistic psychology and applied behavioral science. With a skilled consultant who is professionally recognized and recommended, these problems and concerns can be handled effectively. A few professional organizations are concerning themselves with ethical and responsible use of behavioral science as it is applied to training processes. One of these, Association for Creative Change, has made significant progress in developing standards for training and accrediting practitioners.

Planning for integration of human relations principles into management and health care delivery systems now accounts for a very small percentage of state budgets. As yet in Georgia, no long range planning has been done, though the importance is recognized. The first step is to evaluate thoroughly what has been done thus far and use that data to determine applicability to community health in the 1980's.

The prognosis for the future of our society at home and as a world community looks bleak unless much new emphasis is placed on intentional behaviors. Health care delivery systems must be developed that recognize and act on the human values espoused by Bill of Rights, Public Law (PL) 93–641 and the World Health Organization concept of health.

Jimmy Carter's approach to government, both in Georgia and in the Presidency, indicates that he values input from his grass roots constituency and intends to support and act on the values that have been discussed in this paper.

Farsighted health professionals are coming to a world view of the human family as a world community, interdependent and interrelated as people and as societies. They are recognizing the earth as one giant organism which is struggling for health and for which prevention of social disease is as necessary as treatment of the symptoms.

Purpose, Definitions, Concepts, and General Principles of Nursing

IRENE G. RAMEY, Ph.D., R.N.
Minneapolis, Minnesota

PURPOSE OF NURSING AND DEFINITION OF ITS PRACTICE

The purpose of nursing is to provide to individuals and their families those services which contribute to the maintenance of health, to the recovery of health, or to the facilitation of a peaceful death. Professional nursing is a process which involves assessing the patient/client's status of health, developing short and long-term health objectives, planning and implementing appropriate nursing activities to help the patient/client and his family reach those objectives, and evaluating the effectiveness of those nursing actions. Nursing activities include, not just those aimed at curing or alleviating pathological conditions, but also maintenance of health, prevention of illness, rehabilitation, habilitation, counseling and teaching. The practice of professional nursing includes (1) direct and indirect services to patients/clients and their families; (2) teaching, directing, and supervising other nurses, nursing students and auxiliary personnel; (3) collaboration with other health professionals in assessing the needs of patients/clients, planning for their care, providing diagnostic and therapeutic services, and evaluating the progress of the recipient; and (4) adding to the body of nursing science through research.[1]

DEFINITION OF NURSING SCIENCE

Nursing is a practice profession which exists in response to a social need which implies both a charge and an obligation. The practice of nursing, therefore, necessitates the selection of relevant bodies of knowledge, of appropriate theories, concepts, and principles, and synthesizing them into a relevant, appropriate framework for utilization of the nursing process. Because nursing, as well as many other professions, draws heavily upon theories, concepts and principles developed in other disciplines, synthesis becomes exceedingly important in the assessment, planning, intervention, and evaluating phases of the nursing process. Such theories, concepts, and principles, which are derived from other disciplines, and which were validated with other samples under other circumstances, are being cautiously tested in nursing practice for their validity and reliability. At the risk of over-simplification, a few overall statements about the purposes of some of the other health professions will serve to differentiate the purposes of nursing: the field of medicine diagnoses and treats pathological conditions; the field of dentistry diagnoses and treats pathological conditions of the body structure of mastication; the field of pharmacy compounds medications used in the prevention and treatment of pathological conditions; the field of public health diagnoses and treats environmental problems detrimental to health. As was mentioned above, the field of nursing has traditionally had as its major purpose the provision of services which contribute to the maintenance of health, to the recovery of health, or to the facilitation of a peaceful death. This purpose has, however, greatly increased in scope in recent years with the many advances in all fields and with the movement of nurses into new areas of specialization. However, true to its tradition, and in contrast to medical research, nursing research is concerned less with physical and mental diseases, but is concerned more with the general and specific problems of human being such as pain, anxiety, suffering, and hospitalization and its attendant stresses. Current nursing studies fall into five major categories: (1) building the science of practice, (2) refining the artistry of practice (with attention to health outcomes), (3) establishing structures for optimal care, (4) developing methodologies and measurement tools, and (5) application of findings.[2]   Researchers and practitioners in nursing are concerned with developing a systematized body of knowledge about the life process in man which is obtained and tested through use of the scientific method. There is an open system of knowledge flow among the disciplines and professions which precludes strict territorial claims on areas of knowledge. The various health professions can be viewed as a series of overlapping open systems not only in regard to bodies of knowledge but also in regard to functions and professional goals. With the increased realization of a need for an interdisciplinary approach to health care, and with increasing numbers of options becoming available in health professional education, the delivery of health care will increasingly become more like the skillful orchestration of a symphonic composition, with each health care professional credentialed for his own role and accountable for his actions and their outcomes.

CONCEPTS OF THE LIFE PROCESS USED IN NURSING

The *individual* is a person who, like all living things, has evolved from the primordial matter of the universe and developed into an exceedingly complex electrodynamic unit with both structure (organization) and function (process). Although the individual is an open system, he possesses integrity consistent with his organization and process. Although he is subject to the laws of the universe, he also has the capability for purposefully directing his activities. He functions as a unit, not as a collection of anatomical systems. Awareness of the past, present and future is thought to be more highly developed in human beings than in any other species. Such sentience may be the cause as well as the effect of experience and of cultural beliefs and behaviors which make each individual uniquely different from all others. Certainly, the ability to communicate through written and spoken languages enables humans to build up their past, discuss the present, conjecture about the future, and reach out to touch the lives of others in a way not possible to other species.

Interaction is a concept which denotes the constant dynamic interplay between the *individual* and his environment. Environment connotes the physical universe, inanimate objects, animals and other individuals either singly or in various constellations or groups. The individual is continually affecting or being affected by his environment; he is an open system with a continual input and output of energy in progress. The interaction may vary in frequency, and in magnitude over time. The individual breathes the air and inhales the oxygen but exhales carbon dioxide and water vapor; he walks, sits, or lies on various substances and deforms them but is likewise deformed by pressure; he is bombarded by cosmic rays, X-rays, electromagnetic waves, light waves and sound waves, but stops, deflects, and creates some himself; he interacts with animals and with other people, and they change the courses of each others' lives. Therefore, the individual is not just a passive recipient of forces external to him to which he only adapts, but he himself is also an active force in the universe.

Self-regulation is a concept which has also been termed as homeostasis, or homeokinesis[3] by other authors. Self-regulation is usually thought of as an individual's attempt to achieve equilibrium and stability, a static rather than a dynamic concept. However, since human beings have evolved phylogenetically through millions of years into their present systems of structure and function, and since each individual develops ontogenetically from conception to death, it seems more appropriate to think of self-regulation as being a process enabling an individual to achieve an ever increasing complexity of organization, permitting continual evolution into a higher form of life. Since all forms of life can be considered examples of negentropy, such a view would explain the individual's tensions, motivations, and drives. As an individual interacts with his environment, changes occur in him, and his internal regulatory system which controls anabolism and catabolism serve to reestablish his functioning within those parameters of upper and lower deviations which he can tolerate. A perfect balance never exists. When an individual is unable to maintain that pattern which is unique to him, we term him "ill". The rate of recovery from such a deviation varies from individual to individual, and it varies within the same individual at various times during his life span.

Unidirectionality refers to the movement of the life process in a single direction from conception to death. It is on-going and irreversible. Regression cannot occur. The changes which have occurred in an individual cannot be eradicated. Repetition of behavior or functioning which occurs must be viewed as a new event taking place within the context of the total constellation of the individual and the environment at that particular time (as differentiated from a constellation which existed at a previous time). Changes occur in an individual as a result of the normal sequence of development for which he has been programmed by his genetic inheritance. In addition, changes occur because of his interaction with his environment. The rate at which the changes occur vary from one individual to another, and they sometimes may vary within the same individual as he passes through different phases of life.

Rhythmicity refers to the patterns of functioning of an individual. Biorhythms may occur in circadian, lunar, seasonal or annual patterns. It is believed that rhythmicities are due to either exogenous or endogenous factors, i.e., external to the individual and imposed by the environment, or innate to the individual in the sense of a built-in biological clock, even though the latter is not fully understood at this time. In addition, socially imposed rhythms which are related to holidays, work weeks, vacations, or other important remembered events, become very meaningful to an individual. Rhythms vary in frequency and in magnitude, and must be studied on a longitudinal basis so that the impact of disturbances can be analyzed.

GENERAL PRINCIPLES OF NURSING PRACTICE

1. The nurse relates to the patient/client as a unique individual who must be viewed within the constellation of his psychosocial and geophysical environment.
2. The patient/client participates in the nursing process to the extent that he is able.
3. The role of the nurse expands and contracts depending upon factors such as the patient/client's needs for health care services and their availability, as well as the nurse's own unique knowledge and abilities.

The general principles enumerated here are applicable in any setting in which nurses provide services to patients/clients. Laws or principles which may be used to guide nursing practice in specific types of patient's problems are being developed through the research process.

References

1. Ramey, Irene G. Meeting today's challenges to nursing service and education. *Nursing Forum*, **VIII**, (2): 160–175 (1969).
2. Gortner, Susan R., Doris Bloch, and Phillips, Thomas P. Contributions of nursing research to patient care. *J. Nursing Admin.*, **VI**, (3): 22–28 March, April, (1976).
3. Rogers, Martha E. *An Introduction to the Theoretical Basis of Nursing.* Philadelphia: F. A. Davis Company, 1970. p. 63.

An Analysis of the Nursing Process

JUDITH A. PLAWECKI, R.N., Ph.D.
Minneapolis, Minnesota

Nursing, one of the "helping" professions, has as its purpose the attainment, maintenance and/or restoration of the most optimal level of health for each individual. This purpose is not unique to nursing; it is shared with the other health-related "helping" professions. The unique contribution of nursing is expressed in its committment to and concern for the total individual. In order to achieve its purpose, nursing care must be deliberate, systematic, organized and individualized.

The concept, nursing process, has been generally accepted by nurses as the fundamental process involved in their provision of high quality individualized nursing care. It is accepted as central to all nursing activities because it is applicable in any setting, within any philosophy and/or frame of reference and allows for flexibility, creativity, and individuality in the practice of nursing.

Essentially, the nursing process is a systematic method of problem-solving applied to the nursing situation. Naturally, some writers define the nursing process in slightly different ways. In general, nursing process is defined as an orderly, systematic approach to determining a client's needs or problems (current or potential), delineating a plan of nursing interventions, implementing the plan, and evaluating the results of the interventions in light of the client's original needs or problems.

The effective use of the nursing process requires a total integration of the nurse's intellectual, interpersonal, and technical knowledge and skills. Intellectual processes involve the nurse's knowledge and skills in areas of problem-solving, critical thinking, decision-making, and prioritizing. Interpersonal knowledge and skills reflect the nurse's ability to communicate, listen, convey interest, show compassion, share knowledge and information and obtain required data. Technical knowledge and skills involve her utilization of those procedures, methods and devices necessary to bring about the desired outcomes.[1]

For the purpose of analysis, the nursing process is usually broken down into four distinct phases. These phases are assessment, planning, implementation and evaluation. These phases represent an attempt to facilitate the presentation and analysis of the total concept. In reality this is an artificial separation of actions which cannot be separated.

ASSESSMENT

Assessment is the first phase of the nursing process. According to Yura and Walsh, assessment is the act of reviewing a situation for the purpose of diagnosing the client's problems.[2] The purpose of the assessment is to collect, organize, analyze, and interpret information related to the client's state of health.

The assessment phase begins with the systematic nursing history and ends with the formulation of the nursing diagnosis, a clear concise statement of the client's needs or problems which will require assistance from nursing. Peplau has clearly summarized some of the steps involved in the process of assessment. She indicates that the thought processes include (1) sorting and classifying of data; (2) comparing of data; (3) applying concepts and determining relationships; and (4) summarizing of synthesizing data.[3] It is imperative to remember that a systematic assessment is essential to identify the nature and extent of the client's existing problems, to determine the nature of his potential problems and to rule out nonexistent problems. This type of an assessment procedure provides the nurse with an invaluable base line of information for determining changes in the individual's needs over a period of time as well as a means for evaluating the effectiveness of the nursing interventions which have been utilized.

PLANNING

Planning is the second phase of the nursing process. This phase begins with the determination of the specific nursing diagnosis and terminates with the development of an individualized specific plan of care. The process of planning requires concerted collaborative efforts between the nurse and client, since it is during this time that realistic goal-directed plans are made to assist the individual to cope with his unmet needs and/or diagnosed problems. The purposes of the planning phase are to: (1) prioritize the client's diagnosed problems; (2) differentiate between problems that could be resolved by nursing interventions from those that could be addressed by the client or those that could be more appropriately handled by other members of the health care team; (3) develop the immediate, intermediate and long term goals of the nursing interventions; (4) identify alternative nursing interventions that would meet the stated goals; (5) select from the alternatives the specific nursing interventions that would best meet the objectives; and (6) develop the specific individualized plan of nursing care. Yura and Walsh state that the nursing care plan is the blueprint for action. It provides the direction for implementing the plan and the framework for evaluation.[4] A total nursing care plan provides a centralized source of information about a client, his needs, and his nursing care requirements. It reflects a knowledgeable, creative, and intellectual nursing activity.

IMPLEMENTATION

Implementation is the third phase of the nursing process. It involves action and is the phase in which the nurse initiates and completes the interventions necessary to accomplish the predetermined nursing care goals. This phase begins with the development of the nursing care plan and ends when the nursing actions and recording are completed. Yura and Walsh indicate that the challenges facing the nurse during this phase are to be able to skillfully integrate and coordinate the activities of the health and nursing team members, provide direct care to the client, and to appropriately delegate responsibility for this care to other nursing personnel according to their backgrounds and abilities.[5]

The implementation phase includes more than the provision of the plan of care. It involves the continual assessment, planning, and evaluation activities of the nurse and her client. The plan of care has to be continually monitored, modified, and revised in accordance with the changing responses of the client during the nursing interventions.

EVALUATION

Evaluation is the fourth and final phase of the nursing process. It follows the implementation of the nursing care plan. Carrieri and Sitzman define evaluation as the continuous process through which appraisal of the effectiveness of the previous steps in meeting the patient's needs is provided.[6] The purpose of the evaluation phase is to appraise the client's behavioral changes which are the direct result of the nursing interventions. Evaluation must be expressed in terms of achieving the expected outcomes or goals of the nursing actions. It is the natural completion activity of the entire nursing process because it ascertains the degree to which the nursing diagnosis and actions were correct; it identifies the strengths and weaknesses in the entire problem-solving process and it indicates where the plan of care must be reassessed, replanned, modified, reimplemented and reevaluated.

The nursing process is a dynamic, changing, and cyclic activity. It is adaptable to any setting, appropriate for any point of reference, and compliments any philosophy of nursing. The process is flexible, demands ingenuity and creativity, as well as a great deal of nursing knowledge and skill. The nursing process is a logical framework for functioning which is applicable, meaningful and relevant for the novice practitioner and the experienced clinician since its primary function is to facilitate the development of an individualized nursing care plan for each client.

References

1. Yura, Helen and Walsh, Mary B. *The Nursing Process: Assessing Planning, Implementing, Evaluating*. Second Edition. New York: Appleton-Century-Crofts, 1973, p. 69.
2. *Ibid.*, p. 26.
3. Peplau, Hildegard E. "Process and Concept Learning." In Burd, Shirley and Marshall, Margaret A. (Eds.). *Some Clinical Approaches to Psychiatric Nursing*. New York: The Macmillan Company, 1963, pp. 333–336.
4. Yura and Walsh *op. cit.*, p. 93.
5. *Ibid.*, p. 30.
6. Carrieri, Virginia Kohlman and Sitzman, Judith. *Components of the Nursing Process: Nursing Clinics of North America*. Vol. 6, No. 1 (March 1971), p. 121.

Characteristics of Advanced Level Nursing Practice

A. MARILYN SIME, R.N., Ph.D.
Minneapolis, Minnesota

The concept of level is an important issue in nursing at this time. There probably is no other discipline which has as many levels of *formal* preparation as does nursing. One may identify the following as different levels of preparation for nursing practice: vocational, associate degree, diploma, baccalaureate degree, master's degree and doctoral degree.

Over the past few years these levels of preparation have been categorized in various ways. Associate degree and diploma preparation were at one time described as preparing a "technical" nurse practitioner, while baccalaureate preparation was for a beginning "professional" nurse practitioner. This distinction has fallen into disuse largely due to undesirable connotations many feel about the term "technical." Another categorization of these levels of preparation has been in terms of the licensure examination which may follow. Vocational nursing preparation entitles the graduate to take the examination for licensure* as a Licensed Practical Nurse (LPN); associate degree, diploma, and baccalaureate preparations for licensure* as a Registered Nurse (R.N.). More recently, a number of state nursing organizations have taken the position that vocational and diploma preparation be phased out and that the baccalaureate degree be established as the level of preparation required for entry into the profession.

The nature and number of levels of formal preparation for entry into the practice of nursing, therefore, are currently under discussion and redefinition.

For purposes of contrast in the subsequent discussion of advanced nursing practice, the nature of beginning level practice will need to be described. Without getting into the issue of what *ought* to be entry level preparation for nursing, the beginning level practitioner will be characterized.

At the beginning level, the practitioner has knowledge of the more or less clearly articulated frame of reference which characterized her nursing program. The beginning practitioner can assess nursing situations and plan nursing care according to that frame of reference. She has a body of knowledge which includes physiological and psychosocial principles underlying nursing care. Within the frame of reference of her basic program she is able, therefore, to analyze nursing situations and apply appropriate care measures.

*Assuming the preparation meets the regulations established by each state.

Advanced level will be used here to refer to that level of practice expected from a graduate of a Master's degree program in nursing. I will define the characteristics I would expect from a practitioner at the advanced level.

CONCEPTS OF ADVANCED LEVEL NURSING

Over the years, nursing has defined advanced level in varying ways, but these ways generally have had a common basis. The way in which advanced level practice has been defined has generally been on the basis of some characteristic of the content studied in a graduate program. By content I mean the facts, the principles, and the tools relevant to the field.

One way in which advanced level practice has been defined has been on the basis of a circumscribed, "specialized" area of content. Such specialization has been thought of as concentration on a particular area *within* nursing. Thus, we have had Master's programs offering degrees in various specialities such as Medical-Surgical, Psychiatric Mental Health, and Public Health Nursing. My position is that a circumscribed, specialized area of study within nursing does not in and of itself prepare for advanced level practice. We have examples of specialization at the baccalaureate and post-diploma levels.

Another way in which advanced level practice has been defined has been in terms of the addition of a *new* area of knowledge or competence—or, an expansion of the role. Earlier, such new preparation has been in teaching, administration, or research; more recently, for example, in areas such as physical assessment. Such an "addition" to nursing knowledge does not necessarily prepare for advanced level practice.

A third way in which advanced level practice has been defined has been in terms of the quantity or volume of content studied. We have developed programs in which we maintain one would get greater depth or breadth of knowledge in nursing. The amount of knowledge one has does not, in and of itself, define an advanced level practitioner.

Each of these ways of describing advanced level preparation or practice has been in terms of content—that the content studied and knowledge gained is greater, more specialized, or different from that of the beginning level. Although specifying content studied does define the kinds of problems a practitioner will address, it does not clarify the issue of level.

CONCEPT OF STRUCTURE

Specifying content studied or knowledge obtained is describing a collection of elements—elements from a body of knowledge in a discipline. However one may define the content of nursing or of a nursing speciality, it is organized in some way. There is structure to it. The body of knowledge in any discipline has structure. A structure is the interrelationship of all the parts of a whole. With reference to a body of knowledge, the parts can be thought of as concepts. The interrelationship of those concepts has to do with how they are organized. The concepts may be organized according to descriptive kinds of categories to form a whole—a taxonomy; or, concepts may be related to other concepts into statements called propositions, and, further, those propositions related to other propositions to comprise a different whole—a theory.

To know structure is to know how things are related. I think the concept structure is very important to a concept of advanced level practice. I believe the characteristics of advanced level practice emerge from the practitioner's perspective of structure. In order to achieve advanced level practice, the practitioner must develop a conceptual structure for her own practice, and look for structure in clinical phenomena.

IMPACT OF A PERSPECTIVE OF STRUCTURE ON PRACTICE

I mentioned earlier that beginning level practitioners learn the frame of reference underlying their basic program. I believe an advanced level practitioner needs to develop her own structure for practice. To develop a structure for nursing practice, one must analyze the elements underlying one's own practice and combine these elements into a pattern or framework. I would expect at least these questions to have been considered:

- What is my overall goal for nursing practice and what are the components of that goal?
- What sort of specific patient outcomes or behaviors are consistent with that overall goal?
- What nurse actions have probability of achieving such patient outcomes?
- What assumptions about man do I hold and how do these assumptions relate to the goal and nurse actions?
- Are my nurse actions and goal consistent with my own value system?

Since the advanced level practitioner has consciously developed the structure of concepts and attitudes underlying her own practice, she is aware of all the parts and relationships of that structure; she is aware of the impact on her practice from a modification or change in that structure; she would be able to articulate the conceptual structure underlying her practice.

As a result of developing a conceptual structure for her own practice, I believe her practice would have these characteristics:

- nurse actions are consistently directed toward a measureable goal.
- since the relationships between antecedents and consequences have been

critically analyzed, the predictability of a given nurse action achieving a specified outcome can be more or less stated empirically.

- judgements of an ethical or moral nature are consistent with and based upon an articulated value system.
- criteria for evaluating effectiveness of nursing care are explicit and comprise adequate evidence of effectiveness.
- the practitioner can evaluate her own practice; external validation is not required.
- the practitioner has a basis for judging the relevance of new ideas and technologies for nursing practice and deliberately chooses to modify or add elements to practice.

The second perspective I attributed to an advanced level practitioner is a searching for structure in the clinical phenomena observed—that is, a deliberate searching for relationships among the factors and forces affecting nursing. Searching for relationships among clinical phenomena requires a focus of attention on what is not known as well as on what is known; attending to an increasingly larger number of variables; meticulously recording important data; and, through systematic observation, detecting relationships among variables. From this perspective emerges additional characteristics of the advanced level practitioner:

- The nature of the outcomes achieved with patients are more substantial, more holistic. Since the practitioner is searching for structure, for relationships, the nature of man is no longer distorted by dividing him into arbitrary chunks which are focused on in isolation. Rather, outcomes are sought which reflect the pattern, rhythm, and harmony of human behavior and experience.
- There is an active contribution to the body of knowledge underlying practice. The contribution to knowledge need not involve formal research designs. The contribution may consist of base rates of a number of important clinical factors (for example the incidence of fear before surgery) and the identification of characteristics of patients who seem to benefit from a particular treatment plan.
- There is continuing communication with colleagues concerning nursing knowledge.

In summary, the advanced level practitioner can develop and reshape the conceptual structures underlying her own practice, can achieve substantive patient outcomes at a specified level of predictability, can evaluate and judge the relevance of her practice and the relevance of new ideas and technologies for her practice, and does develop and communicate nursing knowledge.

NEED FOR ADVANCED PRACTITIONERS

Nursing is currently in a state of flux and is trying to identify and define itself. Forces within nursing are in disagreement over basic and advanced preparation in nursing. Forces outside nursing are enticing nursing into functions which haven't been adequately considered and logically related to the purpose and direction of nursing care.

Given the current state within nursing and the fact that rapid changes are occurring in health care delivery and in the demands of the public, I believe it is particularly important to continue to increase the number of advanced level practitioners in nursing. We will need nursing leaders who have a gestalt for their own practice. Such leaders, in my opinion, will be less likely to respond to pressures, changes, and arguments *singly*, and more likely to have a clear view of the implications of such proposed changes to the fundamental nature of nursing. It is such leaders who will, indeed, shape and direct the future of nursing.

Minorities and Underrepresented Groups

W. Scott Sprinkle, M.P.H.
Atlanta, Georgia

The Civil Rights Act of 1964 brought to the nation's consciousness the existence and needs of the minorities. Subsequent amendments to the Act in 1972 and the growing publicity associated with ERA (equal rights for women) further raised the level of the public's awareness of underrepresented groups.

Of paramount importance to health care organizations, passage of PL 93-641, in addition to other provisions, advocated for the rights of consumers. This legislation promoted consumers of health services to activists in the health planning-decision making process. News events in the late 60's and the early and mid-70's, that carried forward the focus on individual and group concerns and needs, vied for the public's attention: Watts, Attica, Watergate, rights to privacy, informed consent, malpractice, forgotten children, deinstitutionalization, sterilization, abortion, right to life, ordination of women, homosexuality, etc.

The impact of group dynamics on the health care delivery system has been felt. These "change-makers" have enabled a rebirth of the public agency's responsibility as a "change-seeker". Our mission in health cannot be limited to only changing disease patterns, but should also encompass the broader goal that includes the responsibility for promoting and protecting human rights.

As a user of the public's funds, a large health provider organization such as the Georgia Department of Human Resources carries a two-fold responsibility concerning minorities and underrepresented groups.

The first responsibility must be for its own human community. This responsibility must be accepted as both a legal and a moral mandate. The legal mandate is embodied in such Federal legislation as: "The Civil Rights Act of 1964, as amended in 1972"; "The Age Discrimination Act of 1967"; and "The Rehabilitation Act of 1973". The moral mandate is from the people. It is often expressed in the agency's own mission statement. As an example, the Georgia Department of Human Resources' mission is to: "Provide and advocate for those human services which promote the well-being and self-sufficiency of all Georgians in a manner that recognizes human dignity."

A public agency, such as the Georgia Department of Human Resources, which brings to focus the State's mental and physical health effort along with its social services programs, should embrace this first responsibility willingly. In addition, it should, by its own effort, demonstrate a leadership role to other governmental and private organizations.

By the same token, health program units within such a Department should be expected to be the most vocal advocates for the rights and dignity of the minorities and underrepresented groups. The Division of Physical Health, one of two health divisions within the Georgia Department of Human Resources, has defined its mission "to improve the quality of health to all Georgians." Health is used in this mission statement in the broad definition of not just the absence of disease alone, but as a multifaceted concept of social, emotional and physical well-being. The mission statement can be viewed as a recognition and an acceptance by this Division of its obligation to minorities and underrepresented groups within its own human community.

The "Affirmative Action Plan for Equal Employment Opportunity" subsequently adopted by the Division of Physical Health reaffirms this commitment through five broad goals. These are:

1. To guard each person's dignity and encourage a positive self-image in every aspect of race, color, sex, marital status, religion, age, national origin, political affiliation, physical handicap or job status;
2. To free the Division from any defective systems that effectively discriminate against people because of their race, color, sex, marital status, religion, age, national origin, political affiliation, physical handicap or job status, or other irrelevant factors to the job to be done;

3. To assist, encourage, and motivate employees to enhance their abilities to realize more fully their human potential and compete for other and higher level positions within the agency;

4. To achieve within the Division of Physical Health representation of minorities and women proportionate to the available work at all pay levels; and

5. To actively recruit from all segments in the population and work force, including those who are educationally and economically disadvantaged, and offer opportunities for employment to people at all skill levels.

However, the achievement of these goals requires positive actions to eliminate any vestige of discrimination in employment and to assure equal employment opportunities for applicants and employees alike.

For the job secure majority to make such a commitment to change requires the group to recognize its own need for change, determine its ability to change, and establish a climate for change. After recognizing its discriminatory blind spot, the next priority is a need for a human relations training program with a focus on racial bias.

This training experience for the Division of Physical Health has produced two major benefits directly related to its action plan. It improved existing communication and established new communications, as well as personal ties, between individuals. It also allowed individuals to become aware of their own corporate and individual prejudices. Subsequent training workshops, which were requested by the original participants, are continuing this learning experience. Several action steps by the Division of Physical Health are attributed to this training. Examples include:

1. Monitoring employment to detect and eliminate discriminatory practices;
2. Studying promotion practices to assure open opportunities;
3. Utilizing of individuals to their best skill levels; and
4. Reprioritizing of resources.

The importance of an agency assuming the role of "change-seeker" within its own internal human community cannot be overstressed. Until the agency fulfills this first commitment, it cannot assume its second responsibility, that of being a "change seeker" in the external human community.

A health services agency must reach out to the external community of service users. A public agency, in particular, must relate to groups of individuals who have joined together to represent a common need. The agency should not only recognize such groups, but should actively seek their concerns. The viable health care agency will not establish a defense against group expression, but will encompass group representation.

As an example, health consumers, prior to the late 60's allowed themselves to be "cared for" and did not see the need for forming a group voice. They allowed the health provider to determine their needs and to decide the delivery systems through which these needs would be met. From their perspective, the health care provider tended to remove decision making from the consumer. Clients were talked to, but not talked with. The nurse mothered them, the health educator organized them, the physician decided for them and proclaimed to them, the environmentalist regulated them, the institution stripped them of their dignity, and the administrator thoroughly confused them. Georgia's New Health Outlook places primary health decision responsibility on the individual. It recognizes that it is the individual who will determine his own lifestyle and, therefore, his health status, and not the health provider.

The minorities and underrepresented groups of the 60's and 70's, blacks and women, were and are demanding change in order to acquire human dignity. The minorities and underrepresented groups in the 80's, the consumers, will be demanding change in the accessibility, availability, and quality of health care.

Unless the needs of all minorities and underrepresented groups are heard by the health provider organization, the agency cannot know its human community. Knowing its human community, and incorporating the interests and concerns of all its members, can increase the effectiveness of health services for both providers and consumers.

Patient Attitudes

RITA LINDENFIELD, Ph.D.
Professor and Social Worker-in-Chief
Clark Institute of Psychiatry
Toronto, Canada

Doctors are to be applauded for the strides they have made in pinpointing diagnosis and in developing more effective methods of treatment. But what about the attitudes of patients and relatives to diagnosis and treatment?

After careful examination and extensive tests, the doctor often is faced with the unenviable task of telling the patient and the family that the patient has a serious illness. Hopefully this illness will respond to a regime of treatment. In this aspect, they will have to work together. The doctor is compassionate, carefully outlines the regime and arranges to see the patient again. The patient and the family thank him politely—and go home.

Home seems cold. It seems different. It *is* different because someone has a serious illness. No longer are the commonplace activities commonplace. A pall has descended.

In the meantime many feelings surge about: sadness, fear both of the known and the unknown, anger, guilt, shame—a host of feelings. As a result, individual family members, or the family as a whole, demonstrate a wide range of reactions which, in turn, give rise to a wide range of behavior. As a social worker I am concerned about these attitudes and the subsequent behavior for many reasons, not the least of which is the possible negative effect on the patient's illness and the treatment.

After the initial numbness of shock one of five reactions likely will occur. One or another of these may predominate and be consistent, or the reactions may fluctuate with now one, now the other, gaining ascendancy. Five of the most common reactions are:

A Conspiracy of Silence

There is an unspoken understanding that the diagnosis, the treatment plans are not to be mentioned, not referred to by anyone to anyone, even within the family. Friends, employers, teachers, telephone and are given the agreed-upon innocuous information: "Oh, she is doing well. Yes, we are all fine. Thank you." Calls directly to the patient are met with the same wall.

Over-Protection

The family rush in to protect the patient in an unrealistic fashion. The patient can do more, can participate more, in normal activities. However, the family hovers over the patient, quick to set limits. The patient may welcome this tent of protection, or reject it. If it is the latter, consternation will reign.

Denial

The doctor is wrong! If the patient had that illness, symptoms would have shown up long ago. And so recently the patient looked well and was very active. Perhaps they should take her to another doctor. After a rest she will be fine. The patient "forgets" to take the medication which so graphically recalls what the doctor said.

Isolation

The patient and the family accept the diagnosis and make no secret about it. Although immersed in sadness they follow the treatment to the best of their ability. Meantime they withdraw from friends and the extended family who want to, and might, offer much solace.

Overactivity

Sometimes the patient becomes very passive, but one or more of the family fly into a flurry of activity. They are quick to tell others, seek advice from a great variety of lay people and others, frantically search for literature about the illness, and structure the patient's daily activities. Meantime the patient may remain passive or become involved in an equally frantic attempt to dispel anxiety.

These reactions are natural and normal when the patient and the family are first told the doctor's findings. People do need time to absorb the shock, to realign their forces. But if such attitudes are prolonged they, very effectively, may sabotage all the efforts of the doctor and the treatment prescribed.

We must pay attention to these attitudes; we must work with them. This is not to imply that we try to change them overnight. Indeed in the rare instance it may be better to accept them as temporary essential crutches. Only a careful assessment based on sound knowledge will guide such a decision. However in most instances, if these attitudes persist, the patient and the family may become at best, reluctant recipients of the ministrations of the doctor; at worst, they may avoid such ministrations. But most patients and families *want* to cooperate with the doctor. Failure to do so suggests that their attitudes have crushed their will!

Improved Pregnancy Outcome

Jules S. Terry, M.D.
Atlanta, Georgia

Having a baby in the United States today is a more hazardous process than it is in at least 13 other countries in the world. In the United States, interest in our health care has been concentrated almost entirely in the system for delivery of health care, ignoring other preventative factors which have far greater impact on the well-being of the pregnant woman and her child. These factors include biological, environmental, and lifestyle components which directly influence all of the facets of our society and contribute in large measure to the poor perinatal statistics which prevail in this country. Such statistics forecast a directly proportional high incidence of mental retardation, since the high incidence of teenage pregnancy results in a high prematurity rate, with a large number of mentally retarded babies.

The United States lacks comprehensive prenatal care services which stress early and adequate care. There is almost a complete lack of genetic counseling services, although 16% of conditions that lead to mental retardation are based on chromosomal abnormalities. The elimination of rubella is possible through an aggressive immunization program of our young population. Also, in this age of universal availability of Rh immunoglobulin, it is inexcusable for a patient to carry an Rh sensitized pregnancy. Incredible as this is, there are many Rh negative women who are not routinely given Rh immunoglobulin after the termination of pregnancy, regardless of the duration of pregnancy.

Although we are aware of the many factors that enter into improved pregnancy outcome, our concentration has been in the development of a system for delivering health care which will serve only to take care of the problems produced by lack of prevention. The currently developing system for delivery of perinatal care is a regionalized system which provides several levels of care of mother and baby. Next must be addressed a prevention program aimed at improving the future potential of those yet unborn.

An essential element of the perinatal health care delivery system is the early identification of the high risk patient, preferably in the first trimester. The perinatal period is the optimum time for preventive intervention. A woman ideally should come to pregnancy when she is physically and socially mature, has been immunized against rubella in childhood, and is in good nutritional status, so she can undertake a pregnancy which is wanted, and for which she is prepared. Significant reductions in maternal, fetal and neonatal morbidity and mortality can be achieved if mothers and babies at high risk are identified early and if the technology and expertise of modern obstetrics and pediatrics are readily available.

Guidelines for the regional development of perinatal health care services to improve the availability and utilization of perinatal care for high risk women and infants have been developed by the National Commission on Perinatal Health. These state that the development of a regional system for delivering perinatal services requires more than simple differentiation of the hospitals delivering various levels of services. Attention must be given to coordination and communication among all components of the system. These components must comprise a functioning system through planning and decision making, culminating in a plan which has improved the outcome of all pregnancies as its only objective. A statewide perinatal plan for the improved outcome of pregnancy should result.

Planning for delivery of perinatal health care services must involve the family, the community, the community hospital, and the intermediate and regional perinatal centers. None of these entities can plan separately, for all of them are interdependent. It is anticipated in the planning that the area to be served will be determined both geographically and by the number of live births in that particular area. One of the by-products of this planning may be to identify hospitals which should terminate their obstetrical and newborn services.

To provide the highest quality of perinatal care available, the perinatal health care delivery service must include levels of care, financing, screening procedures for identification of high risk, education (both professional and lay), evaluation of the effectiveness and efficiency of the program, and consultation and referral. If the system deals with the components of lifestyle and environment, education of young people must be included so that they may become aware of the potential hazards of becoming pregnant at an early age. The advantages and disadvantages need to be clearly outlined so that they can make an objective decision.

The environment of poverty and social deprivation foster ignorance of and reluctance to enter into the health care delivery system, with many poor people unaware of even how to approach the system. Health professionals must help improve pregnancy outcome by making high quality health care available, affordable, and accessible.

With the identification of the high risk maternal patient at the local level, it is essential that a functioning consultation and referral network system be in place. This will allow the patient to be referred to the appropriate level of services for optimum care required in her case. Such a system must evolve through the participation of the small community hospitals, and the larger intermediate care hospitals. No one of these three components can stand alone. The consultation and referral service should also have provisions for emergency and non-emergency situations. Additionally, a communications link must be set up at the varying levels of care within agencies in the community. For this the use of a "hot line" telephone which makes consultation available on a 24 hour basis is optimal.

Levels of care are an integral part of the regionalization of perinatal care. There are three levels of care designated as primary or community level, secondary or intermediate level and tertiary or the regional perinatal center level. The primary level is provided through local health providers and the community hospital, which care for uncomplicated maternity patients and well or minimally ill newborns. These providers and hospitals should be capable of managing unforeseen complications of labor and delivery. It is essential that a relationship with the intermediate and the regional perinatal centers be established involving direct communication, consultation, and transportation. The community level should be particularly involved with the provision of prenatal care, the early detection of potential problems, and directly involved in the development and implementation of health education and interconceptional care.

Intermediate or secondary level hospitals should provide care for the management of uncomplicated maternity and newborn patients, as well as most high risk obstetrical patients and moderately ill infants which do not require ventilatory assistance or pediatric surgery. These hospitals are usually in larger population centers, and will probably deliver the largest number of patients. In these facilities a higher degree of sophistication, both in personnel and equipment, is to be expected. The newborn and obstetrical facilities should be a separate entity and should have the competency to perform emergency procedures quickly. These level two hospitals should have an ongoing active relationship with the community level one hospitals, as well as with the regional perinatal center. There should be a system for transfer of at least the newborn between the various levels, depending on the type of care needed by the infant at that particular time.

A level three hospital, or regional perinatal center, manages all types of high risk obstetrical and critically ill newborn patients, as well as its local caseload of uncomplicated maternity and newborn patients. Capability in the intensive care nursery should include prolonged ventilatory support as well as the ability to perform specialized diagnostic and therapeutic procedures such as cardiac catheterization, open heart surgery, and other major pediatric surgery. Of course, support services of laboratory, X-ray, blood bank, anesthesiology, and others must be available without delay on a 24 hour basis. The staff and the facilities in these organizations should be of the highest caliber and sophistication. The level three hospital has the ultimate responsibility for the consultation-referral mechanism. The level three hospital, also, will be the chief utilizer of the newborn and obstetrical transport system, and as such, should address itself to its requirements.

All levels of providers of perinatal health care should be responsible for perinatal health education, and their skills in this essential component should be upgraded on a systematic basis. It is also the responsibility of all three levels to educate the professional community, both medical and nursing, in the various aspects of perinatal health care. All those involved in delivering perinatal health care services should be obligated to provide community education, especially for young people. Objectives of these community efforts should be to teach the physiology of pregnancy and its effects on the immature body. Moreover, the higher incidence of premature delivery, subsequent mental retardation of the child and the higher incidence of obstetrical complications in adolescents must become common knowledge. Then, the far reaching social, psychological, educational, and economic consequences of early childbearing should help change values and create appropriate lifestyles.

The cost of perinatal care in actual dollars is high. It is important, therefore, that the value of this care be stressed and that legislators and others appropriating funds understand that appropriations for perinatal care are costly. In the long run, however, if there is a reduction in the incidence of mental retardation there is a great saving. The average stay in an intensive infant care nursery in the state of Georgia is ten days at $400 per day which comes to $4,000, while the care for someone who is mentally retarded comes to $20,000 per year with a life expectancy of 40 years. It is clear to see where the savings are realized.

Optimally, then, pregnancy should occur in a physically mature woman, whose nutrition has been adequate. She should have been previously immunized against rubella and should be in good health. Biological factors which might influence pregnancy, such as Rh problems or familial genetic disorders, should be identified so that the patient can make an educated, objective decision. Pregnancy should be a wanted and healthy addition to a two parent family. The consequences of increased obstetrical risk, increased incidence of prematurity, and increased incidence of subsequent mental retardation as well as the social, economic, psychological, and educational sequelae of a teenage pregnancy should be common knowledge to all adolescents and young adults. If our values and lifestyle can be modified so as to reflect the foregoing principles, then the improved outcome of pregnancy in this country will be assured and the need for sophisticated, costly perinatal care proportionately diminished.

Teenage Pregnancy Prevention

FRANK M. HOUSER, M.D.
KATHLEEN McNEAL, M.A.
Atlanta, Georgia

The high national rate of teenage pregnancies has become a growing, serious problem for those concerned with adolescent health. About 10% of the nation's teenagers conceive each year, and 6% actually give birth. Teenagers in the United States have child-bearing rates among the world's highest.

Last year in Georgia, there were 22,716 teenagers between the ages of 10–19 who delivered a child; this was fully 25% of all live births. This figure ranks the state 4th highest in fertility rates for the nation.

Pregnant teenagers are high risk for complications at birth, and low birth weight infants are much higher among adolescents. Prematurity is a primary factor causing mental retardation, cerebral palsy, and other chronic health problems.

The problem of teenage pregnancy has many implications in an adolescents' life—economic, educational and emotional. It is the most common cause of school dropouts among teenage girls. They are more likely to have to work to augment the family income, and many remain financially dependent on their parents and society, with little opportunity to acquire vocational skills. The emotional adjustment to a newborn child and the ensuing full time responsibility is fraught with anxieties and frustrations. Estrangement from family and peers, and the struggle to redefine identity in new roles may be two important reasons for the high incidences of divorce and child abuse in teenage parents.

Although birth control has been available to adolescents without parental consent for several years, the incidence of unintended births is escalating. Fifty-nine percent of all sexually active teenagers are not usually using any method of birth control; with sexual activity usually unplanned and sporadic; many find birth control too calculating. Others have erroneous facts about sex, and about how to obtain and use birth control methods. Teenagers frequently avoid local family planning clinics for fear of being recognized by acquaintances, especially in rural or suburban communities.

Pregnancy for some adolescents provides an escape from an unpleasant home or school situation. A child may represent the only component in a maladjusted teenagers' life that offers unconditional love and loyalty. Unfortunately, the experience may only serve to delay the vital development of fulfilling mature intimacy.

Adolescent sexual activity, in the past, has been portrayed as representative of the minorities and the poor, but present evidence from actual fertility statistics indicates that teenagers from higher income and nonminority groups are sexually active in earlier ages. The average age of beginning menses is now at 11 years of age, perhaps attributable in part to environmental factors, such as improved nutrition.

To reverse this trend of ever-increasing adolescent childbearing, we must continue to broaden communication with all adolescents about their goals, feelings, attitudes, and needs.

One model now being used in two counties of north Georgia is a Teen Pregnancy Prevention Program, primarily focused in the school. The program is provided by the county health department and the local school boards. It is presented to all students in grades 7 through 12. The initial program component is a sound/slide presentation which describes the clear risks of adolescent pregnancy from medical, emotional, and socio-economic perspectives. This is followed by group discussions led by a nonjudgmental, well-informed health department facilitator.

The challenge of teen pregnancy prevention is complex. Given our national statistics, the socio-economic future for a pregnant adolescent and her child is not encouraging. In the promotion of the philosophy of an ever-improved quality of life, prevention and responsibility must continue to be our emphasis and commitment.

Primary Health Care

FRANK M. HOUSER, M.D.
RUTH MELBER, R.N., M.A.
Atlanta, Georgia

Public Law 93-641 gives legal basis to the premise that health care is a basic human right of every American. The essentials described in the law to assure this right are accessibility, availability, affordability and quality of care.

These essentials are not elements of the existing medical care delivery system. Today's system is characterized by spiraling costs, manpower maldistribution, and limited access, with virtually no quality assurance. In addition to these problems within the medical care system, the disease patterns with which the system must deal have changed from primarily infectious to primarily chronic in nature.

These complexities, which affect our entire societal structure, require a new outlook on health. Such an outlook must include consideration of the effects of the environment, individual's lifestyles, human biology, and the medical care delivery system. These components are necessary to the development of a health care system for the 1980's.

The foundation on which such a system should be built is a comprehensive program of primary health care. Primary health care provides ambulatory services which address local health needs in terms of current disease patterns and which are accessible, available, and affordable. To be relevant to current disease patterns and local health needs, primary health care services must include preventive and supportive activities as well as management of chronic diseases. Other services offered would be health education, genetic counseling, and environmental awareness, in addition to diagnosis and treatment of acute illnesses.

Studies have shown that approximately 80% of health services can be effectively managed by trained personnel other than physicians and dentists. Nurse practitioners and physician assistants may well be the primary health providers in the 1980's. The appropriate utilization of these personnel would be a less expensive method of primary care provision than the existing model.

Availability, accessibility, and quality must be addressed by the development of innovative models of service delivery which provides a single entrance into the system. Efforts are currently being made in Georgia to modify the system based on these directions. Three models are in various stages of development.

In Atlanta, Grady Hospital and the Fulton County Health Department have developed three primary care centers located in high population, low income, urban neighborhoods. These centers are staffed by nurse practitioners, physicians, dental hygienists, nutritionists and allied health personnel. Acute medical care, health education, dental services, contraceptive methods, nutrition counseling, and on-going management of chronic diseases are among the services provided. Medically, the centers utilize Grady Hospital specialty clinics and in-patient services to assure comprehensive care.

Efforts are underway to develop primary care services in medically underserved rural areas. In five such underserved counties, primary care centers have been established within the existing public health system. These centers offer the services of a nurse practitioner functioning under medical protocol, with medical backup and referral to medical resources. It is recognized that these centers do not meet all of the health needs in rural communities but they have made health services more accessible and available than previously. Current planning recognizes the necessity for linkage of these centers to additional health resources.

In one area, outside the large urban centers, a program is being developed to serve a county of 75,000 persons. The role of county health department is being redefined. An ambulatory care clinic is being developed as a core service. The primary provider in the clinic will be a nurse practitioner with backup care provided by a local internist. The clinic will be open during afternoon and evening hours. Coverage will be provided at times the clinic is closed by the salaried E.R. physicians in the county hospital. The hospital emergency room will also serve as the referral source for clinic patients requiring hospitalization.

As an integral component of the local health department, the clinic will, in addition to acute care services, provide a system for management of chronic diseases, health education, nutritional counseling, dental services, family planning, immunizations, well-child care, and prenatal care.

This concept was developed jointly by the local health department and the county medical society. The medical society formally endorsed the program to assure the essential linkages to the remainder of the area medical care delivery system.

The development of methods of quality assurance for services and continuity of care is essential to the continuing planning process.

The Psychobiology of Human Motivation

SIR HAROLD HIMSWORTH, M.D., F.R.S.
London, England

Viewed from the standpoint of their behavior, the different species of animals range from those, like the tiger, that operate alone, to those, like ants, that function as organized communities. In this spectrum, man occupies an intermediate position. He is neither entirely an individualist nor entirely a social animal. He is, at one and the same time, both. In consequence, his motivation is subject to two conflicting sets of forces. On the one hand are those deriving from his egotistical traits impelling him to satisfy his personal inclinations; on the other, those arising from his social, prompting him to cooperate with his fellows and to subordinate his own self-seeking to the common welfare of the group to which he belongs. This is the dilemma that lies at the root of the problem of human motivation. Accordingly, in any approach to this subject, it is to the way these two conflicting tendencies reveal themselves to the conscious thought of the individual that we should direct our attention.

As William James pointed out, many years ago, it is the peculiar property of any proposition that relates to the satisfaction of a biological need to appear to the individual as self-evident. Of course, a man will take care to ensure he has enough to eat. Of course, if he is under the threat of attack, he will take steps to provide for his defense. Of course, he is attracted by a pretty girl. But needs only arise in the context of situations. A man's response to any particular set of circumstances depends essentially, therefore, on the idea he has of its implications. Given this, the form of conduct he habitually adopts to deal with situations of this kind springs immediately into his mind and, under the compulsion of his egotistical proclivities, he feels impelled to give effect to it. As William James said, for the individual concerned, the connection between the thought and the deed in such cases is "absolute and *selbstverstandlich*, an a priori synthesis of the most perfect sort needing no proof but its own evidence."[1]

In principle, similar considerations apply in respect of those ideas men hold in virtue of their needs as partly social animals. But here there is a difference. Ideas that relate to the satisfaction of an egotistical need are directed explicitly to their purpose. Those that relate to a social, only at one remove. And it is here that the now almost forgotten contributions of Wilfred Trotter are of such signal importance.

The end to which man's inherent social proclivities are directed is to enable a species, the members of which are only partly social animals, to form integrated communities and so realize their collective strength. That, however, they can only do to the extent they think alike on matters of common concern. Unless, therefore, man had become possessed, in the course of his evolution, of inherent traits that disposed him to imbibe, and accept as his own, the beliefs and customs of those with whom he was associated, his species would have been forever precluded from developing on social lines. Given such traits, however, as Trotter pointed out, any idea that comes to be held in common by a group of men will, *ipso facto*, acquire an authority over their thought processes, so that it becomes, in their estimation, "an axiomatically obvious proposition" which it would be foolish or wicked to doubt.[2] Thereby it graduates from the status of a concept to be considered on its merits, to that of a belief which it is morally incumbent upon them to accept. It is this that lies at the basis of the phenomenon men call conscience. It is this that accounts for an individual's susceptibility to public opinion, his fear of ostracism, his acceptance of social authority, his capacity for altruism and for the fact that men in different groups can hold, in all sincerity and conviction, such divergent views on one and the same subject. In any consideration of human motivation, or the contingent subjects of ethics, law and politics, it is, in consequence, to the pervading influence of this factor in human affairs that we should pay particular heed.

More than a century ago, Charles Darwin drew attention to the potential importance of man's inherent traits as a social species for any understanding of his conduct.[3] At that time, however, the study of animal behavior was in its infancy and there was not the background of knowledge to allow this line of thought to be followed up. But now the situation is changing. As a result, we are increasingly coming to see that similarities in conduct between men and lower animals may have a significance for us comparable to those in anatomical structure and physiological function with which we have long been familiar.[4]

References

1. James, William. *Principles of Psychology.* MacMillan, London, 1890, (republished 1910). Vol. II, p. 386.
2. Trotter, Wilfred. *The Instincts of the Herd in Peace and War.* Ernest Benn, London, 1916.
3. Darwin, Charles. *The Descent of Man.* John Murray, London, 2nd ed, 1887, (first published 1871). Chapter 4.
4. Wilson, Edward O. *Sociobiology: A New Synthesis.* The Belknap Press of Harvard University Press, Cambridge, Mass. and London, England, 1975.

Public Acceptance of Non-M.D. Providers

RASHID BASHSHUR

Professor, Department of Medical Care Organization
School of Public Health
University of Michigan
Ann Arbor, Michigan

Public attitudes toward non-M.D. providers of care (i.e., physician assistants and nurse practitioners) were investigated in a sample survey that was completed in 1975 in a rural community in Maine. The study had several objectives, but only those pertaining to public acceptance of non-M.D. providers are reported here. The overwhelming majority of the population of this community is white, and a substantial proportion of them are poor. Over the last 100 years the area has experienced a small net decline in population. Medical resources have also been limited, and hence attempts were underway to experiment with various approaches and innovations aimed at redressing problems associated with inadequate health care manpower and facilities. The use of non-M.D. providers was included among other innovations, such as the use of interactive television, group medical practice, and multi-site delivery settings. A crucial issue with regards to all of these innovations is their acceptance by the public, hence the problem addressed in this paper.

In the survey the following explanation was given to the respondents:

> Some health programs and doctors today use specially trained persons called physician assistants, Medex or nurse practitioners to help in caring for people . . .

Subsequently, the following question was asked, the responses tabulated against membership status in Rural Health Associates (RHA), the innovative organization:

> From a patient's point of view, how do you feel about receiving some medical care from such a person?

A grouping of the specific answers given by the respondents revealed slightly over one third of the sample to be in favor of the idea, the remainder about

This research was supported by a grant from the RANN program of the National Science Foundation.

evenly divided between those accepting the idea with reservations and those rejecting it (Table 1). The basic independent variable was "membership status," and it identified the *subscribers*, those persons who were covered for services under capitation arrangement with their premiums fully assumed by HEW (formerly OEO), the *users* who received their services on a regular basis from RHA on a fee-for-service basis, the *eligible non-users* who, by virtue of limited income, qualified for enrollment but had not done so, and the *non-eligible non-users* who neither qualified for enrollment nor utilized RHA on a regular basis.

In this instance the subscribers and the users of RHA were significantly more favorable to the idea than the non-users. This trend can be interpreted to reflect the positive influence RHA had on people's acceptance of the non-M.D. provider in their own care. A more direct test of this hypothesis is provided in Table 2.

A distinction was made between acceptance of non-M.D. providers and preference for them over an M.D. for specific diagnostic functions. For each diagnostic function, respondents were asked to indicate whether they "would let them [non-M.D. providers] do it," and for those who answered affirmatively, "which would [they] rather have: M.D. or non-M.D.?" Data were also gathered elsewhere on whether the respondents had been treated by a non-M.D. provider. The percentages shown in Table 2 differentiate between those with and those without a previous experience with a non-M.D. provider, and indicate the percentage distribution of their responses according to their acceptance of and preference for specific providers for each diagnostic category. It is odd to note that referral to an M.D. was the least accepted function for the non-M.D. providers. But, as expected, taking blood pressure and medical history were viewed favorably. The positive influence of previous experience is shown in all categories,

TABLE 1. Acceptance of Non-M.D. Providers by Membership Status.

Attitude Toward Receiving Care from Non-M.D.	MEMBERSHIP STATUS					
	Subscriber	User	Eligible Non-User	Non-Eligible Non-User	Status n. a.*	Overall
Accept	46%	43%	22%	28%	17%	35%
Accept with reservation	24	24	30	33	40	28
Not accept	16	29	37	32	26	29
Conflict response; don't know	14	4	11	7	17	8
Total	100%	100%	100%	100%	100%	100%
Number of cases	55	193	27	120	42	437

x^2 = 33.73
d.f. = 12
sig. = .001
contingency coefficient = .27
*n. a. = not ascertained

TABLE 2. Acceptance of Non-M.D. Providers for Specific Services
by Type of Service and Experience.

Type of Service and Previous Experience	WOULD ACCEPT NON-M.D. AND PREFER:			Would Not Accept	Total (%)
	Non M.D.	M.D.	No Preference		
Take Medical History					
With Experience	19	42	35	4	100
Without	22	34	27	17	100
Make Routine Physical					
With Experience	10	41	18	31	100
Without	8	29	9	54	100
Sew up Cuts					
With Experience	12	36	17	35	100
Without	7	29	10	54	100
Examine Sore Throat					
With Experience	10	35	29	26	100
Without	11	32	18	39	100
Take Blood Pressure					
With Experience	26	31	29	4	100
Without	25	32	28	15	100
Refer to M.D.					
With Experience	9	29	22	40	100
Without	6	22	15	57	100

particularly among those who held firm negative views. The major shifts reflecting experience are shown in the negative response "would not accept," and to a lesser extent among those without preference. Nevertheless, the preference to have an M.D. was prevalent in all diagnostic categories, and no more than one quarter of any of the groups preferred non-M.D. providers. In brief, these figures do reveal a substantial public receptivity to the extended clinical role of the non-M.D. provider.

This receptivity is generally enhanced with experience. Hence, efforts aimed at increasing the supply of non-M.D. providers in rural areas must take into account the slow rate of acceptance of change and the requisite experience to justify acceptance. Rural residents have indicated that they are likely to accept an innovation only after it proves itself.

Voluntary Infertility

GEORGE THOMASSON, M.D.
ANDREA JACKSON
Atlanta, Georgia

Pregnancy should be a positive experience resulting from a loving family relationship. The positive effect on the community unit sharing in this relationship is a traditional value understood by our society. In other instances problems for individuals and the community arise as a result of someone becoming pregnant. Everyone in our society benefits from the former situation, and likewise, everyone suffers from the latter. For example, not only are the mother and father involved in the circumstance of teenage pregnancy as well as their respective parents and family members but the community as well shares in the consequences with school, social and cultural issues which arise. If the child is retarded or disabled, the additional burden born by society in providing supportive and remedial programs is significant.

Today it is apparent that many factors involved in the issues related to pregnancy are lifestyle considerations. Do some teenagers become pregnant intentionally, and why? Do they understand the subsequent effects on future lifestyle factors? How do these situations relate to the problem of child abuse? These are a few of the situations which individuals influence by personal decisions.

The community shares in decision-making as it provides the environment in which decisions are made and results dealt with. Why is the birth rate declining proportionately more in the middle and upper socioeconomic strata than the lower? How do we deal with social and cultural standards inhibiting utilization of these programs? How widespread is the concern that family planning is based on a sinister concept such as genocide? Are these significant factors in community acceptability in only limited areas? How do we provide the community an opportunity to participate in decision-making to provide a favorable environment for individual decision-making? Why is the responsibility placed on the female partners to deal with contraception? Many of these issues require a broadened perspective involving behavioral psychology, anthropology, and sociological skills.

Many of the technologies currently and potentially available bear directly on biological and research considerations and will require continued prospective studies. What is the risk to a teenager of becoming pregnant? What is the risk to her child? What problems occur when older women become pregnant? How do personal health factors affect the mother and her child? What effect do existing physical or mental disabilities have on the outcome of a pregnancy? What are the adverse effects of hormonal contraception? How is this affected by longevity of use? Is there a genetic risk in hormonal contraception? What needs to be done to develop more practical male contraception? The health care system in our country is manifestly involved in planning, providing, and evaluating programs and methods of voluntary infertility. With the exception of spermicidal products and condoms, every current contraceptive method and all planned methods are provided by the medical care system. How does this limit availability? What method of funding could more realistically underwrite the cost of this system? How do providers' attitudes facilitate or inhibit the use of any contraceptive technique? How do we keep providers knowledgeable about the most appropriate matching of techniques and personal needs. How can quality of care be ensured and realistically monitored in a diverse population utilizing a fragmented delivery system? The health care system must address these issues as well as those in basic research for effective decisions to be reached about providing accessible, available, and effective systems in voluntary infertility.

The Quality of Life

ROBERT H. BECK
Regents Professor
University of Minnesota
Minneapolis, Minnesota

Two critiques of the American quality of life can be looked to for their paradigmatic adequacy and possible benefits to be had from taking them seriously. One of these critiques is an adaptation of Plato's thought on physical and mental-moral health of the individual and the body-politic. The other is drawn from the thought of Mao Tse-tung, Chairman of the Communist Party of the People's Republic of China.

These two quite different criticisms do agree on two points, that is, (1) the American ideal of progress is one of unrestrained consumption and (2) there is anything but a harmonious relationship between our chief social classes. The good Chinese man or woman, as the protypic Platonic or Greek good person, is restrained and dedicated to the welfare of his/her society.

Having recognized this area of overlap the two critiques have to be taken up separately. First there is the matter of goals. Plato sought what his predecessors had honored for a very long time, that is social harmony. To reduce his argument to its minimal essentials, social harmony would only come to be when each person did what he/she was suited by his/her nature to do. Those whose natures were similar would do well to band together, thereby forming social classes of those with like temperment. In the Platonic class-structure there would be the *hoi poloi* at the base of the pyramid of power and responsibility. Above them would be the guardians and at the apex of the triangle, the philosopher-king, whose closest colleagues made up the powerful Nocturnal Council described not in the *Republic* but in Plato's last dialogue, the *Laws*. In this class-structure the lower class could be thought the workers of the world, the traditional hewers of wood and drawers of water. These people held opinions but were guided by little knowledge and had no wisdom at all. The guardians, too, had more courage than wisdom but they at least were educable beyond the rudiments and it was from their ranks that one looked for the person truly characterized by a love of wisdom, the philosopher, who should be king. All would be arranged in orderly fashion and the philosopher-king, closeted with the Nocturnal Council, could be counted on to guide wisely.

Mao Tse-tung substitutes the upper-echelons of the Chinese Communist Party for the philosopher-king and calls for a perpetual struggle against self-seeking and the threatened dominance of an elite, a bourgeoise. The face of struggle without end seems to bespeak the dynamic but, in fact, Mao Tse-tung's civic design is static. Leaders come and go but the ideal norm is as unchanging as Plato's ideal Republic. For Plato, as for Mao Tse-tung, wisdom is not widely held. As we know, the Chinese people seem to share in power more effectively than those who would populate Plato's city-state. But that sharing in "crown power," as the British once described the tie between England and such independent nations as Canada, Australia and New Zealand, is more for show in the trappings of ceremony than in reality.

How might one respond to these critiques? There is little question today of the self-indulgence of those with the ability to be self-indulgent, whether this be seen in terms of individuals or nations. Our best hope for enhancing the quality of life is through education. Families and other institutions cooperate in informal aid; of great importance but held apart from the formal education through schools and collegiate education. It is the latter to which our attention is given. Specifically the assignment that schools in a democracy have is to prepare a large number of young people who develop the habits of cooperating, displaying a social and esthetic conscience in their everyday civic behavior, and willing to contribute to the well-being of others even while developing themselves and their careers. It goes without saying that this training should accustom students to communicating, the latter being essential for cooperation.

The alternative is the leadership of a small minority, an elite however much this miniuscule fraction denies that it is a minority that leads, that sets the fashions, the standards and, ultimately, the laws. Although modern society has seen the emergence of a bureaucracy such as is not envisaged by Plato, I would argue that the essential critique to be made is rather like what I have found in the Platonic dialogues, this last, of course, is to be supplemented by the writings of Mao Tse-tung and other spokesmen for the aspirations of the Chinese Communist Party when attempting to inspire the peasants, workers and military. This Chinese critique can represent the essence of the Marxist-Leninist view, in whatever national guise it appears.

Death

ROBERT FULTON, Ph.D.
Professor and Director
Center of Death Education and Research
University of Minnesota, Minneapolis

Death occurs in all living things and marks the end of an individual life process. Death is imminent in life and is part of the continuous transition through which all living organisms grow and decay. In this respect, death may also be viewed as a set of changes that occur in the life cycle of an organism.

In any complex organism three kinds of such changes can be distinguished: First, certain groups of cells die and are replaced through normal biological activity throughout the life of the organism (necrobiosis). Second, certain cells or part of an organ or tissue may die due to blockage of the blood supply or other reasons without fatal consequences for the organism (necrosis). Third, there is the cessation of all vital functions (somatic death).

We have traditionally defined a person as dead (somatic death) if the heart and lungs have ceased to function and if there is failure of the eye to react to light or the skin to pain. However, surface hair and skin and bone cells may continue to grow for several hours after death while the cells of the brain cease functioning within a few minutes. In fact, the problem of interpreting these changes, that is, the problem of deciding when death has actually occurred, has been complicated by recent advances in medical technology. New procedures make it possible to define a person as dead if his brain measures no electrical activity (as recorded on an electroencephalogram) for a period of 24 hours or more. In other words, brain death is said to have occurred when the normal function of the brain has been irreversibly destroyed even though the heart and lungs may still be functioning through the use of heart-lung machines.

The efforts to establish a new definition of death reflects the struggle of humankind to understand and cope with death. Over the centuries we have sought not only to find the causes of death but we have also attempted to understand its meaning. Theologically, death has been explained in terms of God's Will and its cause as a result of humankind's fall from Grace. In modern secular times, death is explained less by Divine Will and more by chemical, bacterial, nutritional, or other biological causes while its meaning continues to be debated.

It is not too long ago that the majority of deaths that occurred were those of children who died from such infectious diseases as diptheria, pneumonia, or diarrhea. While this continues to be the case in underdeveloped countries, people in modern industrial societies die specifically from such degenerative ills as diseases of the heart, stroke, or cancer. Today it is primarily the elderly who die and they die in public institutions under the care of medical specialists. This change in who dies and where they die has had a significant impact on society. For one thing, it means that a young person has limited experience with death and dying. In fact, the present generation of young men and women could be described as a "death-insulated" generation—that is, the chance is only 5 out of 100 that a death will occur in a family before a child reaches adulthood. This statistic reflects a profound change in mortality patterns. For example, more than 11% of the American population (or 21 million people) are now over 65 years of age. The growing population of elderly persons, coupled with changing family patterns and religious beliefs has changed our responses and affected our attitudes toward death and dying.

Death typically is an experience of separation and loss for the survivor. This loss is experienced personally as grief and has been expressed historically through mourning rites and funeral customs. The deaths of an increasing number of elderly people today are experienced by the survivors as minor social and emotional events. This reaction seems to be due to society's increasing stress on the value of a person's utility or function because the elderly are often less central to the lives of their families and communities; moreover, they are viewed as having "lived out their lives." Deaths of children or young adults, on the other hand, are seen as unexpected and unjust and generally elicit a more profound grief reaction.

Traditionally people confronted death within the structure of a set of religious beliefs that gave it a meaning apart from the world of men and nature. Today death is explained more as a normal biological function. However, some find death a threatening and annihilating prospect and choose to deny or avoid it. Others see death as the ultimate challenge and seek, through medical science, or other means, to delay the aging process or reverse death itself. For example, the cryogenic movement attempts to avoid death entirely through freezing the body with the hope that in the future life can be restored to the frozen corpse. The emergence of the new definition of death—brain death—is a different challenge to death; nevertheless like the cryogenic movement, it raises profound medical, legal, and moral questions. Advocates for this new definition of death argue that it benefits society by providing opportunities for the saving or prolonging of a person's life through an organ transplant. Critics of the concept warn against a quick acceptance of this definition because of such unresolved questions as who should decide which definition of death to use? And at what point in time is brain death irreversible? Modern medical advances have made the concept of brain death a reality and it is employed today in organ transplant procedures, but its use has raised many social policy questions that remain to be answered.

Still others are prepared to face and accept death but wish to remain in control of their lives and to determine the time of their deaths. The proponents of euthanasia (the "good death") argue that the right to death is equivalent to the right to life and that people should be permitted to choose when and how they die. It is also their conviction that individuals should have the right to dispose of their own bodies as they wish. In this regard, the Uniform Anatomical Gift Act, which has recently been enacted in all 50 states, provides that the deceased has this right—to dispose of his or her own body. This act reverses a tradition that allowed the survivors to care for the body, a tradition that has been in effect in the Western world for almost 1,000 years and which is known to be practiced since before the Christian era.

The consequences of these shifts in our thinking have yet to be fully realized or indeed fully understood. What might be said at this time is that change in our society continues and nowhere may we expect it more than in our attitudes and responses to death.

Death and the Black

Dr. Stacey B. Day
Sloan Kettering Institute For Cancer Research
New York, N.Y.

Curtis A. Herron
Pastor, Zion Baptist Church
Minneapolis, Minnesota

DR. STACEY DAY: Pastor Curtis A. Herron, you are a black and I am a physician and we are sitting discussing together attitudes towards death, primarily of blacks, although we have agreed to try to relate as many of our observations as possible to people in general, both black and white, as they may be seen through the eyes of a black. Definitively let me ask you the first question. Do you consider yourself a good sounding board for the feelings of the black community—and would you say that you could offer as good an insight into the death attitude of blacks as, say, could be obtained from a similar discussion with a black M.D.?

PASTOR HERRON: I think so. And I think so because I am probably more in touch with the formation of attitudes about life and death than an M.D. would be. I am called upon to attend deaths probably as often as one M.D. would. Moreover, because I have an entry into homes that a doctor does not have, I can go when I am not called because the people are my parishioners. I council them and I am often called to advise them on all the common interests of life, such as working on the job, marriage, sickness or death. These kind of things a doctor is not often called upon to attend, he does not get to know people in all these areas—or if so, not nearly as deeply as I come to know them.

DR. STACEY DAY: Then in a way, I could say that you reflect many aspects—social, psychological, cultural, even political thinking that might be raised in the black community? You will have the potential for a great insight in your comments on the black community?

PASTOR HERRON: I think so.

DR. STACEY DAY: Do you think that in questions of death and attitudes towards death, a white physician is culturally able to satisfy the needs of the black patient who is dying?

PASTOR HERRON: I say that would depend upon his sensitivity towards black people and their attitudes toward death. If he is sensitive, his color will have nothing to do with whether or not he is capable. It is his sensitivity that is important. And I have found that generally white physicians are not very sympathetic in their treatment toward the black patient and especially in terminal cases.

DR. STACEY DAY: That is very important. I have traveled in many countries, including Africa, and I have always found a so called "cultural attitude" toward death. The American situation is somewhat confusing because we have, I guess, blacks and whites as distinct "cultures" yet *of* the same culture. For a doctor at least, I would hope, the primary concern would be the patient irrespective of his color. But you have suggested to me the *necessity* for a white physician to be sympathetic to the black patient. Now I am going to ask you point blank: "Do you feel that there is some sort of implicit distinction (not necessarily meant) in the attitudes of physicians toward a black patient as opposed to their attitudes toward white patients?"

PASTOR HERRON: I have not been in the presence of white doctors treating white patients, so I can't comment on that, but I would certainly suppose they would need to be sympathetic toward *anyone* who is to die. What I am saying, and what I have said, is that generally speaking, I do find that white doctors are not as sympathetic. I am not talking about this in a purely humanistic way—I am talking about their lack of understanding. I don't think they understand the attitudes of black people—their fears, suspicions, their ideas about death, and therefore they are not as feeling in their ministering to the patient as they might be if they had such understanding.

DR. STACEY DAY: Do you think that the black person has a different attitude toward death, toward religion, to God, than the white person in America? Are the fears of the black different?

PASTOR HERRON: Yes, I think that what is different about black religion and white religion is *what is emphasized*. The conception of God is one thing to the older black person and another thing to the white person. One conceives of God according to the situation in which he is in and according to his needs for that God. Black people have been an oppressed people. They needed a God who was a great deliverer, a savior, a great Messiah, a God to be worshipped, a God who would help—perhaps not so much a God to be worshipped as a God who would come and who would help in time of need.

DR. STACEY DAY: That seems important to me. Does that mean that the attitude toward death of blacks is different from that of whites?

PASTOR HERRON: Well working on your question I would say that black people have different ideas about God and these ideas would affect their attitudes towards death. Their attitudes about situations in which they live, also affect their ideas about death. If you have been living in a very bad situation, terrible physical situation, death is not as bad—not as terrible a thing to you, to that person, as it would be to a person who has had it very well in life.

DR. STACEY DAY: I am inferring from what you say, and I might be wrong and you will have to correct me, but I feel from what you say that death can be accepted more easily, possibly by a black person than by a white person. Is that so?

PASTOR HERRON: That does seem to be true but there are other variables, other things that enter into consideration. I do think, however, that the end result, *with all the variables*, is that black people have a very religious way of dying and they approach death with assurance. Older blacks do. Now this is not true of young blacks because young blacks are not coming out of the same situation. They have not been involved in the same kind of church or religious atmosphere as older blacks have evolved in. Older blacks that I am talking about are blacks over 30.

DR. STACEY DAY: Do blacks fear death?

PASTOR HERRON: Yes, there is a great fear of death in blacks, especially older blacks. There is also awe and reverence involved in this fear, and there is what the Christians call the blessed assurance that God is going to save them even in death. The reason that this would be different from whites is because black religion has been a religion of hope. It has been a crises religion, because a God was needed to bring people through their many crises that were continually appearing in their lives. Survival was a religious thing for black people and so God was constantly coming and saving.

DR. STACEY DAY: Does this thinking occur in the Spirituals—what I call the spirituals?

PASTOR HERRON: Yes, it is seen in them. It is most exemplified in the Negro spirituals.

DR. STACEY DAY: Would a white psychologist have anything to offer a black?

PASTOR HERRON: If he can understand where the black is coming from, if he understands the religious emphasis of the black situation, but I imagine that if he is approaching the black purely from an academic point of view, then he could not.

DR. STACEY DAY: Inasmuch as a white psychologist probably could not understand Hindu culture unless he had been in India and examined its background and worked with Hindus. Unless one is familiar with blacks, at first hand, living in America alone doesn't give any priority of understanding of the way blacks may approach life or death?

PASTOR HERRON: It gives him priority in the sense that he is on the scene and if he wants to be sensitive he can. He can see and he can observe, but he cannot know by academia, by being in an institution.

DR. STACEY DAY: That is important. What role does the black family play in the support of each other in the process of grief and mourning, following the death of a loved one?

PASTOR HERRON: The older blacks are very philosophical about death. While there is a great fear, it is not as great as white propagandists would have us believe. But we are trying to avoid the political field. There is a great deal of fear in death relatively but I don't know whether blacks fear death more than whites. I do know that blacks have a theological approach to death generally, and it enables them to accept death in a way. Sometimes their attitude is resignation, and even better than resignation, some find real hope in death because the world situation has been so bad for them. They may say; "Man I'm going to go to Heaven one day," and even though they don't want to stop living, they know that when they die there is going to be "a better world."

DR. STACEY DAY: The family therefore, in your thinking does support each other?

PASTOR HERRON: Yes, mourning is not the terrible thing in death. For the mourning ones, grief is a piercing and painful thing, but not so terrible that they cannot handle it within a reasonable time.

DR. STACEY DAY: Therefore, it could be said in a general way that their mourning might be a *happy mourning*?

PASTOR HERRON: In times past it was. Even now as a minister familiar with funerals I attempt, and am often successful, at trying to develop the funeral into a kind of celebration. Especially when we feel that the person who has died has been a loyal and faithful servant of God. There is no need for sorrow, you see, and so we attempt to testify to our young people that here is one "who has made it." He has paid off. He has done well. And so this is a happy occasion and we try to even bring victory songs—Christian songs—into the funeral for the preface.

DR. STACEY DAY: Can I say I am reminded of Louis Armstrong's jazz immortal rendering—"When The Saints Go Marching In, I Want To be There in That Company." It is almost a sort of victory celebration. As if death could almost be joyful, as if one were almost marching into heaven in a ceremony which is a happy occasion rather than a sad one. In many cultures we see this. In the Jewish testament one may read that "death is a wedding." For the arabs in erstwhile years, riding into death was a glorious and noble ride into the gates of paradise. From what you have said, I would conclude that you feel that in general blacks have a healthy attitude toward death?

PASTOR HERRON: That is a value statement. Whether it is healthy or not I don't know. I think blacks are people who will do anything that they can to survive.

DR. STACEY DAY: Then they are really like whites.

PASTOR HERRON: Yes, maybe even more so. Their history points out that blacks have a strong desire to survive and will go through hell to survive, so that death isn't something that one volunteers for, it is not something that one looks forward to.

DR. STACEY DAY: At a white's death, the family would appear to be important. I have a feeling, which may be incorrect and you must correct me, that culturally one of the effects of slavery was that it broke up black society in terms of family units. When the blacks were slaves for example, a father could be sold into bondage and separated from his wife and children. The whole system of slavery was inconsistent with developing family units. Certainly this was a century or so ago, but have social conditions so changed so as to permit "reconstitution" of the black family, socially and psychologically. Is my thesis right or wrong to begin with?

PASTOR HERRON: It is historically right but it is not altogether right. I think studies show that there is a closer family tie among blacks than among whites.

DR. STACEY DAY: Can you convince me?

PASTOR HERRON: Yes. Black children are not nearly as disrespectful to black parents as are white children to white parents. Black parents need their children more than white parents need their children—hence black parents hold on to their children longer than white parents hold on to their children. Black children need their parents more than white children. And so the family holds on to one another more. Maybe it is an economic thing, I don't know, but for some reason they hold on to each other longer and more tenaciously than do whites.

DR. STACEY DAY: Is that an objective assessment? Do you have support for that thinking?

PASTOR HERRON: Objectively, from the point of view of economics. Black parents need their children. When you need somebody you hold on to them. Black children need their parents.

DR. STACEY DAY: Let us consider the death of children. How would blacks relate to the death of a child?

PASTOR HERRON: I couldn't imagine blacks relating any differently than whites to the death of a child. It is a very painful and piercing experience to have to go through. I think the hardest thing for black people, the hardest deaths, the most piercing and painful depths that I have seen, experienced, and administered, have been those in the case of children.

DR. STACEY DAY: How do black children react if they know they are dying?

PASTOR HERRON: I have not come up under the same kind of religious background of "When the Saints Came Marching In" in *young people*. Young blacks have become more like whites, not very religious and not very theological in their approach. Therefore, when they face death, they do so with nonchalance or irreverence or they approach it with great fear and with no hope.

DR. STACEY DAY: No hope?

PASTOR HERRON: No hope because they have no God to guarantee or to give them assurance. God is not as real to them as he was real to their parents.

DR. STACEY DAY: Now one of the questions I would like to ask is a philosophical question. There has been a tendency among Americans for blacks to group in center city communities and whites to reside in the "suburban city". Social problems arise and new potentials present. As would be true for those born in an Irish slum or on the streets of Calcutta, children born in the ghetto obviously have a *greater potential* for disease and early death. Thus it might be said that center city people are faced with an earlier death or an earlier possibility of death. Is it true first in your thinking and secondly, if it be true, how could you face it?

PASTOR HERRON: It is true that they are faced with earlier death. They see more violence, they experience more crises than whites who are not in the center city. I don't think that they even think about it. It is a way of life and they are not aware that it is very different from the way anybody else lives. I have had that experience. I was very poor but I never thought of myself as being poor, because I didn't know how rich everybody else was until television was invented. Then I discovered how poor I was, but I didn't know before so I could live with it. I think the rage that has happened in the last few years happened because television so enlarged poor people that they became aware of the fact that they were disenfranchised, deprived, and all of that. But so far as death is concerned, I don't think that people think about it. It is a kind of thing that they deal with when it comes to them and they don't anticipate it. I don't think that young people handle this too well. As for older people, I have administered to many people and watched them die and older people die bravely, courageously. They die with strong testimonies and they die with a kind of assurance and a certainty that God is going to deliver them. But young people, as a rule, do not have this kind of courage. They may die nonchalantly, they may give their lives. Young blacks in this day will give their lives only to prove that they are men.

DR. STACEY DAY: In what way?

PASTOR HERRON: It is difficult to go into this sort of political thing. When blacks finally came to realize how disenfranchised they were and how deprived they were, and how emasculated they had been by the dominant culture, then they immediately began to get themselves psychologically free of the dominance. They wanted to prove that they did not need the majority's approval in order for them to be men and so they ran out into the streets; they ran without weapons against police and against the national guard and against the power structure and were often cut down. They knew that they didn't have a chance and they did this because there was, in them, an important drive to prove that they are men. I think this is related to the fact that blacks were attempting to demonstrate to themselves and to the world, that they were men.

DR. STACEY DAY: But isn't this a so called "heroes' death." I would look upon this as sort of a black equivalent of the heroes' death. Isn't this also common to whites in a sense that they might not themselves see.   Remember Tennyson's Charge of the Light Brigade.

> "Their's not to reason why
> Their's but to do and die."

PASTOR HERRON: It may be common to the white, but I am saying it is for a different reason and I think that it is important for physicians to know and be able to distinguish the reason. They may appear to be the same; a young black may be dying in the same way that a white is dying, but he is more than likely dying for a *different* reason.

DR. STACEY DAY: That is very good.

PASTOR HERRON: He may be dying in the same place but he may be dying for a different reason.

DR. STACEY DAY: I understand.

PASTOR HERRON: Psychologically, he is doing something to prove to himself this is what the culture has done to him, what the dominant culture has done to him. It has placed a burden on himself to make him strive excessively to be a man.

DR. STACEY DAY: Can you visualize a future situation coming, let's say, when life ends on a common plane for all men in the U.S.?

PASTOR HERRON: I want to say yes, I think so. But if I said that, then I would have to decide also that in time the church will become less powerful. The influence of the church will have become less powerful in the lives of black people than it is now and I think that is something that I don't like to think about. Yet I suppose it is true. Yes, I think so.

DR. STACEY DAY: Now from what you have said, and everything you have said seems reasonable, would it be reasonable to believe that because of the difficulties in the living of black people, the attitude to life they thus form would better fit a black M.D. to be more sympathetic or empathic when attending the death of a white patient, than has been the case you cited earlier of a white physician in attendance upon a black patient?

PASTOR HERRON: Generally a black person who has gone through medical school and has become a doctor, has developed attitudes that will not be very different from whites. Yet he has the connectional relationships to understand blacks; but having been in a white institution all his life, he has no problem in understanding whites. The truth is that blacks understand whites much better than whites understand blacks. This has been necessary for blacks to understand whites in order to survive. But it was not necessary for whites to understand blacks and so they did not have to, nor have they ever tried to. This is why a black physician could work well with whites, but a white physician may or may not be able to work well with blacks.

DR. STACEY DAY: I want to ask you whether a black physician educated in essentially a dominant white culture, on his return to a black community, practices through the thinking process of a white man. Is he sort of a black man with a white mind?

PASTOR HERRON: Certainly this depends upon the person, upon the sensitivity of the black person, the black M.D. Some of them try to forget whence they have come—many do forget. Others become very busy and rise above the struggle that blacks are very conscious of all the time. There is a struggle going on between blacks and whites or between blacks and the white power structure. We live with this awareness. Our children go to school with this awareness. Whites are not aware of it, but blacks are. The black M.D. who comes out of the white school rises above it and says; "I am so busy healing humanity that I really don't have time to be part of that struggle." That is even provincial for he would say that is his way of rising above it.

DR. STACEY DAY: As a physician, can I ask whether you really feel that we can keep the political situation out of this? Death has social and psychological and other implications which derive from the society in which we live. This society is propelling us into political diversions. The point is, do you feel that it is inevitable that even in the situation of death and physicians, there must be political overlay?

PASTOR HERRON: I imagine that I have to believe that politics is a dominant factor in almost every aspect of life. Therefore, there would be at least some overlay, even in the way blacks approach death, as perhaps in the way whites approach death, and the way a physician if he is white or if he is black, manages a black or white patient. I would say that politics has some overtones to this. You are speaking of politics in a very, very broad sense.

DR. STACEY DAY: Very Broad.

PASTOR HERRON: And in that sense, actually what you are speaking of is whether a physician is a humanist or whether he is not.

DR. STACEY DAY: I strongly believe in humanism. I would go further and say this, that don't you think if we educated ourselves as physicians and our medical students to face our patients as people, if we educated ourselves to be empathic and sympathetic and to face the problems of life and death as people, rather than white people or black people or Indian people or Japanese people, we would see the person who is dying as a person rather than as a black person or a white person. Do you think this possible or not possible?

PASTOR HERRON: I think generally that, as time goes on, an M.D. will be safe to approach a black person as if he were not white or black, just humanistic. Make a humanistic approach to him. But for a black person over 30 or 35 at least an M.D. would need to, should be aware of, that this man has an excessive fear of death, and that he is theologically oriented, and sees that there is hope in death, and it would just be good for him to know that.

Could I suggest to you also that a very helpful approach for white M.D.'s when treating a dying black patient over 30, is to remember he more than likely has come out of a very religious background and has a theological approach to life and death. It would be well to attempt as soon as possible, if the person has one, to bring his minister into the picture and to talk with his minister about whether or not it would be well to tell the patient. A minister should have had such contact with the patient that the minister could say it would be all right to tell him or "No, I don't think you should tell him." It is a religious thing, you know.

DR. STACEY DAY: This reminds me that we, in the U.S., live in a two generational society. The older generation are moving farther and farther into old age and the younger generation is almost totally disassociated from the older generation.

PASTOR HERRON: I would say that.

DR. STACEY DAY: So in that one area, let's say there is an understanding between blacks and whites.

PASTOR HERRON: I agree with that. We have a common thing happening to us.

DR. STACEY DAY: That is a very important thing. Now we touched briefly on mourning and the mourning process. I have been told that blacks seldom bury their dead—that they in fact cremate the dead, whereas generally speaking whites bury their dead. Is this so?

PASTOR HERRON: I don't think that is true at all. In my life time, and I am forty years old, I don't know anyone who has been cremated.

DR. STACEY DAY: That is an important point.

PASTOR HERRON: I don't know of anyone who has been cremated. I have decided for myself that I would choose to be cremated, as an economic factor, because it would not cost my family as much as it would for a funeral. I think we spend too much money on funerals.

DR. STACEY DAY: What value do you see in a funeral?

PASTOR HERRON: I see a lot of value in a funeral. It is the time when we Christian people come together to celebrate and to demonstrate that here is life that has won, a life that has been well lived, and this is the way life could be lived and now we come together to remind each other of the promises of God, that, even if a man dies he can still live.

DR. STACEY DAY: In a psychological sense, grieving and mourning would have a role in your thinking?

PASTOR HERRON: We can soothe the mourning ones and we can give them assurance that this is not the worst thing in the world or the worst thing that could have happened in this life.

DR. STACEY DAY: Would you tell a black patient that he is going to die?

PASTOR HERRON: It would depend upon the patient. I would need to know how he approaches life. What his life is like. I feel more like talking with an older person about death than I would with a young person.

DR. STACEY DAY: Would you tell a black child that he is going to die?

PASTOR HERRON: It would depend upon his orientation to his religious background. If he had the tools of religion to emphasize that there is a great hope for him in death, then I could talk with him about it. I could talk with him about the symbols of another life like angels and heaven and a good God, a loving father, and all the good things that there could be for him.

DR. STACEY DAY: Are black children in the habit of having pets?

PASTOR HERRON: Yes.

DR. STACEY DAY: As much, do you think, as white children?

PASTOR HERRON: No, but they have pets.

DR. STACEY DAY: If a pet dies, do you think that would help a black child face the reality of death?

PASTOR HERRON: I have seen black children have funerals for their pets.

DR. STACEY DAY: What sort of funeral? Could you describe one?

PASTOR HERRON: Well they were pained. It was really that they were emulating an adult funeral, I suppose. They were pained. But it was a sort of a mock funeral. It wasn't really something with which they were deeply involved. And then I have seen them when the whole family was involved in the funeral for the pet. You know it was a mock, it was something that was done as an informal thing, that was done in a back yard or on a hill someplace you know . . .

DR. STACEY DAY: It was told that children play cowboys and Indians and one will bang-bang the Indian dead. Then the Indian falls down and he gets up again immediately. Would you also find a parallel in black society. Do black children play cowboys and Indians and have the same fantasy images?

PASTOR HERRON: When I was a child I did, but then that was a long time ago. They probably play Black Panthers versus the power structure now. I don't know what games they now play.

DR. STACEY DAY: That is interesting. Do black children see Santa Claus as black or white?

PASTOR HERRON: Since the black awareness has really come into being, a child would paint the face of Santa Claus black and with white hair.

DR. STACEY DAY: Santa Claus is a very old man!

PASTOR HERRON: Yes he is very old. They will paint the face of Jesus black also.

DR. STACEY DAY: With respect to older black people, are they shocked about death being sudden or violent?

PASTOR HERRON: I think it depends on how close they are to the person who dies. I think the difference is that there is a kind of resilience in the older Christian that is not found in the young person. For example, you tell a black patient that he has cancer and it would knock him off his pants if he is a Christian. He just goes all to pieces for a couple of days or a couple of weeks, maybe a month or two.

DR. STACEY DAY: The white patient also.

PASTOR HERRON: Yes, but a difference. What I am saying is a person with a religious background will bounce back. He has the resilience to bounce back. After he has gotten over the shock, he knows then his faith sort of takes over as if it were a subconscious thing, just pervading his mind. It takes over and then hope begins to come into him and he remembers the promises that are real to him that *death is not all*.

DR. STACEY DAY: Then the physician could use this to work with the patient? There seems to be a difference between violent death and death, say, of cancer.

PASTOR HERRON: It is hurting, it is painful, and it is shocking, but there is hope for the religious person. It is not there for the nonreligious person.

DR. STACEY DAY: The grieving and mourning, say, when somebody who had cancer or a terminal patient finally dies, is there a sense of relief? When finally the suffering has ended. But the person who has had a violent death, how do they approach that? That's not the same . . . .

PASTOR HERRON: There is not the same relief.

DR. STACEY DAY: I am told even for an older person who died violently on the street, it is no longer being joyful the way you were saying before?

PASTOR HERRON: Celebration! It would be more difficult to pull off the celebration thing for a sudden death.

DR. STACEY DAY: It is, how shall I say it, not in your hands! God called the person who had cancer and he died—there is nothing any human individual could have done. But on the other hand, were it a violent death, at the hands of another human being, it could have been prevented. It could have been a tragedy that should not have happened. Would that effect be part of what affects the mourning?

PASTOR HERRON: Yes.

DR. STACEY DAY: Let us deal with euthanasia. What, in your opinion, is the attitude of the black community toward euthanasia? Is there a possibility at this time of blacks voluntarily asking for or choosing the moment of death?

PASTOR HERRON: No, no! Blacks are extremely guarded about life. They will do anything to live and they never want to die nor do they want to see their loved ones die. You must be aware that blacks, of all races, have the lowest incidence of suicide.

DR. STACEY DAY: No, I didn't realize that.

PASTOR HERRON: Life is an extremely precious thing, in spite of the fact that it is also a hurting thing.

DR. STACEY DAY: That might be corroborated by the painful nature of life in underprivileged or underdeveloped nations–suicide is uncommon in some Asiatics–India for example. Your feeling is that, at this time, euthanasia doesn't enter into black thinking–at this time?

PASTOR HERRON: It enters into the person who is dying. The person who is suffering would like to die, but he would probably not agree to death. It enters into the minds of the relatives who are attending but they would, more than likely, if they are older people, not agree to it. If they were younger people, they might agree to it.

DR. STACEY DAY: Can you consider voluntary euthanasia–a patient asking voluntarily that his life be terminated?

PASTOR HERRON: Not among older blacks. Among younger blacks, yes. I would say yes.

DR. STACEY DAY: In terms of this two generational society this is "the decade of death." Euthanasia is coming in the next 10 years and that soon, as in the last American war, the young generation will put a gold star in their window saying "we gave"–meaning that they gave their parents, their grandparents and everybody else.* The implication is that they sent their parents away to die. This younger generation of whites would appear to isolate their elderly in nursing homes so that the old white is going away to die in relative isolation and not in the family setting. They may die sometimes slowly and sometimes neglected in a broader sense. Would you think this is true of the black community?

PASTOR HERRON: It is not true. Blacks hold on to their older people. It may not be a healthy thing that they do but they still hold on to their older people. They can't give them up in the same way, with the same ease, it appears to me.

DR. STACEY DAY: I would interpret this, then, that probably few elderly blacks are sent away to nursing homes to die. One might find a greater number of older blacks still within the family setting?

PASTOR HERRON: Yes.

DR. STACEY DAY: And therefore there is a great difference here socially between the black community and the white community. In terms of death?

PASTOR HERRON: Yes, I would say that. But I would say that the trend will be toward becoming very much like whites and probably in 20 years, there will not be much difference in attitude.

DR. STACEY DAY: This suggests to me that it is an economic factor.

PASTOR HERRON: It may be.

DR. STACEY DAY: When the economic threshold of the black rises, you might adopt those attitudes which we identify with the white people.

*Attributed to Professor Robert Fulton, Professor of Sociology, University of Minnesota.

PASTOR HERRON: Yes, unless the black movement to retain what has been a black thing, becomes continuous with great strength, and black people begin to cherish and hold on to their traditions. If they do, then they may stem the tide of assimilating or becoming like the whites. And there *is* this possibility.

DR. STACEY DAY: Another problem. Another statement has been raised that death and sex are similar in this society in that they are both taboo. One doesn't want to talk about them and this is one of the reasons why we had this discussion—to get people to talk about death and face death. Is death taboo in that sense among the black?

PASTOR HERRON: You mean that they don't want to talk about it?

DR. STACEY DAY: Yes.

PASTOR HERRON: I will probably preach three sermons a year about death. But then I seize upon every occasion at a funeral to talk about death also. It isn't the most talked about thing, but it is talked about and older people give a lot of thought and talk about death. I don't know about young people, I just don't think they think about it. They may talk about it, but they don't do a lot of thinking about death.

DR. STACEY DAY: Would I be right in saying that throughout our conversation I feel very strongly that you, like whites, have drawn a great sort of differentiation between the two generations. Between the old generation and the new generation that you have suggested to me has revelations of the sort of continuity of the family as a unit in the blacks that it has lost in the whites. That would be one of the differences we discussed. But in general, blacks and whites see this great divide between the older generation and the so called new generation. It would appear, in a general way, that the new generation has more problems to face, in both life and death, than the older generation appears to have. Do you have any way of dealing with that?

PASTOR HERRON: Well, I try to have our church act as a counter culture. I try to emphasize the traditions of blacks that were good and to encourage the young people not to be caught up in the culture of the majority, but to reflect deeply upon what was good about the blacks and hold on to it.

DR. STACEY DAY: I am very much worried in a very personal sense by this gulf, this sort of divisions of communities. I understand historically, why the black people must seek identity. I do understand that. I understand, I think, the purposes surrounding the changes you are undergoing right now and I think these are important in raising what you call, (I don't like the phrase), the dignity of blacks. I understand that. But it seems to me that black studies and all these expressions of blackness we see about us, are wrong. This is how it seems to me personally and I would have to say to be honest, that I don't like them either from the white side or from the black side. It seems to me that if I could have a divine right, the goal would be to bring about social changes that are valid for *all*.

This is, after all, the United States of America, presumably one country. There obviously are differences of degree, but still, shouldn't the total goal be a oneness rather than such as black movements. I don't mean to be rude, but if I were to say that black studies are all nonsense, that we have to change the social faces of our communities, not our races, but change the society in which we live to get the best results, would you respond to me angrily or would you feel that I am wrong or would you feel that actually through the black movement, perhaps you can attain the goals I have described?

PASTOR HERRON: The goal for thinking blacks is the same as you have described, which is one humanism, a one humanity. And so blackness as a movement is not an end, it is not the goal, it is a means to the end. It is important for black people to affirm themselves as human beings, before they can become a part of the main stream of humanity, and blacks must, in order to affirm their humanity, compensate for what has been taken away from them. Their humanity has been taken away from them. So they must first love themselves, and in order to love yourself the slogan "black is beautiful", and all of these things, these slogans, were designed to overcompensate, to bring the child from being nothing to *being*, to make him feel superior to the whites. But in the long run, common sense will bring him back to the medium where he will discover that you are not really better than the whites, but as good as the whites. But you have to overcompensate, to reach down to bring him up, you see. So this is an intermediate strategy, one used to reach the goal which is one humanism. We can not begin to deal with white culture until we can believe in ourselves and love ourselves. There is self-hatred among blacks and it was taught us by whites because we saw ourselves through white eyes. We only thought of ourselves as being whatever they thought we were. We thought of beauty as being based upon what whites thought was beautiful, and we thought of goodness and most values as being only those values of whites, and this was a very damaging thing for black people. An outsider would have much difficulty in understanding that until it was explained to him. So I can understand why you think like you think, and it is natural for you to see that since the goal is oneness, that therefore, we must not talk about being different. But if you could understand that we are different, maybe we are not inherently different (perhaps there *are* even inherent differences), as you know physically, there are some differences between black people and white people. We are different because culture has made us different. We are different for the other racial, ethnic, or inherent reasons that we are different, and there is no need for us to go around talking about how we are no different, when we are different. It is a truth. But we are not inferior, and we have to make ourselves know that, and it is going to take time for black people to finally understand they are not inferior.

DR. STACEY DAY: So this is sort of a growth process.

PASTOR HERRON: Yes, it is an intermediate strategy.

DR. STACEY DAY: Therefore, I as a nonblack, have to be aware of this growth process as much as a black and therefore, I can only understand in terms of reference which you put for me. After you having told me your explanation, I have to mend my original question. Otherwise I don't grow myself, and therefore there is an understanding required of me to see your point of view; but, together, people can move in this way to understand that it would be a growth process. Now before I close, could I ask you one more question? That would be to ask you your definition of life, and although it is personalized, your reason for living.

PASTOR HERRON: A definition of Life? Well, for me there is existence and there is life; and I understand life to be a great stream that is going someplace, that has a destiny. It stands for something; that means something; and it is going ultimately to God. Moving toward God. Life stands for something and its meaning comes from God, you see. And its purpose? The purpose of life is to serve this God and we understand He is at the end of life.

DR. STACEY DAY: Thank you so much. Is there anything you might like to add to our discussion—any thought you would add?

PASTOR HERRON: I am very much concerned how whites handle the subject of death. I would hope that they could become a lot more humanistic than they are. I really don't think that they are very humanistic.

DR. STACEY DAY: Towards blacks?

PASTOR HERRON: White physicians towards black patients.

DR. STACEY DAY: Do you think the white physicians attitude toward the white patient is different from the white physicians attitude toward the black patient?

PASTOR HERRON: It is my experience. I would have to believe that white people treat whites differently than they treat blacks.

DR. STACEY DAY: Pastor Herron may I quote you on our discussion?

PASTOR HERRON: You are free to quote me.

DR. STACEY DAY: Thank you very much.

B. Rural Health Perspectives

The Need for Development of Rural America

CONGRESSMAN TOM BEVILL
Remarks to Clearinghouse for Rural Health Services Research
University, Alabama

The most effective way to increase the ability of rural people to purchase health services is through the improvement of economic conditions. Consequently, any program aimed at improving the health facilities of rural citizens must be directly tied to efforts already underway which advocate maximum economic development of rural America.

Probably the two major problems confronting rural people in terms of health needs are the shortages of health manpower and the inability of many people to purchase adequate health care. While health services for rural citizens have improved markedly during recent decades, they have by no means kept pace with advancements made in other, primarily urban, areas of the country. To improve health services for rural people in the years ahead, it will be necessary to expand the supply of health workers, to improve the organization and administration of health systems, and to improve the ability of people to purchase health services.

A report researched by the Library of Congress advanced a fourfold rationale for increasing health services in rural areas. First, improved health services are an essential community service and also essential to economic development.

Second, improving health services creates jobs and generates income for rural communities.

Third, adequate health services assure rural people of the same access to these services as other people in the country enjoy.

And finally, there is the obvious rationale that adequate health services will lessen the harmful impact of illness, and in many instances, prevent it.

It seems the most acute need in rural health services is the need for doctors. A Gallup opinion survey revealed that rural citizens view the shortage of physicians as the single most important health problem facing rural communities today. In that area, legislation commonly referred to as the rural health clinic bill has been introduced in both the U.S. Senate and the House of Representatives. If approved, this proposal would permit Medicare reimbursement for small-town health clinics that use nurse practitioners or physician assistants. Currently, Medicare requires continual on-site presence of physicians at these remote clinics even though the areas in which they are located are characterized by doctor shortages. The thinking behind this legislation is aimed at helping millions of rural Americans who are presently being deprived of Medicare benefits to which they are entitled. Unless some changes are made to the present system, several health clinics across the country may be forced to close due to the lack of Medicare reimbursement to cover the costs of providing health services to senior citizens. If this issue is not properly addressed by Congress and the Executive Branch, the result would likely be that rural citizens will not only find themselves with a continuing shortage of doctors, but with a shortage of all health workers as well.

As was initially pointed out, improvement of rural health services is directly tied to the total economic development of the thousands of miles of sparsely populated land in this country. The federal government, both at the executive and legislative levels, has and must continue to aid in this economic development. As a Representative of a rural Congressional District, I am aware of the economic problems which confront residents of these areas daily. I also monitor closely the manner in which our federal government is attempting to help rural residents cope with some of these problems. The Appalachian Regional Commission (ARC) offers a good example of the assistance to which I refer.

During the past 12 years, probably no federal program has benefited the southern and eastern poverty regions more than the Appalachian Regional Commission. The Commission came about as a result of the Appalachian Regional Development Act of 1965, which Congress adopted as part of an effort to revitalize the stagnant economy of Appalachia. When the program was initiated, the Appalachian region was characterized as one of the poorest areas in the country. Today, thanks largely to the efforts of organizations like ARC, the region is on the move economically.

Stretching from southern New York to northern Mississippi, the Appalachian region encompasses the area surrounding the Appalachian Mountains, the oldest mountain chain in the United States. In all, ARC serves parts of 13 states. During the years since the commission was conceived, one of the most striking changes in the region has been the reversal of its population trends. For many decades prior to the 1960s, the region had consistently lost people to other areas of the country. Significant in this loss of population was the outmigration of young people because of the area's lack of job opportunities.

In the early sixties, the Appalachian region was losing about 122,000 residents every year. In the late sixties, the figure had dropped to 90,000. Then at the beginning of this decade, the trend reversed itself and the number of people moving into the region began to outnumber those leaving. Over the last five years, there has been an immigration of 60,000 persons annually to the region. Between 1965 and 1973, the region gained more than one million jobs in major industrial employment. The region's growth rate is now comparable to national figures and the Appalachian economy continues to show signs of expansion.

Another indication of the growth can be seen in the area's per capita income figures. During the past 10 years, the region's personal income has risen by 89% while the national figure was rising by only 81%. Countless other statistics point to a stimulated economy in the region and much of the credit for this must be attributed to the effort of the Appalachian Regional Commission.

Continued economic stimulation of this nature in rural areas will have an impact on all phases of the rural lifestyle as we have traditionally known it. And if we continue to meet with success in these efforts to improve the entire spectrum of rural life, we can certainly expect to see the health services improve vastly in rural America.

That is certainly one of my foremost goals in the United States Congress. It is likewise a goal that must be attained if we are to carry on and improve a lifestyle that is uniquely American.

Health Care and Health Resources in Rural America: Trends, Projections, and Federal Initiatives

JOSEPH BALDI
Public Health Advisor
Office of Rural Health
U.S. Public Health Service
BCHS, DHEW
Rockville, Maryland

INTRODUCTION

Within the past decade, and particularly the past five or six years, a great deal of attention has been focused on the health status and health care problems of rural America. The interest of a growing number of Americans in rural communities and lifestyles, likely stemming from the country's disillusionment with an urban-based War on Poverty, Vietnam War, and Watergate crisis, has contributed to a growing knowledge of rural areas and their problems, including those related to health care. Senators and Congressmen, state and local political leaders from predominantly rural states, and President Carter, have frequently talked about the Federal government's and the nation's long-time neglect of rural communities throughout the country, and the need to begin addressing major rural concerns.

Statements and proposals from national, state, and local leaders, and the media that cover these statements, have resulted in the passage of major Federal health legislation, and in the creation and strengthening of a growing number of rural lobbying groups, state and local health coalitions, and organized communities. For example, the Congress has in recent years passed the Emergency Health Personnel Act of 1970, the Rural Development Act of 1972, and in 1974 established the Health Underserved Rural Areas (HURA) program. Moreover, organizations like the Appalachian Regional Commission and the newly established National Rural Center, Rural America, and the Rural Hospital Task Force of the American Hospital Association, and various state and local rural health coalitions are becoming effective representatives and lobbyists for rural health interests.

Although recent trends are encouraging numerous rural health concerns and problems remain and they must be attacked. Before addressing these concerns and discussing trends and recent developments in more detail, it would be useful to study a few demographic and health statistics that help describe rural America and some of the health problems faced by these communities.

Based on 1970 U.S. Census data, nearly 54 million Americans lived in rural areas. The data listed below reveals that residents of these areas have a greater number of factors and conditions that contribute to poor health than their urban counterparts:

- There are a greater percentage of poor (17.6% vs. 10.4%) and aged (28.7% vs. 17.6%) in rural as opposed to urban areas.
- 56.7% of all substandard housing units are located in rural areas.
- There exists a much larger percentage of rural communities as opposed to urban communities that lack adequate potable water supplies.
- Health care needs of the poor and aged, and of rural residents in general, are greater than for the remainder of the population, as reflected in higher death (11.4 deaths per 1,000 population in rural areas vs. 8.6 deaths per 1,000 population in urban areas) and morbidity rates. The infant mortality rate in rural areas (1971 statistics) of 20.0 deaths per 1,000 live births was substantially higher than the 15.8 figure for urban areas. And the percentage of people with chronic health conditions which limit mobility are higher in rural (13.1%) than in urban areas (10.9%).

Moreover, access to health care, both geographic and financial, is a greater problem for rural residents:

- Approximately 40% of all persons who are 30 minutes or more (travel time) from their regular source of care are rural residents.
- Federal health expenditures per capita are lower in rural ($80) than in urban areas ($92) despite the large number of poor and aged in rural communities (1974 statistics). Medicare and Medicaid expenditures per eligible person in California versus Mississippi offer the most dramatic contrast.

	California (predominantly urban)	Mississippi (predominantly rural)
Medicare	$526	$277
Medicaid	$339	$ 32

In summary, despite encouraging trends and developments, rural areas of this country are poorer, have a disproportionate number of health problems, and experience greater geographic and financial barriers to quality health care. It is appropriate now to begin looking at major health problems in rural areas, examining what conditions created the problems, and what potential exists for ameliorating or resolving them.

MAJOR HEALTH PROBLEMS IN RURAL AREAS

Major Problem 1: The Lack of Physicians and Other Health Care Providers in Rural Areas

For the first seven decades of this century, and particularly since World War II, the numbers of physicians in rural areas has continued to decline. The growing shortage of primary care physicians (i.e., general or family practice, internal medicine, Ob/Gyn, and pediatrics) has been particularly disturbing. The trend extends beyond physicians and includes dentists, nurses, pharmacists, and other health professionals as well. Between 1967 and 1972 the number of primary care physicians in rural areas declined by another 7%, while the number in urban areas remained constant. As late as 1975, 561 of 672 (81.3%) of counties and service areas designated by the Public Health Service as medical Critical Health Manpower Shortage Areas (designated whenever the primary care physician to population ratio is less than 1 to 4,000 population) were rural. The population in these areas totalled 9.2 million people.

Physicians have been leaving rural areas or starting their practice in urban areas primarily for the following reasons: (1) increasing numbers of medical students and interns have been encouraged to enter specialty areas, and specialization usually requires association with major urban medical centers, and, (2) the increasing admission standards and cost of medical education favors medical students from wealthier families, the vast majority of whom are from urban areas and who are likely to practice in urban areas after they complete their medical studies.

Fortunately, major developments have taken place in recent years which begin to address the critical medical and health personnel shortages that pose major concerns to the citizens of so many rural communities across the country.

First, the Vietman War required the training of large numbers of paramedics, who upon returning to the United States, sought employment in the emergency and medical fields. The Medex and Primex training programs were developed in the late 1960's. Once the effectiveness of these para-professionals was adequately documented, the Comprehensive Health Manpower Training Act of 1971 became law and provided the impetus for the training of new para-professionals, the nurse practitioners (NPs) and physicians assistants (PAs)—effectively creating a new health profession: the physician extender (PE). Beginning in 1972, Federally supported programs for the training of NPs and PAs, Medex-Primex, and certified nurse midwives grew rapidly. By the end of 1975, the Public Health Service had spent approximately $75 million for PE training and research and nearly 6200 PEs had graduated from training projects (3,000 NPs, 2,500 PAs,

300 medex, and 400 Primex). The Health Professions Educational Assistance Act of 1976 provides the vehicle for continued Federal support of these important health professionals.

The Federal government's active role promoting the development of the PE profession has resulted in most states amending medical and nurse practice acts to allow NPs and PAs to practice. Clearly this represents a major breakthrough for those attempting to address problems related to health manpower distribution. However, existing Federal and private payor reimbursement policies serve as a barrier to PEs fully complementing physicians in health manpower scarcity areas.

Medicare (Title XVIII) currently reimburses for services of NPs and PAs only in those practices where a physician is continually on site. This policy has a direct effect on the State Medicaid (Title XIX) programs which in 34 states do not reimburse for NP and PA services. Two major bills have been proposed to the Congress (S. 708 sponsored by Senators Clark and Leahy, and H. R. 2504 sponsored by Congressman Rostenkowski), that would allow for reimbursement of NPs and PAs in medically underserved rural and urban areas. A compromise version of these bills is expected to pass both houses and be signed by the President by the end of 1977 or early 1978.

A second major factor affecting health manpower distribution in rural and urban shortage areas is the development of the National Health Service Corps Program. The Corps, which was created by the Emergency Health Personnel Act of 1970, recruits and places physicians, dentists, and other health professionals in communities designated as Critical Health Manpower Shortage Areas (CHMSAa). By late 1976, a total of 600 Corps assignees (405 physicians, 93 dentists, and 102 PEs) served 331 CHMSA communities. The modification of the Emergency Health Personnel Act and expansion and tighter management of the Health Professional Scholarship Program will sharply increase the number of Corps personnel to more than 2000 by 1979 and 3500 by 1981.

Another factor which has influenced a fair number of primary care physicians to practice in rural areas is the Federal government's support of medical schools that have, particularly since 1965, developed family practice residency programs. The Comprehensive Health Manpower Training Act of 1971 has provided the vehicle for support of these residency programs. In 1976, $15 million dollars were made available to support the development and administration of family practice residency programs.

The abovementioned developments will certainly impact favorably on rural areas, but the health manpower problems of poor, isolated rural areas cannot be expected to disappear. The Corps scholarship program can perhaps at best platoon health professionals in the most underserved areas. For even in 1976, only 37% of all Corps personnel extended beyond two year assignments, and few remained to enter private practice in their communities. The use of NPs and PAs, many of whom are from areas not far from where they choose to practice, may offer the best solution to the health manpower shortage problem in numerous rural areas.

Major Problem 2: Obstacles Confronting Rural Hospitals

Hospitals represent major health resources in rural areas particularly since rural areas have limited health resources, and because the closure of rural hospitals frequently means that the associated physicians will leave these rural communities.

The growing number of Federal and State regulations pose major concerns for rural hospitals. Rural hospitals must meet multiple and frequently conflicting building, staffing, and reimbursement requirements to comply with Federal, State, and local regulations. A recently completed Hospital Association of New York State study, entitled *The Report of the Task Force on Regulations*, graphically depicts the problems. The report examined the regulatory process and pressures in New York and found that there are 164 regulatory agencies which impact on hospitals, 96 of which were State agencies, and 40 were Federal. The proliferation of Federal and State requirements undoubtedly contribute to the problem of rapidly rising hospital costs challenged recently by President Carter and Health, Education, and Welfare Secretary, Joseph Califano. Moreover, the various codes and regulations do pose major management problems, and in certain instances, may result in the closure of rural hospitals.

In 1976, the American Hospital Association created a Rural Hospital Task Force to begin dealing with these issues, the most disconcerting of which include: (1) facilities and life safety codes, (2) requirements for highly specialized staff or duplicative staff (e.g., separate director and staff needed for acute care and for intermediate care (ICF) or long-term care (LCF) facilities), and (3) inadequate reimbursement rates, especially for non-acute care. Legislators, including Senator Laxalt and Congressman Baucus introduced bills in 1976 that would have addressed some of the major problems faced by rural hospitals. Neither of the bills were passed, however.

Major Problem 3: Concentration of Federal and State Health Expenditures in Urban Areas

In recent decades Federal funding, including funding for health care, has been concentrated in urban as opposed to rural areas. Most recent statistics show $540 per year per capita spent on the urban poverty population, versus $512 per year per capita spent on the rural poor. The most dramatic contrast in Federal health expenditures, already cited in the introduction to this article, exists between the predominantly rural state, Mississippi, and the predominantly urban state of California. Overall Medicare and Medicaid expenditures also show a significant urban bias: Medicare $409 per eligible person in urban areas vs. $365 per eligible person in rural areas, and Medicaid, $180 per eligible person in urban areas vs. $117 per eligible person in rural areas. It is evident that urban states have a better Medicaid package primarily because they have a larger tax base and more resources to match Federal Medicaid funds.

Moreover, the emphasis within HEW and other Federal government agencies has been on urban problems for the past three decades. For example, the major operational health programs administered by the Public Health Service, Maternal and Infant Care, Children and Youth, Community Mental Health Centers and Community Health Centers programs have emphasized service delivery in urban areas.

Only recently, with the passage of the Rural Development Act of 1972 (PL 91-410), did major legislation reflect the Congressional concern that rural areas had been overlooked far too long by the Federal government. PL 91-410 recognized that the improvement of rural areas requires improvements in several sectors including employment, housing, and health. In the President's Fifth Annual Report to Congress on Rural Development, there was commitment of significant programmatic expenditures and assistance to rural areas. The proportion of Federal outlays allocated to rural areas increased slightly from 34.6% in fiscal year 1972 to 35.1% in fiscal year 1973.

Two new Federal health care efforts were established in 1975 to address the problems of rural areas without adequate resources to meet their health care needs. The Health Underserved Rural Areas (HURA) program was established by the Senate Appropriations Committee in September, 1974, and Medicaid funds were made available to fund the program. At about the same time the Public Health Service established the related Rural Health Initiative (RHI) effort to integrate a number of Federal health programs (including the National Health Service Corps, Community Health Center, and Migrant Health Center Programs) to improve the delivery of health care to rural residents.

By the end of 1977 the HURA program had grown to more than 80 projects totalling $15 million, and the RHI effort had increased to more than 240 projects totalling approximately $28 million. Grantees included community-based groups, hospitals, private group practices, state and county health departments—the full range of organizations capable of providing primary health care to health shortage areas.

The bulk of these resources were targeted to areas of greatest need, particularly to communities of the Southeast, South, and Southwest. Priority areas were identified as those with health manpower shortages, high infant mortality rates, a high percentage of elderly and poor, and large numbers of migrant and seasonal farmworkers.

HURA and RHI program goals are much broader than the hiring of staff to provide primary health care services. The programs also seek to foster the organization and coordination of limited health resources from a number of sources into a comprehensive health system. Virtually all of the HURA and RHI grantees have established linkages (i.e., formal or informal arrangements) with one or more of the following institutions or programs: hospitals, community mental health centers, family planning and maternal and child health programs, the Medicaid Early and Periodic Screening, Diagnosis and Treatment (EPSDT) program, the USDA Women, Infants and Children (WIC) food and nutrition program, and county and state health department services.

Major Problem 4: Effective Implementation of The Health Planning and Resources Development Act (Public Law 93-641), and the Role of Health Systems Agencies (HSAs)

Given the limited human and financial resources available for health care in rural areas, it is particularly important that those resources be used as effectively as possible. Fortunately, the concept of health planning in this country has evolved considerably.

Health planning got a major boost in 1966 with the passage of the Comprehensive Health Planning Act (Public Law 89-749). Statewide and area health planning agencies were major limitations, principally the limited resources available to fund the Comprehensive Health Planning agencies, the system worked fairly well. The Congress and many of the nation's health policymakers did not believe that the system was effective enough to begin to deal with a continually growing health care budget which had jumped from $38.9 billion in 1965 to $118.5 billion in 1975, and with the proliferation of additional health care legislation and programs.

In 1974, the Congress passed The Health Planning and Resources Development Act (Public Law 93-641) to try to deal more effectively with the above-mentioned problems. The law established a network of Health System Agencies (HSAs) and replaced the Comprehensive Health Planning agencies in this country's most ambitious attempt at national health planning. Congress, in Section 1502 of the legislation, identified ten national health priorities and asked that these priorities be addressed in the development of the national guidelines for health planning. Whenever possible, these guidelines should include goals and objectives that are realistic and specific because the guidelines are to become a mandate (in the form of Federal regulations) for the Health Systems Agencies to allocate their resources to begin attacking specified major health problems.

It is very important to rural areas that the final draft of the national guidelines adequately respond to the first national health priority defined by the Congress as, "the provision of primary care services for medically underserved populations, especially those which are located in rural or economically depressed areas." Moreover, the present legislation, which requires geographic areas to have the minimum of 500,000 population, with some exceptions, tends to ensure Health Systems Agency (HSA) boards dominated by urban members by aggregating geographic areas containing urban centers.

Two proposals currently being developed by the Public Health Service Health Resources Administration would help ensure that rural areas be better represented on HSAs, and that more reasonable funding be made available to support the administration of the least populous rural HSAs, that frequently cover large geographic areas. The first proposal would amend the "equal to" provision within Section 1502 of PL 93-641 to read nonmetropolitan representation on HSA boards be "at least equal to" the percentage of the rural population within the jurisdiction of the HSA. The second proposal would amend the law to increase the minimum budget allocation for HSAs from $175,000 to $200,000 or $225,000. These measures would serve to ensure that rural citizens participate more fully in health planning under the direction of the HSAs.

SUMMARY, TRENDS, AND CONCLUSION

Before concluding this article, it is worthwhile to examine two health related issues and trends that will impact on rural as well as urban areas over the next several years.

Containing the rapid rise in health care costs has become a major issue raised by the Carter Administration. In 1940 health expenditures were $3.9 billion and represented 4.1% of the gross national product, by 1960 the figures had reached $25.9 billion and 5.2% of the gross national product, and by 1980 it is estimated that health care costs will total $180 to 190 billion dollars and represent 10% of our gross national product. Pressure will be exerted by the Federal government to set ceilings on the rise of hospital costs and perhaps even physician fees. The Health Care Financing Administration (HCFA) was created in early 1977 to combine Medicare and Medicaid, the two massive Federal financing programs whose budget totalled approximately $40 billion this year, to simplify the administration and cut back drastically on fraud and abuse, and overall inefficient management. Managing the existing health financing system is essential before the United States can realistically begin to consider comprehensive National Health Insurance. Rural areas and urban areas alike will be expected to tow the cost line and to manage existing health care resources more effectively than in the past.

Slowly, but with increasing impact, more and more Americans believe that the existing health/medical care system is a bottomless pit that can absorb as much money as is appropriated. Considerably above 90% of all health care dollars go to support the existing health/medical care system. Professionals in this country and abroad, like former Canadian Minister of Health, Marc Lalonde, strongly believe that much more emphasis and funding should be given to other factors, in addition to the existing health/medical care system, which, perhaps as much or more than the health system, affect the health status of individuals. Two major factors that for the most part have not received adequate emphasis in this country, are the impact of lifestyle and the environment on health problems and conditions.

Let us briefly examine these factors in the light of health conditions in rural areas. It is generally recognized that the lack of a potable water supply and poor sewerage, more than any other variables, contribute to high infant mortality rates in many parts of the Southeast, South, and Southwest. Moreover, pesticides and industrial chemicals cause major health problems for thousands of migrant and seasonal farmworkers, and employees of mines, chemical/pharmaceutical, and paper industries. In addition, there is considerable evidence that chronic health problems such as cardio-vascular disease, obesity, and diabetes can be controlled by appropriate diet and exercise.

Rural residents are fortunate that towns with romantic names like Dry Run and Picture Rocks, Pennsylvania, and Miner Hill and Hungry Mother, Tennessee are in vogue and that political leaders, lobbyists, and the media have focused on the health problems of rural America as well as on the fascination with rural lifestyles. For decades the migration of people, including health professionals, away from rural areas and the focus of Federal government programs on American cities resulted in a deterioration of rural America. As has been described earlier, the nation has become aware of and concerned about the health status and problems of rural communities. Extensive Federal legislation including the Emergency Health Personnel Act of 1970 that created the National Health Service Corps, and the Comprehensive Health Manpower Training Act of 1971, and the Rural Development Act of 1972, and the establishment of the Health Underserved Rural Areas (HURA) and Rural Health Initiative (RHI) programs have begun to impact favorably on rural areas. Although the many health and human service problems of rural communities will not disappear overnight, the problems are engraved in our nation's conscience and the recent commitments to assist rural areas will likely continue for sometime—particularly while a man from Plains, Georgia occupies the White House.

References

1. *Forward Plan for Health, FY 1978-1982*, U.S. Department of Health, Education, and Welfare, Public Health Service, August 1976.
2. *Health Services Administration Forward Plan, FY 1979-1983*, U.S. Department of Health, Education, and Welfare, Public Health Service, May 1977.

3. *Draft Initial Statement of the National Health Policy Planning Guidelines*, "U.S. Department of Health, Education, and Welfare, Public Health Service, Health Resources Administration, October 1976.
4. The federal initiative in rural health, by Dr. Edward Martin, *Public Health Reports*, 90(4) July–August 1975.
5. Public Health Service Rural Health Initiative FY 1976, *Briefing for the Policy Board*, July 17, 1975.
6. *A New Perspective on the Health of Canadians*, by Marc Lalonde, Government of Canada, April 1974.

The Rural Preceptorship in Medical Education

ROBERT F. GLOOR, M.D., M.P.H.
University, Alabama

BACKGROUND

The Flexner report[1], issued in 1910, which emphasized the place of the natural sciences and biomedical research in the education of physicians, resulted in a movement of medical education into the University setting and led to the development of many of the university based medical centers now in existence. This movement along with the marked increase in medical knowledge and other trends over time has led to specialization in medicine and a movement away from comprehensive care rendered by a single physician. The overall result has been a tendency towards disease orientation and away from looking at the patient as an individual who is a member of a family and citizen of a community; the focus has been on treatment and away from prevention. While the quality of medical care has improved markedly, in more recent years there has come the realization that something has been lost.

In the 1960's, with the development of the first Department of Community Medicine in the College of Medicine at the University of Kentucky, began an emphasis on the need for physicians to look beyond the individual patient to the family and community from which he came. Thus community medicine, or the diagnosis of the health problems of communities and the development of solutions to these problems, became a new specialty. In 1966, the National Commission on Community Health Services[2] pointed out the need for a personal physician who is the center point for integration and continuity of all medical and related services to his patient. Such a physician will emphasize the practice of preventive medicine, both through his own efforts and in partnership with the health and social resources of the community.

He will be aware of the many and varied social, emotional and environmental factors that influence the health of his patient and his patient's family. He will either render, or direct the patient to, whatever services best suit his needs. His concern will be for the patient as a whole and his relationship with the patient must be a continuing one. In order to carry out his coordinating role, it is essential that all pertinent health information be channeled through him regardless of what institution, agency, or individual renders the service. He will have knowledge of and access to all the health resources of the community—social, preventive, diagnostic, therapeutic, and rehabilitative—and will mobilize them for the patient.

Subsequently, the Millis report[3], also in 1966, pointed to the need for changes in the graduate education of physicians, including the need for comprehensive medical care services and the development of primary care physicians. The report stated:

Many leaders of medical thought have proclaimed the desirability of training physicians who are able and willing to offer comprehensive medical care of a quality far higher than that provided by the typical general practitioner of the past. The physician they conceive of is knowledgeable —as are other physicians—about organs, systems, and techniques, but he never forgets that organs and systems are parts of a whole man, that the whole man lives in a complex social setting, and that diagnosis or treatment of a part, as if it existed in isolation, often overlooks major causative factors and therapeutic opportunities.

One of the qualifications of the physician who renders comprehensive care is thorough knowledge of and access to the whole range of medical services of the community. Thus he can readily call upon the special skills of others when they can help his patient. If the full range of medical competence is to be made effectively and efficiently available, it is mandatory that means be found to increase the supply of physicians who are properly trained and willing to serve in this comprehensive role.

What is wanted is comprehensive and continuing health care, including not only the diagnosis and treatment of illness but also its prevention and the supportive and rehabilitative care that helps a person to maintain or to return to, as high a level of physical and mental health and well being as he can attain. Neither the hospital nor any of the existing specialists is willing, equipped, or able to assume this comprehensive and continuing responsibility; and too few of the present general practitioners are qualified to do so. A different kind of physician is called for.

We suggest that he be called a *primary physician.* He should usually be primary in the first-contact sense. He will serve as the primary medical resource and counselor to an individual or a family. When a patient needs hospitalization, the services of other medical specialists, or other medical or paramedical assistance, the primary physician will see that the necessary arrangements are made, giving such responsibility to others as is appropriate, and retaining his own continuing and comprehensive responsibility.

While stressing the need for and advantages of university based programs the report also called for the involvement of affiliated hospitals and local practitioners.

In the same year the Willard Committee[4], stressed the need for "the development of new programs for the education of large numbers of family physicians for the future." The Committee report defined the family physician as one who:

(1) serves as the physician of first contact with the patient and provides a means of entry into the health care system; (2) evaluates the patient's total health needs, provides personal medical care within one or more fields of medicine, and refers the patient when indicated to appropriate sources of care while preserving the continuity of his care; (3) assumes responsibility for the patient's comprehensive and continuous health care and acts as leader or coordinator of the team that provides health services; and (4) accepts responsibility for the patient's total health care within the context of his environment, including the community and the family or comparable social unit.

Again, the involvement of practicing physicians was recommended.

In 1970, the Carnegie Commission on Higher Education[5] stated that

To improve health care requires:
More and better health manpower
More and better health care facilities
Better financing arrangements for the health care of the population
Better planning for health manpower and health care delivery

The heavy emphasis of the report of the Commission on the first of these components of better health care led to the recommendation that there be developed additional health science centers which were to move beyond the Flexner research model in a diversity and mixture of models.

New developments should be toward greater integration with social needs, or toward greater integration with the general campus, or both.

In addition, the report called for the development of area health education centers which would be attached to local hospitals but whose educational programs would be administered by university health science centers. These area health education centers would

... train medical residents and M.D. and D.D.S. candidates on a rotational basis; they would carry on continuing education for local doctors, dentists, and other health care personnel; they would advise with local health authorities and hospitals; they would assist community colleges and comprehensive colleges in training allied health personnel; and, in other ways, they would help improve health care in their areas."

The report favored a closer tie with scientific and clinical instruction and "tying clinical instruction to work with 'garden-variety' as well as 'exotic' patients."

These trends in concert with movement toward the use of teams in health care, the stress on ambulatory care, and a concern for a holistic approach to the patient as embodied in the concept of health given as a principle in the constitution of the World Health Organization[6], "Health is a state of complete physical, mental and social well-being and not merely the absence of disease or infirmity" have led to a reemphasis on the use of the practicing physician, particularly when he is located away from the university medical center, in the education of medical students. When seen in the context of these reports, it is not surprising that the rural preceptorship has become an important part of the education of the medical student in a number of medical education programs. These concepts are also being applied to students of other disciplines but for purposes of discussion here the model will be limited to the education of the medical student in family and community medicine.

THE PRACTICE SETTING

While family medicine and community medicine are separate entities there are distinct advantages in including both disciplines in one clerkship. Since the student has been oriented toward clinical medicine through all his medical education, the use of a clinical base for his introduction into the practice of community medicine makes the transition take place more smoothly. In addition, the preceptor serves as an introduction to the community and serves as an important resource to the student as he studies the community.

Thus, in the model under discussion, preceptors are chosen on the basis of their being:

1. board eligible or certified in family medicine
2. situated rurally away from the medical center
3. widely involved in their communities.

Some spend one day each week teaching family medicine in the university based family practice center. Others are prevented by distance from such teaching. The preceptors are not reimbursed financially for their teaching contribution during the clerkship.

STRUCTURE OF THE CLERKSHIP

Students are assigned to communities of their choice when possible during their senior year after completing the basic clinical rotations. Overall supervision is provided by faculty from the department of community medicine. Direct patient care supervision is the responsibility of the preceptor.

The communities (usually the physician of the local hospital) typically arrange for housing and meals for the student during the seven week clerkship. A previous week has been spent at the university in seminars providing the student with background for the experience.

Once in the community the student is expected to spend about 50% of his time with the preceptor in both clinical and nonclinical aspects of the practice and the remainder in looking at the community health needs and resources and in conducting a simple research project in some aspect of community health, if this is included.

CLINICAL ASPECTS OF RURAL PRACTICE

The medical student observes and participates in the examination and diagnosis of patients seen in the practice of the preceptor. He notes that there is a difference in the types of problems seen here when compared with those seen in a medical center where many of the patients have been referred by primary care physicians for secondary and tertiary care. He then recommends management, consistent with the resources available in the community, for their problems, assisting in making arrangements for referral when indicated. The student is involved not only with the care of patients in the preceptor's office but is active also in the management of hospitalized patients. Frequently he assists other community physicians in the management of unusual patient care problems in the hospital. The emergency room provides opportunity for further participation in the rural practice and the student becomes involved in the after-hour coverage of the medical needs of the community. He is frequently called to the bedside of a previously seen, or new, patient during the night hours to assist in the management of a newly arisen crisis for the patient. Since he is not licensed he must of course carry out these activities under the supervision of the preceptor who ultimately is responsible for the patients.

His visits in the nursing home with the preceptor provide ample opportunity to learn the techniques and judgments necessary in the care of geriatric patients with chronic problems. Home visits, made as appropriate to the practice of preceptor or with a public health nurse provide another aspect of health care in the community in which the student becomes involved. He may participate with the physician in satellite clinics or those conducted by the health department.

In all aspects of the practice, the student observes the manner in which consultants are utilized by the physician for improved patient care, both by means of referral and by use of the telephone. The important roles played by such specialists as the pathologist, radiologist, and anesthesiologist are demonstrated as patients are cared for.

The use of other professional and paraprofessional team members by the physician is seen as patients require the services of those with skills in such disciplines as inhalation therapy, dietetics, physical therapy, social work or one of the many other important fields. In some practices, the student is afforded the opportunity of seeing the manner in which physician extenders, be they physician's assistants or nurse practitioners, function through the use of patient care protocols and standing orders. The importance of such extenders in meeting the accessibility needs of the rural population becomes apparent to these students. The need that these newer types of providers have for backup and supervision is also noted.

By observing the manner in which the physician uses newer technologies to advantage, such as the reading of electrocardiograms by means of a telephone linkage to a distant computer or use of courier services for transport of laboratory specimens to a distant laboratory, the student realizes that the rural physician need not feel isolated from diagnostic aids necessary to his practice.

The all too frequent accident or other crisis provides opportunity for the student to learn of the emergency medical system available to the community and the problems attendant upon the meeting of such a community need.

In all aspects the student is encouraged to look at health in terms of the elements of good health care as given by the American Public Health Association:

Accessibility
    personal accessibility
    comprehensive services
    quantitative adequacy

Continuity
   person-centered care
   central source of care
   coordinated services
Quality
   professional competence
   personal acceptability of health services
   social mechanisms which enforce or encourage standards of quality
Efficiency
   equitable financing
   adequate compensation
   efficient administration

OTHER ASPECTS OF RURAL PRACTICE

A preceptorship with a practicing physician usually provides the only opportunity afforded the medical student for learning the many nonclinical aspects of private practice. He discusses the development of a record system for the office which meets the needs of the preceptor not only for continuing patient care but also for continuing medical education. In some practices he observes the practical use of the computer as a tool in record keeping in the office. Less sophisticated record systems are also valuable as the student observes the complexities of recording, filing, and retrieving the results of the patient's encounter in the practice, whether in the office, hospital or other site away from the office.

Financial aspects observed include the manner in which fees are determined, billing mechanisms, and problems which arise from patients having extremely limited resources for handling unexpected medical expenses. The mechanisms and problems of billing third-party payors become real to the student as he learns more about private insurances, Medicare and Medicaid.

The role of the physician as an employer of a team of professional and non-professional personnel is demonstrated as the student observes the overall management of the office. Through discussions with members of the physician's office staff the student begins to learn office management as it pertains to the ordering of supplies and other activities so necessary to the smooth handling of the daily patient load.

As the student notices the manner in which the physician practices "defensively" the problems of malpractice become real to him, and malpractice insurance and legal liability are discussed with the preceptor. The student is impressed with both the right and the need of the patient to be properly informed concerning the expected results and risks of alternative methods of management for a particular problem.

As the student participates with the preceptor in hospital staff meetings he learns not only of the privileges afforded the physician by the hospital but also the responsibilities incumbent upon staff members of such institutions. Similarly, he learns of the role of the medical society in the community and state and, most particularly, the responsibilities of the preceptor as a member.

The needs of the private physician in keeping informed of advances in the sciences of health care becomes apparent to the student as he observes the informal and formal means utilized by physicians in meeting these needs. The requirements for relicensing, renewal of society membership and recertification by specialty boards are discussed with the practicing physician.

He also is impressed with the peer and public accountability expected of the physician as he participates with his preceptor in medical audit and utilization review procedures in the hospital and learns of the activities of the Professional Standard Review Organization, required by law.

Through the activities of local planning groups of which his preceptor is a member, the student observes the planning process as carried out by health systems agencies or development boards. He is introduced to the goals and problems of the public health agency for his area since his preceptor is likely to be a member of the board of health.

A most important aspect of the preceptorship involves the role of the physician in the community. The student attends civic club functions with his preceptor and observes other involvement of the physician in community affairs through the school, church or other modality. The potential for family problems as a result of the demands upon the physician both as a professional and community member becomes real to the student and he learns the means which the physician, his wife and his children utilize to avoid or meet such problems. By spending time with the physician and his family away from the demands of the practice the student learns that rural practice need not mean a choice between a profession and family life.

COMMUNITY HEALTH NEEDS AND RESOURCES

In addition to understanding the role the preceptor plays in activities outside the office, the student gains a knowledge of the community in which he is located. He reviews the basic statistics available for the county, making comparisons with those for the state and adjacent counties. He discusses differences with the preceptor and other community leaders. He looks at special reports which may be available. He begins an inventory of the health related activities which, though not those in which the preceptor is involved, are available to his patients. He interviews those working with the local mental health group, the health department, visiting nurse association, and other similar providers of direct health care services. These interviews frequently reveal other sources of services to the student. In effect, he is taking a history and performing a physical examination of the community. The statistics are comparable to the laboratory and other special studies which may be done on the individual patient. His social history of the community includes interviews with the political leaders and other power figures. The family history considers the relationships existing with adjacent communities and other sources of care used by people of the community. He learns of the educational system and other community activities less directly related to health. The review of the industry and employment opportunities in the community provide other facets related to health needs and practices of the people. Thus, supplementing impressions made while assisting the preceptor, the student is making a community diagnosis. His time may not allow him to delve as deeply into certain areas as desired or as necessary to be more accurate in his assessment, but he begins to understand the community collectively. If previous students have prepared similar reports on the community, these may serve as stimulus to explore selected needs or resources in depth.

At times the student designs and carries out a simple research study on a subject of interest to him and the preceptor. The main purpose is educational for the student, so that he may learn how to apply research methodology outside of the laboratory. His study may focus on one aspect of care in the office or be more widely based in the community. It allows the student to develop a working understanding of the tools of the community medicine specialist even as he has learned to handle the diagnostic tools of the family medicine specialist.

Recognizing his limitations and using tact, the student prepares a written report of his study of the community. Frequently this report serves as a basis of action by those in the community who are endeavoring to improve health care services. In fact, the process of gathering the required information may have already started activity toward solving a particular problem or toward wider use of an existing resource. In all this, the preceptor is of help by directing the student to appropriate sources of information, at times personally introducing the student to key people. The preceptor also helps arrange for a formal presentation of the student's findings to a suitable community group.

CONCLUSION

This liaison between the academic physician and the practicing clinician, focused about the educational needs of the medical student, has widened the scope of experience available to the student both in types of patients and disease problems seen and in areas which should be of concern to him as a practicing physician.

At the same time the valuable contributions of the family physician providing primary care are incorporated into the education of medical students. There are of course, advantages to the preceptor, for one cannot teach without learning, and, indeed, he teaches best who practices.

References

1. Flexner, A., *Medical Education in the United States and Canada*, a report to The Carnegie Foundation for the Advancement of Teaching, Bulletin No. 4, D. B. Updike, The Merrymount Press, Boston, 1910.
2. *Health Is a Community Affair*, Report of the National Commission on Community Health Services, May, 1966.
3. *The Graduate Education of the Physician*, Report of the Citizens Commission on Graduate Medical Education, American Medical Association, Chicago, August, 1966.
4. *Meeting The Challenge of Family Practice*, The Report of Ad Hoc Committee on Education for Family Practice of the Council on Medical Education, American Medical Association, Chicago, September, 1966.
5. *Higher Education and the Nation's Health: Policies for Medical and Dental Education*, The Carnegie Commission on Higher Education, McGraw-Hill Book Co., New York, October, 1970.
6. *Official Records of the World Health Organization*, No. 2, Interim Commission, World Health Organization, New York, June 1948.
7. *A Guide to Medical Care Administration*, Vol. I, *Concepts and Principles*, The American Public Health Association, Washington, D.C., 1969.

Rural Health Care:
A Sociological Perspective

R. DAVID MUSTIAN
North Carolina State University
North Carolina

One of the basic beliefs of this society is that all Americans, regardless of social status and earning capacity, are entitled to the best medical care that is available. Over the past several decades, attitudes toward health have shifted from an emphasis on treatment and avoidance to one on prevention. Individuals, regardless of position in life, with physical or mental problems are treated in conjunction with their environment rather than as isolated organisms. Increasing attention is being accorded to the concept of positive health whereby health is a function of both well-being as well as the absence of disease.

Unfortunately, the application of the basic belief expressed above has not been received nor adopted throughout society. Attitudes toward health, availability and accessibility of facilities, location of medical personnel, and the acceptance or rejection of individuals as patients vary quite markedly from region to region. Historically, even the incidence and prevalence of certain diseases have varied by region.

Obviously, need for health care varies over time. Control and treatment of contagious diseases have been affected in most areas with the resultant action being the need to focus attention on degenerative diseases and mental problems.

Benefits of health care are important at societal, family, and individual levels. On-going society functions best with healthy individuals participating and contributing to the system. Good physical and mental health alleviates many family and individual problems.

Place of residence and size of place are major influences on health care. Rural residents usually experience greater difficulty in access to medical personnel and facilities than do urban residents. In addition, the decision of a medical practitioner to locate in an area or an agency's decision to build a facility are influenced by size of place and density of population. Communities experiencing rapid growth or decline face similar problems with health care. These communities must find some solution to the problem of how to recruit personnel and build facilities in a growth period or how to retain personnel and maintain facilities with a loss of population.

One of the most important things for all of us to realize and recognize is that health does not exist in a vacuum. Health, in its true essence, is but one aspect of the quality of life which includes all of the socioeconomic, ecological, and educational factors which make for a satisfactory living situation. The health care system is but one part of the total community system. To effect a change in the quality of health care and thereby effect a change in the quality of life, not only for those needing the care, but for those who are doing the caring, is the overall concern of us all.

To be sure, there is a direct link between the development and delivery of health care services and other community services. The health care delivery system as a part of the total social system of a community must function to meet the needs of people. Health care, as any other service, requires organization, manpower, financing, and facilities. Organization involves both structural and processual factors. Primary interest from a sociological perspective on the structural factor involves the provider-consumer relationships. The processual factor is the means by which health care delivery systems function.

The development, structure, and delivery of health care services are influenced by various community constraints. For example, recruitment of medical personnel seems to be related to community facilities, but other community variables such as median income, median education, or degree of "ruralness" may be more important in influencing location decisions. Thus, major factors and problems which must be confronted in the delivery of rural health services are briefly presented.

AVAILABILITY OF SERVICES

In our large urban centers I suggest that we do not have a problem of availability of either medical personnel or facilities. However, for at least 56 million Americans in rural areas, the availability of doctors, allied health personnel, and facilities are major problems which have current and long-term effects on the level of survival of our people.

Availability refers to whether or not services exist. In the mid-1970's, over 100 counties, which we classify as rural, did not have a physician.[1] Approximately one-half million persons lived in these counties covering over 150,000 square miles. A major factor was the lack of a population base to support a physician. In rural areas, specialized services are even more difficult to make available. For example, in 1974 South Dakota did not have a single neurological surgeon.[2] Wyoming had one and Idaho and North Dakota each had two. In contrast, the City of San Francisco had 32.

Many counties do have adequate population bases to support physicians but are not able to successfully recruit them. Physicians, like other professionals, are not likely to locate in areas where there are few opportunities for colleagual interaction and few support facilities.

Correspondingly, there are many counties and communities which do not have population or economic bases of sufficient magnitude to support medical facilities. Specialized equipment versus basic equipment is even more difficult to make available.

What is needed, then, is consideration of alternative models to improve availability of services. Possible solutions include cooperative ventures by communities in securing personnel and facilities, acceptance of new health workers, and full utilization of available technology to provide services to all areas.

ACCESSIBILITY TO SERVICES

Accessibility is a problem in both rural and urban areas but not necessarily for the same reasons. In urban areas, problems of accessibility stem directly or

[1] Roback, AMA, *Distribution of Physicians in the U.S.*, 1973.
[2] *Directory of Medical Specialists*, 16th Edition.

indirectly from consumer characteristics and awareness, income levels, and service hours that do not mesh with a consumer's schedule. Similar problems are encountered in rural areas with accompanying problems of distance and transportation. The black pauper in a major urban complex and the migrant laborer in a rural area have similar problems in receiving emergency care.

To foster improved quality of life and improved expectancies of survival, we must turn our attention to programs that are reflective of needs of consumers—whether needs are crisis-centered or routine in nature—and that are accessible at the point of presenting problems.

CHOICE OF CONSUMERS

Another major factor of importance is the degree of choice that a consumer has in choosing either a physician or a facility in many areas. With few physicians and many times only one hospital in an area, the situation is one of a monopoly. Potential consequences may include limited care because of large caseloads or lack or specialties and high economic costs because of little competition.

The relationships between medical personnel and patients are crucial in terms of delivery of care and assurance that patients will complete prescribed health plans. It is of particular importance in the small, rural community that relationships of trust be developed. It is also important to recognize the importance of families and social networks in the delivery of health care. Involvement of family members and friends may be a major issue in the individual's decision to seek help and to receive help.

SOCIETAL GOALS

Let us briefly note the impact that societal type can have on the delivery of health care services and in turn on the quality of survival. In our industrial society, basic economic activity revolves around the production and distribution of goods. Primary emphasis in an industrially-oriented society is correlated with high quality of care for a selective portion of a population.

In a service-oriented society, health care services are more widely distributed throughout the population but economic responsibility shifts, Governmental structures must increasingly question the delivery of services in terms of economic support.

The major question's answer may reside within the general populace. The question: What health care do I want and am willing to pay for? Phrased in another way, is health care a must in my perception of quality survival? From a systems perspective, the major question may be: Do we want to deliver health care services to the total population or should we concentrate on special populations such as the aged, the rural, or the disadvantaged?

MAJOR MODELS OF DELIVERY SYSTEMS

Various models have been proposed for the delivery of health care services. The first model which I will describe is the current, dominant, and historical model of record. It is based generally on the free enterprise model whereby providers of a service operate in the market to provide services that will be obtained and purchased by a group of private consumers. Regulation is minimal from the societal level and internal control is effected through professional affiliation. Minimal control of economic factors, i.e., fees and charges, is exerted from systems external to the medical one.

A second model is based on the idea that organization of services with a regional or area focus can best effect the delivery of health care services. This model requires considerable bureaucratic organization and a major setback seems to be in administration of the program due to the need for cooperation and coordination between administrative, service, and consumer systems.

A third model, perhaps still in the testing phase, is the development of competitive health care systems. The fundamental issue is whether areas have the population and economic bases that would support several health systems. In rural areas, economic support may be difficult. Joint or cooperative ventures may be the needed form of organization.

ORGANIZATION OF HEALTH CARE SYSTEMS

The major point to be made here is that health care in the future is most likely to originate within area or regional systems. Major assets of the regional emphasis are (1) coverage of areas of sparse populations—the concept of availability, (2) deletion of duplicate programs, (3) provision of specialties, (4) sharing of economic ventures, and (5) professional support systems.

The physician as an isolated medical system presents many problems. He cannot be all things to all people. Coverage cannot possibly be complete. Professional colleagueship is missing. Group or joint practices provide a minimal professional setting.

Similarly, leaders in small rural communities need to consider alternative solutions in terms of facilities and equipment. The provision of basic care with support of specialized facilities and equipment of nearby urban areas may be the most viable alternative for sparsely populated areas.

HEALTH PERSONNEL

There seems to be a general consensus in many circles that a critical issue in the quality of health care is the need to have more physicians. Various programs have been implemented to augment the training of physicians in medical schools.

In addition, major efforts have been implemented to aid the personnel issue through the development and training of new medical professionals such as physicians' assistants and nurse practitioners.

Early evidence suggests that the consumers have accepted these medical extenders. Emphasis should be turned to coordinating these roles with traditional roles. Of equal importance is the need for legislative bodies to carefully examine these roles and to enact the necessary legislation so that these extenders can perform their functions as members of the total health care team.

COMMUNICATION, TRANSPORTATION, AND TECHNOLOGY

Communication systems and technological innovations will play increasingly important roles in the delivery of health services and in some instances may be the basis for the survival of many individuals. A communication network, particularly in rural areas, permits professional exchange, specialty consultation, and health team cooperation. Some of the feeling of isolation from colleagues is removed with communication and the primary care physician has the means of interacting with specialists when the need arises.

Rapid transportation coupled with communication provide a rudimentary degree of availability and accessibility to our smallest communities. Planning and implementing emergency services are viable means of providing help in rural areas. Communication and transportation innovations make facilities and specialized personnel accessible to the majority of all residents.

Let us consider the impact that computer technology can have on our survival. Today's computer systems enable us to store massive amounts of material, synthesize data bits, create data profiles, and produce outputs rapidly. A computerized filing system gives the physician the capacity to compile symptom profiles and identify patients who may be high risks for given health conditions. Such use has tremendous potential for preventive health care.

THE PEOPLE'S VIEW

With an increase in the quality of life, individuals generally focus greater attention on their health status. To ensure the quality of our survival, each individual will have to turn his attention to a perspective of prevention and maintenance. With an expanding population, we must direct our energies at prevention programs rather than responding to crisis situations. While we can do much through education programs, ultimately the individual actor must assume increased awareness and responsibility. With increased awareness, each consumer will monitor his health state and seek care.

In the future, the quality of health care can be enhanced by the consumer's acceptance of paraprofessionals and cooperation in the best possible use of available and accessible facilities. The recognition that health and health care systems are but parts of total lives and total systems will bring positive results in terms of quality of health care and quality of survival.

Human Services Delivery in the Rural Community: A Developmental Model

DAVID K. BROWN, Ph.D.
Center for the Study of Aging
University of Alabama
University, Alabama

The discrete elements of human interaction in a system of human services delivery have been more or less delineated. They include, for example, the clients receiving the service, lay and paraprofessional change agents in the community, and formal professionals in the "helping" professions. What is perhaps less clear, or at least, more often ignored, is the human and social ecology of the rural community in which services are delivered.

THE ECOLOGY OF RURALITY

The rural community, in a developmental context, has been defined as an informal social system with a traditional folk-like culture. It is isolated both physically and socially, mobility is limited, civic allegiances are extremely localized and form a familial-based nationalism. A strong sense of kinship solidarity raises suspicion toward outside authority, and these kinship and familial connections provide an orientation into which all interpersonal relationships are conventionalized. Furthermore, the recruitment of sociopolitical leadership is particularized through kinship, familial ties and the politics of acquaintance. Finally, traditional life style is not questioned, values are sacred and not challenged, and religious and kinship ties provide the essential guides for behavior.

Having said all this, it must, however, be quickly pointed out that rural communities in the United States are currently undergoing rapid transformations. Indeed the point can be defended that the unfinished state of the development revolution in the United States is currently seated in the rural community. It can be further argued that the seed bed of this revolution lies in human services delivery activities. The economic revolution and the rise of unionism has transpired. Human rights movements, feminism, the rise of the aged, etc., is in its ascendancy. The next critical stage, yet to be played out, rests with the human services delivery network.

SYSTEMS AND ORGANIZATION

The critical task in human services delivery lies in institutionalizing a broad scope of services through administrative structures which are sensitive to rural traditionalisms yet acknowledge the rapid transformation these communities are facing. In other words, the challenge is one motivating and guiding effective service delivery systems while at the same time mobilizing and building community support and legitimacy. The human services organization is required to accommodate to its environment in order to survive while at the same time it attempts to introduce and guide significant changes in the same environment.

It is to be assumed, that the organization—being change oriented—will face a modicum of environmental hostility to innovation. However, its institutionalization means, that eventually, if effective, the innovative programmatic services it introduces will be valued by the community and that the community will become supportive both of the organization and the innovations it represents. Institutionalization of human services thus entails a sociocultural process by which new services, organizationally sponsored and directed, are integrated into the community and acquire a self-sustaining capability.

HUMAN SERVICES SYSTEMS: A MODEL

The broad parameters of a system of human services rests squarely upon points of human contact and exchange between clients and the organization. These pivots of contact which serve to activate and mobilize clients seeking and demanding service priorities in their own behalf and which catapult the organization into change agent and advocacy roles, are suggested as follows:

1. *penetration*—which requires organizational contacts of an extensive and continuous nature, reaching to those most isolated and alienated sectors of the rural community.
2. *information*—or the process of information distribution about extant services and access to them.

3. *belief*—which requires an intensive, ongoing range of public contact so as to arouse and generate consensus over the value and flexibility of service plans and programs.
4. *confidence*—or a level of extended interactions which inspires a feeling that service providers care about the citizenry, that organizational efforts are based on integrity and efficiency and that citizens' actions, wants, and needs are considered meaningful in the system.
5. *action*—the appeal to the aspirations of community peoples for increased human well-being which motivates them to significant action and positive achievement.

CONCLUSION

It has been argued herein that the key to effective human services delivery rests within the human contacts between and among clients and the serving organization. The critical function of the organization is to manage and direct service delivery while, at the same time, mobilizing clients to articulate service priorities. The ultimate effect of this process is to produce an integrating capability of service delivery within the community.

It has been further maintained that a simultaneous development process goes on both within the organization and the community. The organization goes through an evolution process of trying to evolve those roles, authorities, and capabilities that enhance its delivery effectiveness while—at the same time—the community undergoes a transformation of its institutions which enhance its capability to meet human need. Discreet exchanges of power and legitimacy among and between the organization and the community form the nexus around which the development revolution of human services delivery is currently occurring.

Human Services in Rural Areas

LEON H. GINSBERG, Ph.D.
Commissioner, Department of Welfare
Charleston, West Virginia

Rural areas in the United States have suffered significant neglect over the past several decades and this has affected the quality and extent of human services in such areas over that period of time.

For purposes of this discussion, a rural area is one that is not connected to a Standard Metropolitan Statistical Area (SMSA) of 50,000 or more. In other words, as the term is used here, a rural area is a nonmetropolitan area. The distinctions discussed are between metropolitan and nonmetropolitan areas.

When it began, the U.S. was a strictly rural nation and most of its people were involved in agriculture or other rural industries, such as mineral extraction, trapping, and logging. However, during the first 200 years of its history, the nation became less rural and more urban and metropolitan until, at this time, most of the people—some two-thirds—live in metropolitan areas.

The nation changed from less to more metropolitan because of a variety of economic factors, the most important among them being the mechanization of the mineral extraction and agricultural industries, which had been the backbone of the rural economy. When these industries began to decline, many people from rural areas relocated from small towns to the larger cities, where better economic opportunities were available. As the nation became more focused on mercantile matters, manufacturing, and finance, there was less concern and need for a rural population and more need for an urban population to engage in these activities.

However, the significance of the population trends was probably exaggerated. In fact, there were some thinkers in the 1960's who believed that the nation would become almost totally metropolitan by the 1980's because the population trends were in that direction. That did not happen. In addition, the emphasis of the times was on the problems of the inner city and on the problems of metropolitan living. Urban attractions and urban concerns occupied the minds of the people and also occupied the government agencies which provided funds to improve the quality of life in urban areas as well as opportunities in urban areas and in turn, increased the population problem by attracting more and more people from the small towns and rural areas to the cities.

NEW GROWTH IN RURAL AMERICA

However, by the early 1970's the trend was beginning to be reversed. Currently, there is a larger degree of growth in nonmetropolitan areas than in metropolitan areas. That is, although both are increasing, the increase in population in the small towns and rural areas is greater than it is in the metropolitan areas. There are several reasons for the reversal of this trend.

Most important is the fact that many people prefer nonmetropolitan living. In fact, most of those who left rural areas for cities did so not because they made a conscious choice to relocate to cities but because they had to leave for economic reasons. There were simply no jobs for the young men and women in the small towns. Many of those who remained in small towns were young children, older people, as well as the poorly educated, and less aggressive. Presently, however, there are economic opportunities for people in nonmetropolitan areas. Therefore, many more people are staying in small towns and many others are moving into them.

Those who are moving to small towns include many who are engaging in the resurgence of the two major nonmetropolitan industries. Mineral extraction offers better possibilities for employment and economic gain because of the OPEC price rises and the generally increasing value of energy. As energy costs have risen, coal mining and other energy extraction industries have grown in small towns and rural communities. In addition, the decline in the surplus food available to the nation and the increase in the need for food world-wide have made agriculture a stronger industry than it was.

Although both mineral extraction and agriculture are still widely mechanized and require fewer people for more production than they once did, they are both on the increase, particularly compared to other industries in the U. S. Therefore, they are holding more people and attracting more people to small towns and rural areas. Of course, an improved economy in these two industries increases the opportunities for professionals, business persons, and others in nonmetropolitan areas.

Rural areas are also growing because of an increase in the number of people used to staff institutions of various kinds. Correctional institutions, public hospitals, and other facilities that are often located in rural areas need more staff because the public has demanded an increase in the quality of care given in those institutions.

In addition, there has been a great growth in rural areas that are focused on retirement opportunities for older people. Some of the fastest growing counties in the U. S. are retirement counties and older people from metropolitan areas are moving to these retirement areas in large numbers. Some of the retirement areas are in the warmer, southern counties but others are in New England and other areas of moderate to cold climates.

Higher education institutions in rural areas have also grown in recent years. Some formerly small teachers' colleges have become large state universities which attract many faculty members and students and that fact, too, has increased the size of the rural population.

In addition, some large businesses are decentralizing to small towns from metropolitan areas.

POLICY ISSUES FOR RURAL AREAS

There are a number of public policy questions that must be asked and answered about rural areas if some of America's population distribution problems are to be resolved.

Foremost is the concern that life must be improved in rural areas for the benefit of the rural population as well as for those who live in metropolitan areas. If massive dislocations of people from rural areas to urban are allowed to occur once again, it will be to the detriment of both the metropolitan and nonmetropolitan areas. There is significant value in keeping the population distributed between all size population centers. Massive concentrations of population on just a few acres are likely to be detrimental to the total population —both those in them and those outside.

Therefore, there is a need to equalize human services in nonmetropolitan areas with those in metroplitan areas. For example, any reformed welfare plan ought to provide payments of relatively equal size to people in large cities and small towns to prevent the dislocation of people from nonmetropolitan areas to metropolitan areas.

Simple social justice also requires that there be sufficient health, mental health, counseling, and educational services in nonmetropolitan areas. There are also some indications that the problems of nonmetropolitan areas are more severe than those in metropolitan areas. That is, as older people move into retirement counties, there will be a need for greater health services. In addition, years of neglect and maldistribution of resources have led to situations in which the poorest housing, the poorest health standards, and the fewest resources to deal with health and other human problems are in rural areas. Equalization is a necessary objective.

Of course, the same kinds of concerns are true of law enforcement programs. Small community justice is as important as adequate justice in metropolitan areas. Humane law enforcement officers and humane human services programs for those found in violation of the law are important in rural areas, many of which have corrections programs and detention facilities that are relatively primitive when compared to those available in metropolitan areas.

ATTRACTING PROFESSIONALS TO RURAL AREAS

One of the problems of rural areas has historically been that of attracting professionals in fields such as medicine, nursing, psychology, and social work. In the past, it has also been a problem to attract teachers to rural areas.

Several solutions have been used to overcome these difficulties. One of the most important is the National Health Service Corps, which recruits young health practitioners and brings them to rural areas in cooperation with local communities. Some remain in those areas and have had a strong positive effect on the delivery of adequate health services to small communities.

In addition, some professional training programs have begun to develop and implement internships, traineeships, field placements, and other practical, experimental opportunities in rural areas. These have had a positive effect in orienting new professionals to the special problems and attractions of small communities. However, the major solution to the problem of attracting professionals to rural areas is probably simply time and the increase in training programs for professionals. For example, although it was difficult to attract teachers to rural areas a few years ago, the surplus of professionally educated teachers has provided sufficient numbers of teachers to cover the educational responsibilities of most nonmetropolitan areas. The same is becoming true for nurses and social workers and it will become true for physicians, pharmacists, and psychologists. There are simply more people who want to pursue these professions than there are metropolitan opportunities for them. Therefore, it is very likely that time will take care of the problems.

In addition, some rural areas are developing models for effective emergency health services which can transport patients to nearby metropolitan facilities rapidly and which might also be effective in demonstrating the ways persons with a variety of training can help solve health problems. It is possible that most of the health problems in most of the world are problems of health education and health training rather than health professionals. Therefore, some professionals may be able to use themselves as educators and supervisors of physician extenders and other para-professionals in the health fields. Essentially, the models of human services delivery probably need to be adapted to the special problems and resources of rural areas. There are probably too many efforts to simply duplicate the experience of the urban community when it might be more appropriate to find new ways of delivering services to rural areas.

CHARACTERISTICS OF RURAL AREAS REQUIRING SPECIAL ATTENTION

Practitioners in rural areas need to understand a variety of things about those areas and need to prepare themselves in specific ways to work with rural areas effectively.

One of the most important features of rural areas that must be understood is that religion plays a major role in a small community—probably a more important role than it plays in most urban and metropolitan areas. Therefore, the rural practitioner must learn to work with and through religious institutions, particularly churches, and must develop positive relationships with rural religious practitioners, particularly ministers.

That does not mean that rural human services practitioners must be religious themselves. However, it does mean that they must have respect for and must have the capacity to deal with religious institutions and leaders.

In addition, rural communities are more likely to be traditional than modern— more likely to be primary group oriented rather than secondary group oriented, and more likely to appear to be conservative than liberal or radical. Therefore, effective practitioners in rural communities must learn to work within traditional norms. That does not mean that one must identify with the specific values of that society. It may only mean that the practitioner must learn to be respectful of and capable of relating effectively to a traditional value system. It means being aware of and competent in dealing with family orientations. It means being cognizant of the importance of primary groups, such as groups of people sitting around in coffee shops or friendship groups, as important molders of community values and important sources of community participation and communication rather than looking to more formal organizations and the press as the sources of these kinds of processes.

One must also learn to have patience in small communities. Rapidly expecting to change situations that have come to reality over decades is unrealistic for the rural practitioner.

The traditional quality of rural communities also may make demands on the individual behavior and personality of the human service practitioner in the rural community if he or she wants to gain acceptance and credibility in the community. For example, wearing more or less modest and traditional clothing, the absence of beards for men, and a relatively conservative or at least discrete personal life seem to be factors of great importance for effective practice in rural communities.

One must also learn, because of the traditional nature of rural communities, to work with the natural helping networks. One may find that certain kinds of health and mental health services are delivered by individuals, rather than by institutions. One may discover that the public shelter for transients is operated by the local sheriff, rather than by any formal association designed for the purpose. One may discover that the children's counseling services are handled, in reality, by a local minister, rather than a mental hygiene clinic. All of these things influence the ways in which one practices in the small town. They mean that one must build upon the local community's resources and that one must also be very careful to avoid offending local community norms and procedures, which may render a new human services professional ineffective.

Confidentiality is also a problem of significance in rural communities and its impact is somewhat different in the small town than it is in the metropolitan area, where many of the models of confidentiality have been developed and defined. For example, it is hard to miss seeing a local community resident entering a social agency office. Keeping that fact confidential is sometimes difficult. Therefore, home visits and meetings in nonregular places such as restaurants or parks may be necessary in order to protect the confidentiality of some clients in some ways.

Perhaps the biggest problem for some professional practitioners in rural areas is the lack of professional colleagues, stimulation, and supervision. It may be that the most effective rural practitioners are those who like to be on their own, who like to operate independently, and who have the capacity to supervise themselves and to stimulate their own professional growth through the use of professional journals, meetings, and other resources. It is usually impossible for the practitioner in the small town to have the same opportunities for guidance and direction as his or her colleagues in the very small towns. However, for the correct kind of professional, that may be a blessing in that it affords one an opportunity to try new things, to be one's own supervisor, and to test one's self against one's own values, rather than those of a colleague or supervisor. It does mean, however, that one must be personally vigilant about insuring one's continued careful study of literature, attendance at professional meetings, and the initiation of and maintenance of informal contacts with fellow practitioners whenever that is possible.

In all, many of the best human service practitioners find rural human services challenging and pleasant in many ways that contrast markedly with providing services to more urban areas.

Collective Sharing in Rural Health

JOHN B. O'LEARY, M.D.
Professor of Family Practice
University of Minnesota Medical School
Minneapolis, Minnesota

I. THE SOUTHWESTERN MINNESOTA HEALTH COOPERATIVES

Small rural communities around the world share a common problem in that they all have a shortage of health professionals. The traditional approach to this problem has been from the top down—a government or private foundation will try varying amounts of bribery and force to get professionals to move into rural areas.

Senator Dick Clark, Chairman of the Subcommittee on Rural Development of the Senate Committee of Agriculture and Forestry, has said "Our nation has spent millions of dollars, hundreds of millions of dollars, in the past few years to encourage health professionals to practice in medically underserved areas; and, in frankness, these have not been very successful. Therefore, health personnel are still clustered in urban and suburban areas."[1]

In July 1971, a group of six small doctorless towns in southwestern Minnesota decided to band together to see if they could do collectively what they had been unable to accomplish singly; namely, attract health services. It was a long and difficult struggle, but they were eventually successful. They now own and operate two busy clinics. They have done it all on their own without outside support. The clinics are now running in the black and the people are proud of their efforts.[2]

If running two doctors' offices and making money at it were all that health is about, this story could end right here. Even in these days of galloping inflation, it is hard to lose money operating a doctor's office. Fortunately, the farmers in southwestern Minnesota are among the best in the world at applying the advanced technology of agriculture to the production of healthy crops and animals. A group of these farmers participated in a series of monthly teleforums patterned after the community education programs of the Cooperative Extension Service.[3] These forums involved bringing groups of 30 to 40 farmers into contact with the University of Minnesota faculty by means of telephones coupled to loudspeakers. The month preceding each meeting was spent reviewing basic science research literature pertaining to human and animal health so that all participants could take issue on matters of substance. At the end of that year, there was general agreement that the cows and the corn were doing much better than the people because farmers had been engaged in applying *health* research while doctors continued to concentrate on uncontrolled studies of *disease*.

These meetings sparked an interest in the development of a truly comprehensive preventative program which could be a model for other communities to emulate. Barton, one of the early leaders of this group, had studied the barriers to forming cooperative health organizations among southwestern Minnesota communities[4] as well as the intercultural conflicts between health sciences students and a rural population.[5,6] Knowledge of these factors not only helped to insure the success of the clinics, but provided a theoretical background to the development of the Southwest Minnesota Community Health Cooperatives.

II. THE COSTA RICAN MODEL

Dr. Rodrigo Gutierrez, Dean of the University of Costa Rica Medical School, has said that health programs which operate from the top down do poorly but that those systems which are developed by the people themselves can be spectacularly successful.

One such system has been developing for 25 years in the communities surrounding San Ramon, Costa Rica. Spearheaded by Dr. Juan Ortiz, the program now has 42 community health stations and serves 80,000 people. The nurse is the primary care person who comes from, and lives in, the community she serves. The health stations are visited twice weekly by nine teams of health professionals operating out of the "Hospital Without Walls" in San Ramon.

The outstanding features of this system are that it was founded by the people, has spread one community at a time and operates on local money. The community organization is so extensive that every adult is on some committee. There is an excellent communications network involving newsletters and radio. The health record system appears so complete that it would be difficult for any individual to escape receiving needed health services and health instruction.

The nurses at the health stations play a key role in the success of the system. They maintain on-going responsibility for the preventive health maintenance of each and every individual through a detailed record system. This record and reminder system is so comprehensive that every child in the area receives all preventive services, including immunizations and screening for mental and physical development, at all age levels.

Even more important, the nurses share this responsibility for health maintenance with the citizens of each community through local committees which are active not only in medicine but in other areas of social need. For example, committees for youth, committees for schools and committees dealing with alcoholism.

None of this could exist without extensive communication being an integral part of the system. There were classes in such things as nutrition and oral hygiene. There was extensive use of radio in some areas and there were excellent community newsletters. Small wonder that there is so much cooperation and that pride in the system is so great.

Even more important are the free and open lines of communication which allow everyone to share responsibility for decisions that are usually, and foolishly, kept as "medical secrets" in many parts of the world. It comes as a cultural shock to realize that the people of the communities actually participate with the medical staff at the hospital in San Ramon in weekly autopsy conferences. How right and beautiful it is for everyone to share responsibilities in health. When a child dies, not to say what could "medical professionals" have done, but rather what could all of us, as part of a community, have done to prevent the death of our child.

III. IMPLICATIONS FOR THE FUTURE OF RURAL HEALTH

In Milio's[7] discussion of a strategy for changing health-damaging life patterns to health-generating life patterns, she points out that the range of health options available to an individual is determined by the organizations which are geographically accessible to that individual. Lathem[8] has advanced the argument that these organizations must be large and powerful to produce significant changes in health status.

Merton[9] has stated that technological innovations will be adopted and persist only if they fit the structure of a society. The central institution in rural Minnesota is the school. Farmers who live equidistant from several towns in Minnesota invariably identify their home town as the one in which their children attend school. In other parts of the world, other structures may predominate; in Costa Rica it may be the local community political structure and in the Bible Belt of the southern United States it may be the church, but structures exist in every community.

To succeed, a health system must involve all the structures of a rural society because health problems do not exist singly—they are all multidetermined phenomena. As the authors of the Harvard report on barriers to primary prevention, Broskowski and Baker[10] point out:

Approaches to primary prevention must take into account the total ecological systems in which the population resides. Previous work by Kelly demonstrates the necessity of understanding the ecological complexities in planning preventive interventions at the level of broad social institutions, such as the schools.

Regardless of one's theoretical model of the phenomenon in question, effective preventive programs aimed at the total population must certainly be aimed at the primary institutions of society, such as schools and work settings, and will therefore require the involvement of specialists familiar with the special phenomena that exist in large-scale organizations. For example, any adequate comprehensive program of primary prevention must address itself to the work environment in which so much of the population resides.   Sophistication with work environments and the organization factors concomitant with such environments will certainly be necessary.

References

1. Clark, D. Hearings before the Subcommittee on Rural Development, U.S. Government Printing Office, Washington D.C. (1977).
2. O'Leary, J. B. and Barton, S. N. Health cooperatives in rural communities. *Biosci Commun.* 87: 1-13 (1977).
3. O'Leary, J. B. Urban-rural teleforums as an educational tool for intercultural health education. Proceedings 2nd International Conference on Health Care Distribution Systems. University of Miami, November 25, 1975.
4. Barton, S. N. Covert aspects of intercommunity communication: an application of Bion; PhD diss. University of Minnesota (1975).
5. Barton, S. N. and O'Leary, J. B. The rhetoric of rural physician procurement campaigns: an application of Tavistock. *Q. J. Speech* 60: 144-154 (1974).
6. Barton, S. N. and Weber, R. G. Cognitive differences between health science students and a rural population. *J. Med. Educ.* 50: 1120-1121 (1975).
7. Milio, N. A framework for prevention: changing health-damaging to health-generating life patterns. *Amer. J. Publ. Hlth.* 6: 435-439 (1976).
8. Lathem, W. Community medicine: success or failure? *New Engl. J. Med.* 295: 18-23 (1976).
9. Merton, K. *On the shoulders of giants.* Harcourt, Brace & World, 295: 18-23 (1976).
10. Broskowski, A. and Baker, F. Professional, organizational, and social barriers to primary prevention. *Amer. J. Orthopsychiat.* 44(5): 707-719 (1974).

Mental Health Services in Rural Areas

LUCY D. OZARIN, M.D., M.P.H.
Assistant Director for Program Development
Division of Mental Health Service Programs
National Institute of Mental Health
Washington, D.C.

Until recent years, mental health services were not generally available in rural areas except in the immediate vicinity of a state mental hospital. People who could afford private care could obtain it by traveling to the nearest city which had a psychiatrist or private mental hospital. For most people, care had to be sought at a state hospital or, when available, an outpatient psychiatric clinic. Many large rural states have only one state mental hospital. In South Dakota, for instance, the state hospital is located in the extreme southeast so that residents of Rapid City must travel over 300 miles. A similar situation exists in Wyoming where the state hospital is in the extreme southwest only 80 miles from Salt Lake City, Utah but about 300 miles from Sheridan, Wyoming, a population center.

This situation is changing. Following recommendations made by a study of mental health needs and resources in this country,* Congress passed the Community Mental Health Centers Act in 1963 to provide financial support to construct locally-based mental health centers. In 1965, the law was amended to provide support for staffing the centers. Amendments in subsequent years give additional support for services to children and to those with alcohol and drug abuse problems. Congress revised the act in 1975 (PL 94-63) to broaden its provisions and to strengthen local control. The preamble of the law states (Section 302) "community mental health care is the most effective and humane form of care for a majority of mental ill individuals."

Community mental health care, as defined in federal law and regulations, means that a comprehensive range of mental health services (inpatient, outpatient, partial hospitalization, emergency, follow-up, screening and preventive services) will be accessible (economically, geographically and at the times needed), available and with continuity of care as long as care is needed. Each state is required to establish catchment areas each with a population base of

---

*Action for Mental Health, Report of the Joint Commission on Mental Illness and Health, Basic Books, New York 1961.

75,000–200,000 although exceptions can be made for smaller or larger populations. Catchment areas may include part of a city or a number of counties. Thus, catchment areas in large cities may encompass only a few square miles but in sparsely populated rural areas may contain up to 50,000 or more square miles. In order to bring the service as close as possible to those who are in need, all centers have established satellites. In rural areas where centers cover more than one county, the satellites are usually placed in the county seats and may be staffed on a resident or itinerant basis. A satellite may be as far as 165 miles or more from its administrative office.

Delivery of service in a rural area is more difficult to arrange than in an urban area. Distance and space pose problems in reaching the population in need. Service delivery is more expensive both for clients and staff. Travel time lessens time available for treatment. The cost of service becomes higher for each unit of care provided. Rural centers usually serve a poorer population so collections are smaller. In addition, professional staff are often urban trained and prefer to remain in cities. Recruitment is difficult and once staff is obtained, retention may be a problem. Continuing education, now becoming mandatory for certain professionals, is difficult to obtain without traveling to a university or city. Most professionals have been trained in urban settings and many prefer the professional and cultural advantages available there.

Despite the difficulties in providing services in rural areas because of financial and staff limitations, mental health centers have been established in many rural areas. The country has been divided into 1500 catchment areas (3,100 counties). In mid 1977, about 660 centers have received federal support. About 240 or a third are located in nonmetropolitan rural or part-rural areas. A center in eastern Oregon encompasses 13 counties and two Indian reservations. One in northwest Kansas includes 20 counties and the northern Arizona Comprehensive Guidance Center in Flagstaff covers almost half the large State area in five counties and includes two Indian tribal councils as corporate partners. Some predominantly rural States now have centers serving all their catchment areas. These include Idaho, Kentucky, Montana, and Vermont.

The annual report for 1976 from the eastern Montana Regional Mental Health Center describes its activities. The eastern Montana Mental Health Center, with main office in Miles City, serves 94,000 people living in 17 counties which cover an area of 50,000 square miles. The Center was founded in 1967 and is a nonprofit corporation governed by a Board of 16 citizens. Numerous problems during its early years were gradually overcome and now the Center is a viable service unit. About 45 professional and support staff are employed including two psychiatrists with a vacancy for a third psychiatrist.

The Center has three main bases located at Glasgow, Glendive, and Miles City, each a town of about 10,000 and the largest in the area. Three additional sites have staff based there and are located in smaller communities. Several counties are served by itinerant staff for a total of ten service sites.

The Center's annual report for 1976 describes its activities in the past and currently. In 1973, 178 people from the catchment area were sent to the single state hospital 300 miles away. In 1976, 12 people were admitted to the state hospital and 296 were hospitalized in general hospitals in the catchment area. Currently, 53 residents of the catchment area remain in the state hospital, a decline from 120 only 18 months ago.

In 1976, almost 8,000 outpatient sessions were provided to clients, 422 were treated in partial hospitalization programs at the three main locations and 480 emergency calls were answered on a 24-hour service. Thirty percent of the clientele are under 18 years of age.

Consultation and education services are provided to 75 different agencies or groups each month including schools, courts, law enforcement and correctional agencies, clergy, welfare, health agencies.

Several new programs broaden the range of services. An unused wing of the local general hospital in Miles City has been leased and is used for transitional care providing a less expensive alternative to continuing hospitalization for patients who no longer need intensive treatment. A residential group home has also been opened to provide a halfway house for patients discharged from the State hospital. Eight people and a resident manager can be accommodated here.

Several Indian reservations are within the catchment area. Contracts have been made with the respective Tribal Health Boards to supply funding to hire, train and supervise Native American mental health workers who report their activities to the Center and for whom the Center provides backup professional help.

Over 1,500 new admissions received Center services during 1976; 450 were under age 17, 900 had weekly family incomes under $150 a week and another 400 had weekly incomes of under $200. About a third were referred by self, family and friend and 20% were referred by local physicians. Schools and community agencies each referred about 11%.

For the year ending 1976, total revenues were almost $600,000 with State funds contributing almost half, Federal funds somewhat less than a quarter and local Government funds less than 7%. Patient fees, contracts with schools and other agencies, and third party payments contributed the rest. Expenditures totaled $508,000 of which almost 70% covered personnel expenses.

Rural areas which are not served by federally assisted mental health centers may have mental health services available through county or state auspices. These services are usually not comprehensive and are limited to outpatient and and consultation and, in some places, partial hospitalization. Some State mental hospitals provide outreach services and may station staff in distant locations or send out traveling teams. Inpatient care may require travel to distant hospitals and 24 hour emergency service is often not available. Financial limitations and desire for local county autonomy are cited as reasons why rural areas may not seek to provide more comprehensive services.

Individuals and families with adequate means or health insurance often seek private care but in rural areas where privately practicing professionals are not easily available they may seek help at their local mental health centers. About 40% of center clients are self, family or friend referred and another 15% are re - ferred by physicians. About 60% of the centers are located in low median income areas. Most clients are in the young adult and middle age ranges. Predominant problems seen at centers are maladjustments, family difficulties and personality problems with psychological/physiological symptoms. About 10% carry a diagnosis of schizophrenia and 13% are diagnosed with depression. Prompt attention to mental health emergencies and crises on a 24-hour basis often forestalls the need for expensive inpatient care which disrupts family life.

Center clients often have a variety of problems: medical, psychiatric, economic, social, legal, vocational. Since helping resources are limited in rural areas it is highly important that available agencies (welfare, vocational rehabilitation, courts, probation, police, schools, public health nurses, physicians, hospitals) coordinate their services on behalf of clients. Linkages between agencies on a formal or informal basis must be established so referrals can be made easily. This is illustrated by the linkages between mental health centers and the Rural Health Initiative/Health Underserved Rural Areas (RHI/HURA) projects.

The RHI/HURA program is a federally assisted effort to help rural communities designated as health underserved to develop primary (ambulatory) health care systems with linkages to hospital and other specialized health facilities. In mid-1977, 244 such projects had been funded. A study was carried out in 1976 to determine their relationships with federally funded mental health centers. About two thirds had overlapping catchment areas in one or more counties, one third of the RHI/HURA projects were currently using some mental health resource and another third were planning to do so. About 6% were budgeting for mental health related personnel, usually social workers. In one two-county rural area, two nurse practitioners 80 miles from the RHI base and a psychiatric nurse in a satellite 65 miles from the mental health center are seeing clients together as appropriate and are covering emergencies for each other. The psychiatric nurse is on the RHI Board. In another state, a large one-county RHI has contracted with and is paying for services from the mental health center. The center furnishes consultation and inservice training to RHI staff and psychiatric and speech and hearing services to their clients. Another RHI has budgeted funds for a psychiatrist to be hired by the local mental health center who will be stationed at the RHI site which is also a center satellite. In another State, a center satellite has allowed the RHI staff to park their trailer on their property until permanent quarters are found. A RHI physician in another project provides medical back-up to mental health center satellite staff between visits by the center psychiatrist. Plans are under way to station mental health personnel in selected RHI/HURA projects to facilitate coordination and delivery of health and mental health services.

Providing mental health services in rural areas poses special problems but, as described above, it can be done if adequate funding, facilities, transportation and staff are available. Staff ingenuity, creativity and hard work are essential. Local community support and involvement provide the foundation for the efforts described here.

Rural Mental Health— A Community Perspective

ALAN R. GOODWIN, Ph.D.
MICHAEL T. SMITH
Indian Rivers Mental Health Center
Tuscaloosa, Alabama

The community mental health center concept was originally endorsed as a potentially more efficient and economical alternative to institutionalization. The probability that others could be prevented from requiring hospitalization was another attribute. In the late 1970's community mental health centers have broadened to include twelve basic and mandated services, which include:

Inpatient Services	Services for Elderly
Outpatient Services	Screening Services
Day Treatment Services	Followup Care
Emergency Services	Transitional Services
Consultation and Education Services	Alcoholism and Alcohol Abuse Services
Services for Children	Drug Addiction and Drug Abuse Services

This has created much discussion of "the least restrictive environment," "prevention programming," and "outreach services." These issues are particularly important to the development of rural mental health services.

This contrast in services is partially due to subtle differences between what is appropriate behavior in an urban context versus a rural context. For instance, a marginal child abuse incident might be considered a "good lickin" in a rural family. To some extent, in areas of less than 10,000 people, there has been less of an assimilation of the values predicated on congested, media bombarded, and transient city life. The facts that rural families are more of the extended variety, that they produce many of their own essentials, and that there is less out-migration to preserve long-established roles and relationships obviously affects the quality of life. Other issues such as transportation, church influence, the sanctity of the medical doctor, the power of the local sheriff, etc., affect the style of mental health delivery.

Nonetheless, these generalizations do not entirely fit many rural communities. It is the variabilities across locales which make any discussion of rural mental health delivery incomplete. However, experiences in developing mental health services in a three county catchment area, two of which are predominantly rural, suggest that there are two critical variables. The first is to understand the unique features of a given rural community. For example, Indian Rivers Community Mental Health Center serves West Blocton, Alabama, which is an appalachian-like community with considerable mining interests, a relatively small black population, and a strong element of isolation. In contrast, the community of Carrollton, Alabama, with its farming qualities and its old South culture, has a need to accommodate outside resources. Therefore, service "appropriateness" is derived from accepted community standards of behavior and both our prevention and intervention efforts reflect an awareness of these standards.

The second critical variable is an ability to be innovative and flexible. In the two rural counties served by Indian Rivers, the staff has recognized the need to utilize indigenous resources. For example, the existing institutions in the communities such as schools, churches, health facilities, jails, and civic organizations, have been fused with mental health. In this regard, there has been a keen recognition of the political and public relations component of service delivery. A clinician not only wears his "intervention hat" but he also must be the public relations expert. While taking an active interest in the social issues in the community, he must learn how to cautiously communicate his views. He can be progressive but not militant, conservative but not rigid. He runs a strange gambit between being publicly available, while respecting the ethics of his profession, especially confidentiality.

This accessability to the public is the key to initiating a vital program in a small community. With mental health being an "alien" service, visibility helps to build a foundation of trust. For this reason, schools and health departments are utilized as mental health outposts. However, this frequent scrutiny and the knowledge that there is little anonymity in rural areas places the practitioner in a sensitive position. Accountability is a word of mouth system. Although creative efforts may receive scant acclaim, one can be assured that any short-coming will be widely communicated. This obviously reinforces conservatism; therefore, the job of being carefully innovative is a complex one.

Granted that rural mental health is in ways more intricate than urbanized services, it is apparent that the practitioner may require a larger repertoire of skills. A specialist in biofeedback is in less demand than a clinician who can counsel with families, consult with teachers, train paraprofessionals and serve as a consultant to law enforcement and the courts. Additionally, as one serves all of these functions, it is imperative that one realize how values can bias one's impressions of others. A worker must be able to collaborate not only with those who hold similar beliefs, but also with those who approach matters differently. Under these circumstances, the abilities to problem solve with people are frequently tested. Those clinicians who can achieve a degree of objectivity will have the best success in working with the whole community to meet their needs. For the rural mental health worker, results prove to be the primary reward.

The best results in a rural setting, however, are often found with preventive programs. In an area that cannot support reactive mental health services which expensively attempt to eliminate individual problems, a preventive mental health model is often more functional. Clients can be seen more economically, especially when those who might require short term treatment receive it early. Receiving it early, though, is dependent again on making services easily accessible. Utilizing resources such as Headstart centers and schools helps to make prevention possible. Instead of a teacher wondering about a child's behavior but not knowing where to turn, regular on-site visits create an alternative. Additionally, consultation with a psychologist or social worker allows a teacher an opportunity to develop new skills in early detection. When this is achieved, there is a dramatic increase in preventive mental health agents.

Similar innovations in rural mental health are unlimited. For instance, eight youngsters, aged 9 through 11, traveling on a boat for nine days is a unique experience for youngsters from a rural landlock community. Over the last four years, staff members from Indian Rivers have developed such trips for youngsters in the caseload of the Child and Parent Program. The intervention results have been impressive, but the public relations impact has also been just as important. In a sense, the mental health experience of these youngsters and their families has been viewed much more positively because of their inclusion in the boat trips.

Similar results were experienced in other settings. At a garment plant, staff mingled during the lunch break to provide accurate information about mental health services and to clarify some of the many misconceptions about mental health. Misconceptions included the notion that mental health was synonymous with "craziness" and services were only for the indigent. Staff has returned to the garment plant on four occasions and the results have been increased referrals and the initiation of a training program for employee supervisors to more effectively work with their employees who may require referral.

The referral process has been broadened by our outreach efforts. In county jails, groups are now operational and open to all residents who elect to participate. The goal is intervention, but this is also an opportunity to encourage more transient jail residents to seek follow up services upon their release. As to the latter, the most dramatic results can occur if someone enters into an intervention program immediately upon release.

Working closely with the Department of Pensions and Security workers as collateral has facilitated mental health intervention in a number of rural family systems, while confronting the notion of DPS as simply the "welfare agency." One interesting approach has been to work more with the DPS case worker as a co-therapist as opposed to a more formal collateral relationship. For example, a mental health staff member is working in a co-therapy relationship with a DPS worker in an unwed mothers group. This allows mental health and DPS workers to share their sometimes unique skills and have a more effective impact on the population of concern.

This is especially true with the juvenile agency councils which mental health staff members have initiated. These councils focus on improving the communications between human resource related agencies who are concerned with children and adolescents. Included have been mental health related agencies, the Department of Pensions and Security, law enforcement and the courts, the health department, ministers, and educators.

Beyond simple collaboration is contracting. Often, contracts will result from a period of affiliation. For instance, contracts with schools have made it possible to have contact with students and families who never would have sought mental health services. The staff promoted themselves through nonclient volunteer groups, informal discussions with students in the hall, meeting in the teachers' lounge, making impromptu presentations in classrooms, and participating in events like parents' night. While Indian Rivers gained exposure in these settings, the contracts also provided a stable funding source.

The predictability of funding will largely determine the future development of rural mental health. Currently, programs in rural areas do not generate monies to sufficiently meet costs. Therefore, rural programs are dependent upon federal funds for self-preservation, not to mention expansion. The best means of counteracting the steady diminution of grants is to stimulate grass roots support for programs. In rural community mental health centers, community is the big word.

Group Social Competence Training with Clinical Populations

JANE ENGEMAN
MATTHEW CAMPBELL
University of Cincinnati
Cincinnati, Ohio

Personality theorists and psychotherapists almost unanimously emphasize the importance of interpersonal relationships to an individual's physical and mental health. As Argyle[1] states, "It is well known that relations with others can be the source of the deepest satisfactions and the blackest misery. . . . Many people are lonely and unhappy, some are mentally ill because they are unable to establish and sustain social relationships with others." There is, however, much less agreement among workers in the helping professions regarding the most appropriate treatment for individuals suffering from a lack of satisfying interpersonal relationships. This article describes social skills training, a competency based approach which has been used to treat a variety of clinical populations who complain of unsatisfactory social relations.

The task of social skills training is two-fold: identification of specific interpersonal behaviors and attitudes that are problematic for the individual and direct training of the client in the interpersonal skills he or she lacks. These target skills are learned and practiced in the relative safety of a group of people sharing common difficulties. The structure of treatment and the specific training techniques used are designed to foster the ultimate goals of maintenance of the newly acquired competencies and generalization of these skills to a variety of life experiences. The kinds of competencies which are typically the focus of treatment are initiating, maintaining, and ending conversations, expressing feelings, responding to others, and social problem solving. Social skills training is a response acquisition approach which is grounded in the behavioral and social-learning theory traditions.[2,14] Basically this model views both normal and problem behaviors as a result of a person's past learning experiences.

CONCEPTUALIZING INTERPERSONAL DIFFICULTIES

The following abbreviated case description will illustrate the use of competency based framework to view psychopathology.

The authors wish to thank their colleagues in the Social Skills Applied Research Group, Daniel Langmeyer, Ph.D. and John Steffen, Ph.D. for their thoughtful suggestions on this paper.

Janice, a 27-year-old woman of average appearance was referred to the authors by her therapist. At the screening interview, she was tearful and reticent. When Janice did speak, her voice wavered and she did not look directly at the interviewers. She related that her work supervisor had recently requested that she take a leave of absence from her job as a floor nurse because of her depressed affect and her difficulty in giving orders and relating to other staff members. Although Janice had a few close female friends, she never dated regularly. She expressed envy at the ease with which the men and women with whom she worked related to each other. Janice's therapist was concerned that Janice might be too passive to receive benefit from the group.

Our competency based model interprets Janice's complaint as a sample of her life difficulties. Her depression is conceptualized as arising from specific skills deficits such as her inability to assert herself by tactfully but firmly giving instructions to those she supervised, her inability to make light conversation to men, and her inability to give herself reinforcement or credit for what she did do well. Janice's ineffectiveness and passive behavior were not understood as manifestations of her depression. Instead, the social skills deficits were viewed as leading to a life style very low in satisfaction, which in turn led to the feelings of depression and loneliness. In contrast, a more traditional medical model may have viewed Janice's depression as a disease process which could only be treated with long term psychotherapy to uncover the highly inaccessible intrapersonal conflicts causing the illness and with medication to relieve the symptoms.

There is considerable data which suggests that Janice is not an isolated case. Many individuals do not have an opportunity to learn the interpersonal skills necessary for a given role they may be called upon to perform. In terms of developmental patterns, some of these individuals have parents who neither modeled effective interactions nor provided opportunities for the child to shape his or her skills. Others have been isolated in childhood for idiosyncratic reasons such as hospital stays, childless or rural neighborhoods or racial factors. Some individuals have difficulty only when placed in new interpersonal situations such as the role of divorcee or beginning a new job or educational program.

Although there are a number of schema for categorizing interpersonal inadequacies, a review of clinical social skills training programs suggests five main foci for intervention: (1) actual skill deficits such as not having a repertoire of conversational opening lines or using inappropriate levels of eye contact; (2) inability to discriminate the circumstances under which certain types of behavior are normative; (3) inability to take interpersonal risks because of fear of negative evaluation from others; (4) inability to solicit and use feedback; and (5) inability to smoothly time and pattern responses. Typically, interpersonal problems result from a combination of two or more of these factors and a problem in one area usually permeates other areas of functioning.

TREATMENT APPROACHES

Because skills training is an educative therapy, such techniques as didactic instruction, discussion, demonstration, rehearsal and knowledge of results are the primary means for helping clients develop new adaptive behaviors. The specific content of the training is individualized through careful identification of client interpersonal competencies and deficits, both before and during treatment. This assessment may include self and peer ratings of various interpersonal skills and feelings.

Frequencies of social behaviors may be evaluated using questionnaires or diaries. Behavioral samples of interpersonal interactions may be obtained using role play assessment techniques.

A description of a typical group meeting is presented in order to acquaint the reader with the process of an interpersonal skills training group.

> There had been a holiday break from the group meetings and members were asked to report their progress during that time. Joe shared that he was working in a town about 50 miles away for the school quarter. He didn't know anyone there and wanted to contact a woman he had met the previous summer but had many doubts. Would she remember him? What could he say? Would she mind if he called at work? Group members discussed these issues, which helped Joe to decide to call the woman at work and generated some possible conversational openers ("Hi, this is Joe White, remember last summer at the clinic, the guy with the poison ivy.")
>
> Joe role played telephoning the woman with one of the female group members playing the other role. Joe expressed how he had often thought about her since that time, told her what he was doing now, asked if it was convenient to talk and made arrangements to call her at home. After Joe self-assessed his performance the other group members and leaders commented on the roleplay, giving him encouragement for improved voice quality and smoothness and for arranging to call her at home. They suggested that he might have sounded too intense about wanting to see her ("That might scare her.") Joe was asked if he would like to roleplay the same situation again. He said he felt like he could make the call now.
>
> The remainder of the meeting focused upon skills necessary in more intimate relationships—empathic listening, self-disclosure, and setting and increasing limits in a relationship. Group members suggested situations which had been difficult for them which required these skills and these situations were roleplayed. The session ended with videotaped presentation of the meeting's role plays and assignment of outside practice tasks.

It may be difficult for the reader to realize how difficult the simple act of calling someone on the telephone may be. Yet, for the individuals who have been referred for this form of treatment, making a telephone call may be a major obstacle as are many other seemingly everyday interpersonal behaviors.

In order to help the clients achieve the goal of increased interpersonal effectiveness, six techniques have commonly been employed: (1) *Didactic lecturettes and information sheets*. Such instruction is employed to transfer information, to structure the group, and to encourage the expectation that new interactional patterns can be learned. (2) *Dyadic and group discussion*. This component helps generate the desired sense of universality and group cohesiveness. Discussion of alternative modes of behavior also provides an opportunity for clients to learn from each other. (3) *Role-playing, role-rehearsal, and modeling*. Group members enact various social situations which have been difficult for them and watch others do the same. This process and the accompanying feedback help the person become comfortable in new roles and develop a repertoire of alternative interpersonal behaviors. (4) *Verbal and videotape feedback*. The group critiques each performance of a roleplay. The roleplays are often also reviewed using videotape. This feedback serves to differentiate between a high and low quality performance; provides information to help correct low quality performances; and offers positive reinforcement for satisfactory performance. (5) *Outside task assignments*. These assignments which ask the members to practice skills between meetings are particularly important to achieve the goal of transfer of training to the extratherapeutic environment. (6) *Buddy system*. Members may pair with other members between group meetings to assist and support each other with outside task assignments.[6]

Employing these six components, a graded approach to the acquisition of new behaviors and attitudes is used. The clients master simpler behaviors, feel more competent and confident, and progress to more complex behaviors.

We would like to stress that clinical populations do not find these new tasks easy, although clients generally report that groups are both helpful and enjoyable. In order to help clients deal with their often extreme discomfort with trying out new interpersonal behaviors, cognitive approaches to dealing with social anxiety are stressed.[5,18] Substitution of negative self statements ("I'm afraid to introduce myself to him.") with self-reinforcement ("It's good that you want to try.") and self-instruction ("Ask him a question about his opinion on the lecture.") is taught.

Social skills training is typically conducted in groups for a number of reasons. A group provides a safe and supportive social environment in which individuals can experiment with interpersonal behaviors and thoughts. Through discussion and role plays, the members of the group can share an array of social situations and potential responses to these situations. Although there is homogeneity among client problems, each client possesses different individual assets which allow for mutual learning. In short, a group, as well as being a cost-efficient way of using resources, provides resources unavailable in individual therapy. The actual timing and spacing of such groups varies considerably, depending on the population and goals of the group.

APPLICATIONS

Although similar educative methods have been in use in human relations training for many years, the wide application of interpersonal skill groups in the treatment of clinical populations is a relatively recent development. For example, this approach has been used with apparent success with psychiatric inpatients[10] (Gutride, Goldstein and Hunter, 1972), depressives,[12] juvenile offenders,[16] and psychosomatic patients.[20] Similarly, social skills training has been used to increase the interpersonal effectiveness of college students,[13,19] to reduce parent-child conflicts (Twentyman and Martin, 1974), and to improve marital interactions.[4] Assertiveness training, or appropriate expressiveness training[8] is being conducted on a wide-scale basis for clinical and nonclinical populations. Although, conceptually, assertiveness may be viewed as one component of interpersonal skill, in practice, assertiveness training and social skill training are very similar.

Evaluations of skills training programs have indicated their overall effectiveness with individuals with diverse presenting problems. The author's own post treatment and year long follow-up assessments suggest that this treatment approach has helped clients interact more comfortably with others in work, school and informal situations. They report taking more initiative in interactions, socializing more frequently and becoming closer to partners in relationships. Clients have not only indicated they feel more confident about themselves, but they appear less depressed and more satisfied with themselves generally. Most important, they feel that they now have more control over their own behavior, realizing that they can continue to increase their competence through experimentation with new behaviors.

There are important considerations which particularly favor use of interpersonal skills training approaches in community settings with nontraditional psychotherapy clients (i.e., those who are *not* what Schofield[17] designated as YAVIS patients—young, attractive, verbal, intelligent and successful). The directness and concreteness of the approach, the focal nature of skill training and the minimal requirements for initial verbal skill make this an appropriate therapy for the many clients who would prematurely terminate long term introspective psychotherapy.[7]

Skills training programs may be conducted by a variety of helpers from the mental health paraprofessional to the primary care giver. Indeed, while professional consultation may save reinventing the wheel, indigenous workers are often the best observers of the necessary social competencies for survival and growth in a given community. Furthermore, agencies other than mental health clinics and psychiatric hospitals are employing social skills training approaches. Rose[15] reports on group assertiveness training conducted in five different social agencies: a family health service clinic, a county mental health clinic, a church-affiliated family service agency, and a clinic located at a school of social work. Schools and school related facilities are developing early intervention social skills programs[3] which they hope will decrease the incidence of disability at later stages in life.

Despite our enthusiasm for such skill based training programs we do not suggest that the reader view all clinical problems through social competency-colored glasses. There are instances in which acquisition of new interpersonal skills is only one of many facets in the treatment of an individual.[11] There are other occasions when an individual possesses all the necessary interpersonal skills but may require a therapeutic modality such as insight oriented psychotherapy or family therapy to be able to make use of these skills. In screening clients, these differentiations must be made, although the art and science of appropriate client-therapy pairings is only in developmental stages.

In summary, social skills training, with its focus upon identification and acquisition of essential interpersonal behaviors is an approach that can be adopted in many settings with various populations. The proven effectiveness of such programs, the efficiency of the group approach, and the importance of interpersonal skill to emotional well being argue for increased consideration of this alternative therapeutic approach.

References

1. Argyle, M. *The Psychology of Interpersonal Behavior.* Baltimore: Perguin, 1967.
2. Bandura, A. *Principles of Behavior Modification.* New York: Holt, Rinehart, and Winston, 1969.
3. Cowen, E. L. Emergent directions in school mental health. *American Scientist* 59: 723-733 (1971).
4. Eisler, R. M., Miller, P. M., Hersen, M., and Alford, H. Effects of assertive training on marital interaction. *Archives of General Psychiatry* 30: 643-649 (1974).
5. Ellis, A. *Reason and Emotion in Psychotherapy.* New York: Stuart, 1962.
6. Engeman, J., and Campbell, M. *Interpersonal Competence: Training and Assessment.* Paper presented at the International Communication Association 27th Annual Conference, Berlin, Germany, May, 1977.
7. Garfield, S. L. Research on client variables in Psychotherapy. In A. E. Bergin and S. L. Garfield (eds.) *Handbook of Psychotherapy and Behavior Change: An Empirical Analysis.* New York: John Wiley and Sons, 1971.
8. Goldstein, A. Appropriate expression training: humanistic behavior therapy. In A. Wandersman, P. Popper and D. Ricks (eds.) *Humanism and Behaviorism: Dialogue and Growth.* New York: Pergamon Press, 1976.

9. Goldstein, A. P., Sprafkin, R. P. and Gershaw, N. A. *Skill Training for Community Living: Applying Structural Learning Therapy.* New York: Pergamon Press, 1976.
10. Gutride, M., Goldstein, A. P., and Hunter, G. F. *The use of modeling and roleplaying to increase social interaction among schizophrenic patients.* Unpublished manuscript, Syracuse University, 1972.
11. Lazarus, A. A. *Behavior Therapy and Beyond.* New York: McGraw-Hill, 1971.
12. Lewinsohn, P. M., Weinstein, M. S., and Alper, T. A behavioral approach to the group treatment of depressed persons: a methodological contribution. *J. Clin. Psych.* **26:** 525–532, (1970).
13. MacDonald, M. L., Lindquist, C. U., Kramer, J. A., McGrath, R. A., and Rhyne, L. L. Social skills training: the effects of behavioral rehearsal in groups on dating skills. *J. Counseling Psych.* **22:** 224–230 (1975).
14. Mischel, W. Toward a cognitive social learning reconceptualization of personality. *Psych. Rev.,* **80:** 252–283 (1973).
15. Rose, S. D. *Group Therapy: A Behavioral Approach.* Englewood Cliffs, New Jersey: Prentice-Hall, 1977.
16. Sarason, I. G., and Ganzer, U. J. Social influence technique in clinical and community psychology. In C. D. Sprelberger (ed.) *Current Topics in Clinical and Community Psychology,* Vol. 1. New York: Academic Press, 1969.
17. Schofield, W. *Psychotherapy, the Purchase of Friendship.* Englewood Cliffs, New Jersey: Prentice-Hall, 1964.
18. Thoresen, C. E., and Mahoney, M. J. *Behavioral Self-Control.* New York: Holt, Rinehart & Winston, 1974.
19. Twentyman, C. T., and McFall, R. M. Behavioral training of social skills in shy males. *J. Consulting and Clin. Psych.* **43:** 384–395 (1975).
20. Wooley, S., Blackwell, B., Fedoravicus, A., Terry, A., Bird, B., and Billups, V. *Illness Behavior: A Learning Theory Model for Psychosomatic Disorders.* Paper presented at the meeting of the Association for Advancement of Behavior Therapy, New York, December, 1976.

Some Developments to Provide Health Services in Rural Areas

BOND L. BIBLE, Ph.D.
KATHLEEN M. QUINT, Ph.D.
State College, Pennsylvania

Some fifty-six million people live in rural America. They reside in every part of the U.S. from the West and Intermountain region to the Northeast and the South. Between 1970–73 a significant reversal of the out-migration trend from rural areas occurred when the total growth rate of non-metropolitan areas (4.2%) exceeded the growth rate of metropolitan areas (2.9%).

Many people living in rural areas lack goods and services considered essential to quality living. A basic aspect for quality living is access to health care.

The two most basic health care problems of rural areas are medical manpower and organization of services. The most acute needs are more primary care physicians and better organization of primary medical practice. Solving either of these would aid in accomplishing solution of the other. Adequate manpower enables the task of organization to go further than it otherwise could, and group organization of practice can make recruitment and retention of physician manpower easier to accomplish. Of the various methods considered for geographical redistribution of medical manpower to rural areas, the one that may hold the greatest promise is recruitment of physicians to organized groups.

Rural and urban areas differ widely in their characteristics, but generally speaking, in rural areas, the physician shortage is more acute; persons must travel longer distances to obtain health care; emergency health services are more deficient; work-related injury rates are higher and a comprehensive approach to health care delivery often is not present.

THE PRESENT SITUATION

Considerable study has been given to methods for alleviating the maldistribution of manpower; approaches currently under consideration or trial involve both attempts to relocate physicians and other health workers and the development of alternative methods for ensuring availability of health services when needed.

A number of methods to place more physicians or other health workers in deprived areas have been proposed or tried. These include forgiveness of student loans, payment of tuition, practice grants, tax exemptions, or other financial inducements in return for service or a period of service in deprived areas; community development of medical facilities and guarantees to physicians; government-sponsored community health centers, special efforts to recruit medical and other health students from the deprived areas in question; the decentralization of undergraduate and graduate medical and allied health education through rural perceptorships; greater use of community hospitals for intern and resident training; and legislation providing exemption from military obligation in return for service in a medically deprived area.

Although many of these approaches have merit, it has also become clear that, for some remote rural areas, solutions completely different from the traditional "physician in residence" must be sought. In such areas, emphasis may be needed on expanded transportation and communication capabilities, part-time use of physicians and allied health workers, improved biomonitoring technology, use of new physician-support occupations, better understanding of individual health practices, and development of emergency care and self-help methods to ensure rural health coverage. Community organization for health services is crucial.

RURAL HEALTH CARE SYSTEMS

Basic to the development of rural health systems is the initial planning by rural communities. It must be regional in scope, based on economic service areas. It must involve all segments of the community, provider and consumer alike, with support of planning agencies and government at the county, state, and federal levels. It must be elastic and tailored to the geography and the potential resources of the locality. Hopefully, the plans will always provide a "one door" service for all economic levels rather than serving the middle class or becoming an indigent clinic serving only the poor.

Multiple communities in a logical service area will need to plan together to develop health care systems on an area basis to attract appropriate health manpower working in a group to provide home, clinic, and hospital care. Planning for health services on an area basis makes it necessary to think in terms of time rather than distance. Planning must also recognize divergent needs for services requiring new types of health workers, technology, and emergency care practices.

We must emphasize that there is no one, simplistic solution applicable to all medically deprived rural locales; rather, each area will need to develop its own plan, incorporating these approaches most appropriate to their particular needs. Of prime importance is coordination of planning. There is an urgent need for system development with physicians and hospital linkages.

MODELS FOR RURAL COMMUNITY PRACTICE

Numerous experimental models for rural health care services have been developed in this country, and many more are in the planning stages. Some show promise of being adaptable to the needs of multiple communities. Some have been less than successful and should be allowed to quietly fade from the scene.

Several medical schools have initiated rural health services. The University of Florida College of Medicine utilizes medical and nursing students with resident physicians to develop health services in Lafayette County. The program to train former medical corpsmen (MEDEX) is being conducted by medical schools in the States of Washington, Pennsylvania, California, South Carolina, and Utah. There are now 52 accredited programs in 30 states for preparing assistants to primary care physicians. Livington Community Health Services in California demonstrates the effectiveness of cooperation between a medical school (Stanford University) and a local rural community to provide health service when little or none was previously available. WAMI is an acronym standing for the States of Washington, Alaska, and Montana, and Idaho and is an experiment in decentralized medical education. Its objectives are: (1) to increase the number of places in medical school without a major construction effort; (2) to increase the number of primary care physicians throughout the WAMI states; (3) to improve the health of WAMI residents; (4) to expand continuing medical education programs for health care professionals; (5) to broaden experiences of future doctors by using clinical facilities in various communities; and (6) to correct maldistribution. In its brief development we find that a higher percentage of WAMI students have elected family practice (62%) than non-WAMI students (47%). Of the first 14 residents to specialize in family practice, 6 are practicing in rural or sem-rural areas in the Northwest, 2 are military physicians, and 6 have faculty posts in family practice departments of medical schools. WAMI appears to be a solution to at least some of the problems of medical education and health delivery in the Northwest.

A growing use of telecommunications will shrink distances. They enable the health care team not only to extend its services but also to compensate for any numerical shortages in the team. They permit long distance transmission of findings and advice. Rural Health Associates of Maine, for example, uses closed-circuit television for instant consultation between its outlying facilities and the main one. X-rays of patients in the little town of Broken Bow, Nebraska, are transmitted by slow-scan TV to the state university medical center in Omaha where they are interpreted. Good communication, such as two-way radio between hospital and ambulance, is indispensable to good emergency care in rural areas.

Nurse practitioners are being used as the primary care contact at the Hope Clinic in Estancia, New Mexico, in several communities in Idaho, Utah, and Colorado, in several health outreach clinics in the County of Albemarle in Virginia, and in a number of other locations.

It was in Grand Junction, Colorado, that one of the first federally assisted Health Maintenance Organizations had its beginning in 1971. The Rocky Mountain HMO, as it is known, became fully operational as of January, 1974, and had 74 physicians and an enrollment of more than 7,200 after less than one year of service. A rural group practice in Marshfield, Wisconsin, formed a contract practice (HMO) arrangement for interested patients, and the Geisinger Clinic, serving rural counties in Pennsylvania, has a similar development.

ASSISTANCE THROUGH STATE AND REGIONAL HEALTH ORGANIZATIONS

State and regional approaches in planning for delivery of health services in rural areas have been effective in a number of locations. The Rural Health Care Association in Colorado helps small towns evaluate their health care needs and then integrates those needs within the framework of a complete health care delivery system. The complete system includes satellite offices, core offices, urban specialty groups, hospital facilities, and continuing education facilities.

Two of the most successful state-wide health organizations are the Michigan Health Council and the Virginia Council on Health and Medical Care. The Michigan Health Council was organized in 1943 and has today about 100 state organization members. The Council's Placement Service has placed 1,452 physicians and 499 dentists in the State in its 33 year history. The Upjohn Company of Michigan and the Michigan State Medical Society have provided substantial grants for the Placement Service. The Virginia Council on Health and Medical Care was organized in 1946 and has 75–80 organization members with a placement record of over 800 physicians in Virginia during its history.

Other organized efforts to assist rural areas to provide health care services include the Health Systems Research Institute which is an independent, nonprofit organization. The Institute is delivering organized care in several locations in Nevada, Montana, Arizona, Washington, and Wyoming. The Utah Valley Hospital has a rural health project whereby emergency room physicians travel to rural clinics and provide patient care or, when a physician or nurse practitioner is already at the clinic, they offer back-up and 24 hour consultation services. This is a hospital-based program which includes assistance to communities in helping to plan and develop their health care services. Presbyterian Medical Services, originated as a branch of the Board of National Missions of the Presbyterian Church, and operate Mission Health Services in New Mexico and Colorado. They own or administer several rural hospitals, outpatient clinics and nursing outposts as well as an emergency medical service system in a seven county area of New Mexico.

An example of a state legislated organizational approach is that of North Carolina where the General Assembly, on the recommendation of the Governor, created an Office of Rural Health Services. The objective of the Office is to provide assistance to medically needy rural communities desiring access to the medical care system through a primary care program. The five program resource areas provided are: (1) the physicians location assistance program for recruitment; (2) the NHSC liason program; (3) the primary practice rural preceptorship program; (4) the educational loan program; and (5) the rural health centers program utilizing physician extenders including nurse practitioners and physician assistants.

A COMMUNITY ORGANIZED EFFORT

An example of a community health care services organization is the Southwest Minnesota Health Care Enterprises, (SMHCE), a nonprofit, citizen controlled health system designed to serve 15,000 people in two counties. The towns of Lamberton and Jeffers are a part of this system, which explains why today there are two clinics in this area after so many doctorless years.

Six communities joined together to form the system and secured financial support from the Regional Medical Program (RMP). The Department of Family Practice and Community Health of the University of Minnesota School did a survey of perceived wants and resources. A USDA official, who was superintendent of the University of Minnesota's Southwest Experimental Station in the area, was a key leader in community organization.

On the basis of the findings and utilizing new concepts in group dynamics and medicine, a health care system was designed for SMHCE by the Medical School.*

The system used new approaches to solve the problems of practicing medicine in a rural area. Among them were:

1. Creation of an areawide basis for a health system that would eventually serve a multi-community.
2. Utilization of the County Extension Service to involve citizens.
3. A financial commitment from the people in the six communities to fund the system until it could be self-supporting.
4. Establishment of a long-range school and community health education program.
5. Development of the ambulatory care concept that used the hospitals and works closely with doctors already available in the larger region around the six communities.
6. Putting administration of the system in the hands of the board of directors.
7. Primarily this model indicates that health care is embodied in a system of which the physician is an integral part. More than a shortage of physicians, we have a shortage of systems.

*For further reading see: (1) Stephen Nye Barton, Rural health care in southwestern Minnesota: A case study, M.A.P.A. thesis, University of Minnesota, June, 1973; and (2) Stephen Nye Barton, Covert aspects of rural intercommunity communication: An Application of Bion, Ph.D. thesis, University of Minnesota, June, 1975.

Most symbolic of the success of the model is the fact that 11 area towns and 13 townships joined together in March, 1974, to form the Health Services Board of Southwest Minnesota, the first rural joint-powers agreement in the State.

Now that they have created a central group representing all the people of the area, they are able to work together even more effectively in planning a medical delivery system based on their needs as they see them.

SMHCE is attracting a lot of attention because it is working. It works because the people care and have become involved.

One thing is clear from our past experiences in developing models for health care services. Behind all success is generally one man—usually a physician or a small group of men—a mecial society or hospital board—that have the creative intelligence—the vision to see what can be—and the strength to make it happen.

UTILIZATION OF RESOURCES

The principal resource of any area is its people, and the ultimate goal of any program is to raise the level of human well-being. Remedial health services are indispensable to the pursuit of that goal. But in the long run, self-maintenance of health and prevention of disease are at least as necessary.

Careful longitudinal studies are beginning to corroborate the records of insurance companies, the observation of practicing health professionals, and plain commonsense as to the positive relationship between good health habits, better health, and longer life. For example, Dr. Lester Breslow, Dean of the School of Public Health, UCLA, and colleagues conducted studies of 7,000 adults over 5½ years. The results showed that seven simple health habits—three meals a day at regular intervals, eating breakfast, moderate exercise, 7–8 hours of sleep each night, moderate weight, no smoking, and no alcohol or only in moderation—were associated with longer life.

Specifically, a 45-year-old man who practices 0–3 of these habits can expect to live to about 67. A man with 6–7 of these habits can expect to live to 78, a difference of 11 years. The study also reported that the physical health of those following all seven good habits was consistently about the same as those 30 years younger who followed few or none of these practices.

It appears that two generalizations with respect to the effectiveness of current health education and behavior change programs are justified:

First, proponents can point to a number of carefully documented success stories that can be interpreted as having reasonably wide application to similar programs. These involve primarily individuals who already have some strong motivation: patients with a chronic illness or disability, those facing an acute crisis such as surgery or childbirth, or employees whose livelihood may depend on overcoming alcoholism or some other job-threatening condition.

Second, if individual behavior change programs are to be generally effective, they must be accompanied by firm professional guidance and by national policies and mass communication programs designed to reinforce, rather than undermine, the message of health education.

Finally, all of us must recognize that "health does not exist in a vacuum." Health is but one aspect of the "quality of life" which includes all the socio-economic, ecological, and educational factors which make for a satisfactory living situation. To improve rural health, we must also address the totality of deficiences in rural living today.

The Microbiologist: A Rural Perspective

Evan T. Thomas, Ph.D.
Stephen N. Barton, M.D., Ph.D.
The University of Alabama
University, Alabama

The role of the microbiologist in the rural community today is as multifaceted and varied as the organisms involved. Man may suffer from some of the effects of microbes such as illness, food spoilage, and fabric or fiber decay, but microorganisms are essential for a healthy biosphere. It is the microbiologist's role to encourage the benifical functions of the microorganisms while suppressing the undesirable.

The task of regulating all aspects of microbial growth is accomplished by utilizing the physical and chemical changes resulting from biological metabolism. Metabolic research provides the most adequate means of limiting the less desirable characteristics and yet enhancing the production of the useful microbes and their products. All forms of microbes protozoa, bacteria, fungi and viruses are studied. Attempts at regulation vary according to the desired end product, the organism involved, the substrate utilized, or the host attacked. Vaccines, for example, have been developed to produce immunity against many of the common bacterial and viral infections of both man and animals. A few diseases that have been controlled by the use of vaccines include: rabies, lock-jaw, typhoid fever, distemper, brucellosis and small pox. In some cases it has not been possible to use a vaccine and, therefore, antimicrobial agents were developed such as penicillin and the sulfa drugs. Disease control was also encouraged by the introduction of sanitizing procedures and chemical agents that kill contaminating organisms on exposed surfaces. The use of phenol, chlorine, creosol and their derivatives are used in hospitals, in food processing, around the home, and on the farm. Where the above mentioned procedures have not sufficed, other means of disease suppression were devised. The ability to genetically manipulate plants has resulted in the production of smut resistant corn, wheat and others.

Another phase of microbiology has been to explain many of the natural phenomena and then utilize this knowledge to enhance everyday living. The manner in which leguminous plants are able to convert atmospheric nitrogen to plant useable nitrate forms has been known for some time. Today, clover seed suppliers often fortify their products with additional bacteria to exploit maximum advantage of the natural conversion process. Others observed that cabbage could be preserved by placing it in barrels and allowing fermentation to take place. The end product of such a conversion was the first sauerkraut. The modern method of manufacture depends upon adding an appropriate amount of starter culture to the chopped cabbage. Swiss and blue cheeses, wines, vinegars and others each depend upon the addition of known microbes in standard amounts to achieve the desired result. From the old starter dough that many early settlers of this nation used in the making of their daily bread scientists have isolated the yeast responsible for dough rising. Thus we are no longer dependent upon a portion of dough set aside for future baking; we simply add purified yeast directly to the flour and other ingredients to obtain a uniform product.

Research, in part by the microbiologists, is responsible for the conversion of the United States from an agrarian to an industrial society. We are no longer dependent upon locally produced foods for daily consumption. Today food can be transported long distances without spoilage by the utilization of refrigeration. Microbes which break down organic matter are inhibited in their normal metabolism by the lower temperatures. Chilled or frozen foods do not lose their natural flavor or consistency. Another form of preservation was developed by an early microbiologist, Pasteur. He observed that wine could be heated to a predetermined temperature for a given period of time and it would remain potable for long periods of time. The principle developed by this early pioneer is still used today and is a commonly accepted procedure—Pasteurization. Much of our food today has been subjected to heat under controlled conditions so that vegetative microbes are either killed or reduced to a tolerable level.

Our modern sanitation procedures are also the result of microbes at work under controlled conditions. Sewage treatment plants and the home septic tanks are both dependent upon microbes to accomplish the task of breaking organic matter to plan useable forms in the recycling process. The control man exerts over the natural process serves as the key to either speed up or to increase the efficiency of nature's cycle.

The future of the microbiologist-microbe team is bright. We are seeking ways of manipulating microbes to make them more efficient servants of rural people. Agricultural microbiologists are working with microbes to make cattle more efficient protein producers by fortification of silage with highly effective cellulose utilizers. Waste products from various commercial enterprises are being considered as possible sources of raw materials which will be converted by microbes into useable protein. Sewage and garbage are being examined as potential sources of urgently needed energy.

To summarize: the role of the microbiologist in the rural community today is essential for the control and enhancement of life as we know it. Disease abatement, food production and preservation, energy sources, and the recycling of organic matter reflect the keystone, the partnership of man and nature.

Prison Health—A Microcosm of Rural Health

Terence R. Collins, M.D., M.P.H.
University, Alabama

When one discusses the topic of prison health, three assumptions must be made. The first is that health care is a human right. This right cannot be removed by geographic, economic, or personality factors, by court order, administration, or legislative decision. The second assumption is that being in prison is the punishment. Because of this, societal goals should not be to punish convicted citizens by inadequate food, medical care, or social services. The third assumption is that health does not exist in a vacuum. Health is but one aspect of the "quality of life" which includes all social, economic, ecological, and educational factors which make for a satisfactory living condition.

The health problems of rural and prison populations bear striking similarities. Rural populations have been plagued by an out-migration of talented young people, leaving a disproportionate share of low income people living in substandard housing and having inadequate community services. The prison population in recent years has suffered from an immigration of young minority group people with poor education and few talents or skills. Prison housing is often substandard, and supportive social services are nonexistent. The health problems that are present are the same in both groups. The prison population is at a disadvantage in that many of the inmates come from a culture of poverty in which medical or dental care has never been available to them.

Rural areas lack educational opportunities and cultural experiences. Of course prisons lack cultural experiences, but unfortunately in many situations they lack *meaningful* educational opportunities. Many of the "skills" which are taught in the prison setting are not transferable to the society in which they are expected to live and work once released from the correctional setting. Just as there has been a difficult time financing adequate education in the rural areas, the same situation exists in the prison setting for the many inmates who are illiterate.

The problems of environmental health in rural areas are well known. Those of water quality, housing, and pesticide hazards are only a few examples. The same hazards apply in many prison settings because so many of our prison facilities are built in rural areas for "security" reasons. In reality, society admits the need for more correctional facilities but no one wants to have the facility built in their town. One need only visit any large prison facility and you can view the culture of poverty, poor housing, and inadequate or poorly maintained water and sanitation systems. The incidence of occupational work injuries in rural areas is known to be much greater than urban areas. A similar situation exists in most prisons where prison farms are a frequent if not the main source of inmate work opportunities. The problems are aggravated in the prison setting because of low capital outlays for equipment and low maintenance on the existing equipment resulting in a higher rate of work injuries. The prison setting adds another dimension to environmental health, and that is the high incidence of violent injuries which are commonplace in correctional institutions.

If one looks at the health care systems in the traditional way, we might first look at availability of care. The lack of health care personnel in rural areas has been well documented. The same severe shortage is present in prison health settings. The lack of educational and cultural opportunities hamper recruiting of physicians whether in rural private practice or in prison settings. In addition, the working conditions in a prison do not lend themselves to attracting well trained health care personnel. The amount of money paid to prison physicians or physicians in rural health settings is *not* the determining factor in attracting or retaining personnel. Though it has not been studied, it is probable that retention of physicians in prison settings would parallel that of rural health physicians as far as background of the physician and his spouse, and the questionable value of rural exposure during medical school training. As the National Health Service Corps has placed physicians in areas of need, the federal government, at least in the past, had the opportunity of placing physicians in prison health settings by assigning United States Public Health Service physicians who were serving their draft obligations. Now opportunity to obtain this type of physician is limited.

The problems of providing emergency medical services to rural health areas because of the scattered population and lack of centralized medical care facilities is equally applicable to the rural prison setting where other than the primary care which can be provided in the prison itself, rural communities are often the only other choice; thus the prisoners are subject to the same restrictions as that of the rural populace.

Occupational Health: A Rural Dimension

JOEL R. BENDER, Ph.D., M.S.P.H.
Indiana University Medical Center
Indianapolis, Indiana

Few people will question the social benefits of improved health and safety in the work environment. However, this desirable objective is often difficult for the small firm to attain and particularly difficult if the firm is located in a rural setting. Occupational health issues in small businesses and rural businesses, i.e., those firms that are located in an area of 2500 people or less, are often similar, but there are special problems facing rural industries in terms of potential hazards and delivery of health care.

One must recall that nearly two-thirds of the U.S. labor force is located in firms with fewer than 500 employees. In fact, workplaces with 25 or fewer workers employ 30% of the total work force and represent 90% of all the workplaces covered by OSHA. Surveys have indicated a trend toward location of smaller industries in less populated communities. Reasons cited for this trend include: (1) the quality of the environment, (2) a total commitment on the part of these communities toward new industries in the form of tax incentives and financial assistance, and (3) a sense of "belonging" which develops from close relationships between business and workers within the smaller community.

Another important, widely cited reason for the trend away from larger cities that deserves explicit attention is the desire for improved quality of life. Conway Publications has developed an indicator based on this concept.

It is encouraging that the indicator does utilize health factors in the form of the number of physicians and dentists, total hospital beds, and accident death rates. These variables approach some of the issues in preventive medicine for industry but tend to deal with secondary or tertiary prevention. That is, they tend to concentrate on precluding the "progression" of disease or disability resulting from work related pathological processes.

To approach the issues from a primary prevention point of view, one has to recognize the following major constraints: (1) there is a severe shortage of trained manpower interested in dealing with the industrial environment, i.e., physicians, nurses, industrial hygienists, and safety experts, and (2) the diversity and seemingly random location of rural industries makes adequate planning and delivery of occupational health services a difficult task.

As an example, one can discuss the difficulties in meeting the health needs of the agricultural worker. It has been estimated that throughout any given year, approximately eight million workers are exposed to the potential hazards of pesticides. The agricultural work force in the United States also suffers a disproportionate number of occupational injuries and illnesses. The National Safety Council estimates for 1971 indicated that although farm employees make up only 4.4% of the work force, they suffer some 16% of the occupational deaths and 9% of the disabling injuries.

Aside from economic poisons and other chemicals, agricultural workers are exposed to a variety of less well known hazards. The farmer, for example, has his own respiratory disease, "farmer's lung", caused by inhaling dust from moldy hay, oats, barley, corn or beet pulp. There is also bagassosis, caused by inhaling moldy sugar cane; mushroom picker's lung, from mushroom compost; cheese worker's lung, from moldy cheese; and even meat wrapper's asthma, from fumes generated during thermal sealing operations.

There are other occupations which are often by necessity located in rural areas. Those manufacturers involved in lumber and wood products, for example, had the highest injury and illness incidence rates for 1973. This general classification which contains logging camps, logging contractors, millworkers, prefabricated wood construction, and other related operations had an incidence rate of 24 per 100 full-time workers. This is one and one-half times the average for all classes of manufacturers. As might be anticipated, those persons involved in roofing and sheet metal work also had high incidence rates, 27.7 per 100 full-time workers.

Health services are not frequently available to these workers or employees in service industries (auto and electrical repair, education, etc.), wholesale and retail trade, trucking or other businesses commonly found in any rural community. Most rural manufacturing and processing firms are as a result of size not likely to employ an industrial hygienist or physician. A study of industries in a rural north central state indicated that safety experts and nurses were employed in 11% and 9% of the firms with 25 or more employees, respectively. More important was the finding which indicated that the presence of safety experts did significantly lower the distribution of total recordable illnesses and injuries.

Nurses entering the field of occupational health often have little or no previous exposure to occupational health nursing and must develop skills through seminars, corporate training, or area associates. In most instances, the industrial nurse represents the occupational safety and health program for the firm and must take on additional roles such as safety director or assistant plant manager. In addition, the occupational health nurse performs numerous in-plant medical services ranging from emergency medical care to counseling and education.

The rural location and small size of many manufacturing and processing firms does present a challenge to those interested in providing adequate safety and health services. Certainly more work needs to be done to find those aspects of a safety and health program which produce most benefits for the worker. Local physicians should become more aware of the needs of industries located in their community. Cooperative efforts and genuine understanding of occupational health needs by physicians and occupational health nurses could lead to productive efforts toward developing an occupational health team to serve the needs of industry and the community.

Selective training of environmentalists or sanitarians would also be beneficial in providing and strengthening safety and health services to local workers. Personnel administrators often take on the role of safety officer. These people should not only be trained in the administration and record keeping aspects of OSHA but also receive some background in safety and health issues within their specific industry. This individual can then function effectively in the safety committee of the firm. Such a committee should have adequate worker representation and a mandate to improve the status of worker health.

Many firms appreciate and respect the educational approach taken by some compensation bureaus and regulatory agencies. Expansion of this effort would greatly reduce the anxiety of management toward the goals of the Occupational Safety and Health Act to reduce worker injury and illness.

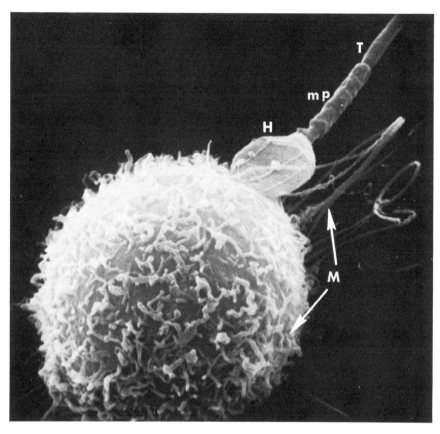

Scanning electron micrograph showing a human sperm in the act of penetrating into a human somatic (i.e., non-egg) cell two hours after admixture. The short and long extensions (M) (i.e., microvillae) covering the surface membrane of the somatic cell participate in drawing in the sperm in much the same manner as occurs during the fertilization of an egg. After fixation, a thin layer of gold was evaporated onto the specimen surface to render it visible in the electron microscope. H = head, mp = midpiece, and T = tail, of the sperm; magnification = 5,725×. (*Photo Credit:* Professor Aaron Bendich)

III

BIOLOGICAL AND
CLINICAL
SCIENCES

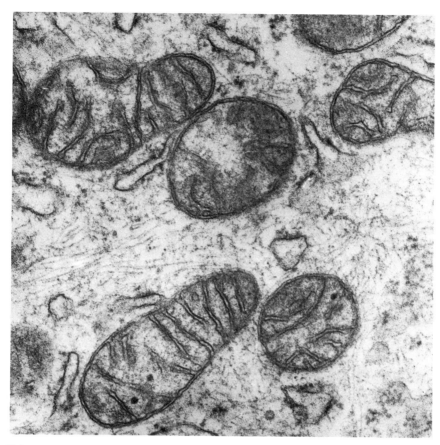

FIG. 1. The internal structure of several mitochondria is easily visualized at this magnification. 41,250×. (*Photo Credit:* Professor Alvin Zelickson)

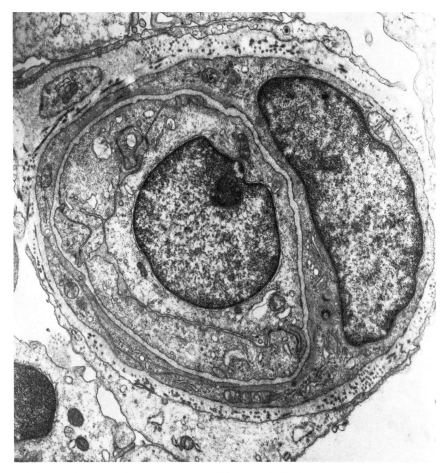

FIG. 2. The detailed architecture of a superficial dermal skin blood vessel is depicted in this micrograph. 12,400×. (*Photo Credit:* Professor Alvin Zelickson)

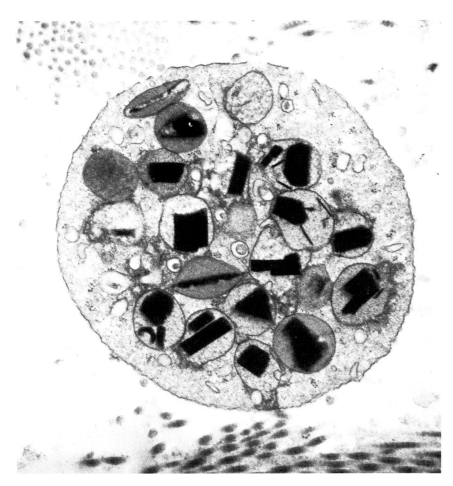

FIG. 3. Eosinophils have a characteristic cytoplasmic granule. 24,700×. (*Photo Credit:* Professor Alvin Zelickson)

FIG. 4. Wart virus groupings are seen in the nucleus of this epidermal cell. 30,600×.
(*Photo Credit:* Professor Alvin Zelickson)

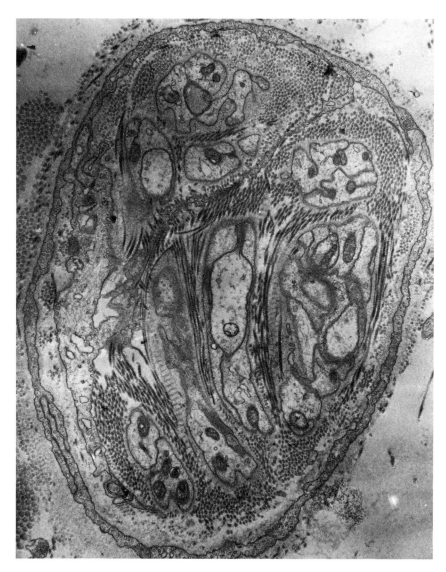

FIG. 5. The ultrastructure of dermal nerve tissue is complex as noted here. 10,850×.
(*Photo Credit:* Professor Alvin Zelickson)

Activated Macrophages as Mediators of Nonspecific Immunity

MONTE S. MELTZER
Bethesda, Maryland

It was established some years ago that injection of animals with certain organisms confers resistance to other unrelated organisms. This was found to be particularly marked in the case of facultative intracellular bacteria such as *Listeria monocytogenes* or *Mycobacterium tuberculosis*. The studies of Mackaness demonstrated that this acquired resistance to facultative intracellular parasites could be correlated with enhanced bactericidal properties of the host's macrophage population. Macrophages with bactericidal properties were described as being activated. Production of activated macrophages was dependent on an immune response by the host and involved an interaction between an immune macrophage and the infecting organism (or some antigenic component).

In the late 1960's macrophages from chronically infected animals were found to be capable of preventing tumor growth in vitro or in vivo. These observations were given added interest when Hibbs reported in 1972 that such activated macrophages were cytotoxic to neoplastic and transformed cells while having little effect on the growth of "normal" cells. These findings suggested that nonspecific immunotherapy of neoplasia might be achieved utilizing activated macrophages.

Activated macrophages have an increased size and ability to spread on glass, an increased metabolic rate, an increased content of lysosomal enzymes, an increased responsiveness to chemotactic stimulae, in addition to bactericidal and tumoricidal properties. The biochemical mechanism of activated macrophage killing of tumor cells is unknown. Indeed more than one mechanism may exist. Several researchers have found that contact between the activated macrophage and the tumor cell is necessary. Others have attributed the cytotoxic action to the production of soluble factors by the activated macrophages. It is conceivable that activated macrophages release a toxin which is most effective when concentrated at the macrophage membrane so that cell to cell contact would give an enhanced effect. Recently Hibbs has reported that lysosomal enzymes are passed directly from activated macrophages into the cytoplasm of the susceptible tumor cell and the cytotoxic action can be prevented by agents that prevent exocytosis of lysosomal enzymes. Lysosomal enzyme content has been closely correlated with the bactericidal action of activated macrophages and similar studies may well hold the key to the understanding of the tumori-cidal action of activated macrophages.

While an important role has been established for the nonspecific action of activated macrophages as a host defense in bacterial immunity, it is less clear how effectively these cells may function in the resistance of an animal to growth of its tumor. Injection into animals of compounds such as endotoxin, pyran copolymer, double stranded RNA or infection with certain bacteria (e.g., BCG, *Listeria monocytogenes*) produce *in vitro* tumoricidal macrophages and also suppress tumor growth *in vivo*. Because of the comparatively weak cytotoxic action of these macrophages *in vitro* (e.g., number of activated macrophages required to kill one tumor cell, the time necessary for killing to result, etc.) it would seem unlikely that activated macrophages are solely responsible for the antitumor action observed *in vivo*. However, the selective cytotoxic action of activated macrophages for neoplastic cells and their ability to respond to chemo-tactic stimulae may confer on these cells an important role in the initiation or potentiation of an immune response to a tumor. The ability to activate the host's resident macrophage population may yet prove to be beneficial to the tumor bearing animal.

References

1. Hibbs, J. B., in *The Macrophage in Neoplasia* (Ed. M. A. Fink), 83–113. Academic Press, New York, 1976.
2. Mackaness, G. B. and R. V. Blanden. Cellular immunity. *Progr. Allergy* **11**: 89–140 (1967).
3. *Immunobiology of the Macrophage* (Ed. D. S. Nelson). Academic Press, New York, 1976.
4. *Mononuclear Phagocytes in Immunity and Infection in Pathology* (Ed. R. van Furth). Blackwell Scientific Publications, Oxford, 1975.

Immunobiological Approaches to Diseases Associated with Aging

ROBERT A. GOOD, Ph.D., M.D.
President and Director
Sloan-Kettering Institute for Cancer Research
New York, New York

INTRODUCTION

As far as we know, man is the only species to be aware of his own mortality. This awareness has led to considerable thought and, in the last ten years or so, much research upon the underlying causes of aging and senescence. A number of theories have been proposed and most seem to take as axiomatic the understanding that a given species, because of its particular endowment of genes, ages and dies in an apparently predefined or prescribed period.

Ten per cent of the U.S. population is presently over 65 and 86% of these people are afflicted with one or more of the chronic degenerative diseases we are all familiar with: cardiovascular disease, arthritis, diabetes, kidney disease, cancer, and certain infections. These diseases are responsible for the decrease in survival of older people. Yet few observers emphasize the other side of the coin: 14% of the aged do not have these diseases. Many old people, as we all know, remain active, alert and relatively healthy until they are very old indeed. Our immediate concern, then, is not so much about the ultimate genetic determinants of old age and the ultimate termination of life itself, but rather it is to reduce the incidence and severity of those age-associated degenerative diseases of the majority of old people and to improve their overall health and activity. Research in immunology has provided clues that immunological techniques may be a way of achieving this goal and we intend to follow these leads as vigorously as we can.

THE IMMUNOBIOLOGY OF AGING

The body's immune system, made up of antibodies, various kinds of white cells, and other molecules of diverse functions, protects us against a wide variety of diseases. Its main functions are to prevent infection from without by viruses, bacteria and other microbes, and to head off cellular anarchy in the form of cancer from within. However, as a result of our own work and that of others over the last decade, the immune system, somehow directed by the genetic program, seems to be ultimately involved in disease processes we associate with aging. The results of this research have all yielded similar conclusions which, oversimplified, can be summarized in three statements.

1. As the body ages, the protective efficiency of the immune system *decreases*.
2. As the body ages, the autoimmune processes (the immune system's propensity to attack the body's own tissues) *increase*.
3. As the body ages, the incidence of cancer *increases*.

The three processes appear to be interconnected and consideration of one usually leads to some accounting for the others.

RAPIDLY AGING MICE

Much of our own understanding of age-associated immunological aberrations has come from research on certain strains of mice. Some inbred mice, such as the CBA strain, tend to be very long-lived and maintain vigorous immunologic functions well into advanced age (about three years, a very long life for mice). In contrast, the New Zealand Black (NZB) mouse and some of its hybrids are very short-lived and develop immunodeficiencies, cancer, kidney and vascular diseases, and immunologic abnormalities early in life. In all mice too, there is a programmed involution of the major lymphoid (immune) organs, particularly the thymus gland, that begins around the time of sexual maturity, and this is followed by degeneration of the peripheral lymphoid apparatus (the lymph nodes and their constituent immune cells). The rate of this decay varies between strains of mice, and among individuals in man, but its onset is generally followed both by decreased ability to protect the organism against infection, by an increased incidence of degenerative diseases mediated by autoimmune processes, and often by the appearance of cancer. This has led us and others to believe that within the thymus function is contained a major control of aging, especially that associated with appearance of diseases of aging. We therefore decided to use a very short-lived hybrid of the NZB mouse (the NZB X NZW hybrid) to study the relation between declining immunologic vigor, autoimmune processes, wasting diseases, cardiovascular disease, and malignancy, and to attempt to find reasons for these processes and ways to delay, prevent or reverse them.

Early in life NZB and NZB X NZW hybrid mice have a vigorous immune capacity, perhaps even more vigorous than the longer-lived strains. However, this vigor rapidly declines: antibody production (humoral immunity) decreases in efficiency, and responses by certain types of white cells, called T-lymphocytes (cellular immunity), become reduced. With this latter type of immunity, decline is evidenced by reduced response of T-lymphocytes to certain chemicals and microbial agents, decreased ability of the animals to reject foreign tissue transplants, including transplanted cancers, as well as an increase in the incidence of spontaneous tumors. In addition, autoantibodies can be detected in the circulation of these mice. These antibodies attack the mouse's own tissues, notably blood vessels in the heart, lung and kidneys. Finally, wasting diseases which appear to be secondarily related to these immunological abnormalities begin to arise. Spontaneous hemolytic anemia, glomerulonephritis (kidney disease), destructive vascular diseases, and spontaneous malignancies are generally the diseases which spell the end for these mice. Similarly in man, autoimmunities, vascular disease, kidney disease, infections, and cancer occur with aging and loss of immunologic vigor.

SUPPRESSOR CELLS AND AUTOIMMUNE PHENOMENA

It may seem paradoxical, but our research has strongly suggested that the increase in autoimmune phenomena is directly associated with the decreasing function of the cell mediated arm of the immune system, probably produced by the involution of the thymus. The thymus is central for directing, through the hormones it produces, the maturation and differentiation of the T-lymphocytes. This lymphoid organ acts hormonally upon immature lymphocytes and differentiates them into one of at least three basic T-lymphocyte types depending upon their prior genetic programming: *Killer T-cells* are responsible for destroying invading microbes or aberrant body cells like cancer cells; *Helper T-cells* assist other classes of lymphocytes, for example B-lymphocytes in their production of antibodies. *Suppressor T-cells* provide a "brake" upon the functions of the other immune processes and thereby keep the normal immune responses from becoming overblown.

Of all T-lymphocyte activities, it is interesting to note that we have found it to be the function of the suppressor T-lymphocytes which declines most rapidly with aging and thymus involution. Without this immunological brake, the frequency of autoimmune phenomena and associated tissue damage increases. In fact, we have evidence which suggests that the interaction of antibodies with a commonly occurring mouse virus, produces immune complexes which bind and damage tissues such as the kidney blood vessels and joints. This damage in turn produces a corresponding antibody and cell mediated attack upon the body's tissues resulting in escalating autoimmune disease. The progressive lack of normally functioning suppressor T-cells as a result of thymus involution would easily allow this process to take place to an increasingly greater degree in the lifetime of the individual.

Of course, research on these mice provides other ideas about the role of declining immune function in the aging process. For example, it may well be that decreased immunological surveillance, which clearly occurs with aging, permits the easier entry of foreign organisms which, through some quirk of evolution, have similar molecular structures as do body cells (so-called forbidden antigens). The immune response to these microbes could become "confused" and attack body cells also having these antigens. Alternatively, others have suggested that certain immune cells which are ordinarily suppressed (so called forbidden clones of cells) are released from this suppression with age and begin to attack the body itself.

All of these ideas point to the need to keep certain immunological reactions under strict control—a function of the suppressor T-cells which declines with age. (Recent information also implicates helper T-cell in the aging process, but we are as yet uncertain how these functions are involved). It seems that if we are to seek means to interfere with age-associated autoimmune processes and diseases, the suppressor (and helper?) function of thymus-derived lymphocytes should be further studied if new knowledge about them is to be utilized against disease of aging.

IMMUNOBIOLOGICAL MANIPULATION OF THE AGING PROCESS

In field studies of human malnutrition, we found that diet had a profound influence upon the body's immunity functions. In extending and expanding these studies to NZB and NZB X NZW hybrid mice, rats and guinea pigs, we found that reduced calorie and/or fat but increased protein intake suppressed the humoral immune (antibody) response but left unaffected, or even increased, the cellular immune response. Fortuitously, we also observed that dietary manipulations significantly reduced the autoimmune wasting diseases in these mice and increased their expected short life span by two-fold. These dieting mice were also better able to reject transplanted tumors and they developed fewer spontaneous cancers. Such striking results pointed clearly to the close relationship between maintenance of cellular immune functions and delay of aging and its associated diseases.

Many of the immune abnormalities of these unusual mice can be reproduced in normal mice by removing the thymus when the animals are born. This procedure artificially produces the autoimmunity, immune deficiency reactions, cancer, and early death seen in the rapidly aging mouse strains. We and others have shown that injecting thymus, spleen cells or lymphocytes, can prevent, and in some cases reverse, the onset of those syndromes in either the mice in which the thymus had been removed or, just as significantly, in the mice strains which age rapidly by themselves. It seems clear that this type of "cellular engineering" has great potential for manipulating the immune processes associated with aging. Further, it has recently been found that hormonelike substances produced by the suppressor T-cells or by the thymus itself have the potential to reverse or prevent autoimmune reactions in much the same way that dietary restrictions or cellular engineering does; we thus also have the possibility before us of "molecular engineering" for use against aging-associated immune phenomena.

OBJECTIVES OF OUR RESEARCH

Nutritional and Cellular Engineering Approaches to Aging

It now seems possible to associate the short and intermediate life span of certain strains of mice with the development of autoimmune diseases and to recognize the relationship of diet to life span, function of the thymus, and autoimmune aging phenomena. Our preliminary studies encourage us to undertake further investigations of these associations. In addition we need to determine how diet and life span implicate certain otherwise noninfectious viruses in the development of autoimmune reactions. We now have the methodology to investigate these questions.

We also hope to further explore the use of cellular engineering to manipulate autoimmune phenomena and early aging in these mice. Since cellular engineering through bone marrow or thymus transplantation has been shown in preliminary experiments to delay or prevent the onset of wasting diseases and death in these animals, it is essential that we further explore these techniques. In addition, it will be crucial to assess the effect upon the aging process of administration of certain hormones produced by the thymus.

SUPPRESSOR CELL ACTIVITY IN AGING

The resolution of the apparently paradoxical situation in aging in which autoimmune reactivity increases while normal immune responses decrease may result from further analysis of the controls on and functions of the suppressor cells. Since this population of lymphocytes is derived from the thymus, thymus involution with aging results in a reduction in number and functions of these cells. The result is an increasing failure to suppress peripheral processes like antibody production with an increasing tendency for autoantibody synthesis and escalation of autoimmune assaults upon various tissues.

We hope to assess the actual changes in the number and function of suppressor cells in aging people and in persons with a genetically determined disease (progeria) in which aging is accelerated. We also will characterize any molecules produced by suppressor cells with a view toward determining their potential use in preventing or even reversing autoimmune phenomena associated with aging. We will attempt to correlate decreased suppressor cell function with changes in output of thymus hormones. Finally, we will begin to evaluate the influence of thymus hormones, suppressor cell products and suppressor cell transplants on aging in mice and other animals.

Present Status of Analgesic Research

ARNOLD C. OSTERBERG, Ph.D.

Group Leader, Neuropharmacology
Central Nervous System Biological Research Department
Lederle Laboratories
Pearl River, New York

The most significant and rapid advances now being achieved in the pharmacology of analgesics involve the mechanisms of opiate-induced analgesic action, the development of tolerance, and physical dependence. Specific opiate receptors, as sites of opiate-induced pharmacological effects, have recently been located and partially identified. It has also been discovered that various mammalian brains, including that of man, contain materials possessing opiate-like effects which are believed to act as neurotransmitters for modulating pain perception. Enkephalin is reported to be a mixture of two specific pentapeptides, one (the more potent) containing methionine and the other leucine at the N-terminal end of a specific sequence of four amino acids. Both enkephalins have demonstrated opiate-like agonist activity using various *in vitro* (mouse vas deferens and guinea pig ileum) and *in vivo* (mouse and rat tail flick or hot plate) test systems. Their activities are reversible by the well known narcotic antagonist, naloxone. The chemical structures and sequence of amino acids have been confirmed by total synthesis and biological assay compared to materials isolated from natural sources.

Since enkephalin is a peptide, it is rapidly degraded and inactivated when administered orally or parenterally to laboratory animals. Even following intracerebroventricular injection enkephalin has a low potency compared to morphine and a duration of action of only a few minutes due to its rapid metabolism. Another polypeptide called anodynin, different from enkephalin, was isolated from human blood and is more potent and longer acting than enkephalin. More recently, a synthetic analog of enkephalin has been reported to possess morphine-like potency and duration of action. Future synthetic peptide chemistry will, undoubtedly, be directed toward means of retaining receptor recognition and affinity with concomitant relative resistance to enzymatic degradation both in the periphery and central nervous system. This represents a rational approach for designing new analgesic drugs. Two larger molecular weight peptides, Substance P, containing 11 amino acids and β-endorphin containing 31 amino acids, have also been reported to be more potent and longer acting than enkephalin. Substance P does not contain the amino acid sequence of enkephalin, but is a slowly acting and potent analgesic following parenteral administration. This suggests that Substance P may be degraded *in vivo* into a new and practical analgesic agent that reaches the receptor in an active form. A yet larger polypeptide, β-lipotropin, containing 91 amino acids and isolated from pituitary gland, has been suggested to be a biological source or precursor of an endogenous opiate-like substance since it does contain the amino acid sequence found in methionine-enkephalin. There is recent evidence that the pituitary may not be the only source, and that enkephalins are probably synthesized directly in the brain.

A report that the level of an enkephalin-like material is lower in the cerebrospinal fluid of patients experiencing pain compared to patients without pain, lends support to the theory of an endogenous opiate-like neurotransmitter for attenuation of pain. In addition, it has been reported that analgesia induced by acupuncture may be reversed by naloxone, suggesting the involvement of endogenous opiate-like factors. This may be similar to the well known phenomenon that pain in one portion of the body may attenuate a preexisting pain in a different area. A technique involving electrical stimulation of certain brain areas has been tried clinically for alleviating intractable pain. It was suggested that low frequency stimulation releases the endogenous opiate-like substance to produce analgesia. Stimulation-induced analgesia, like that of morphine, is subject to tolerance, cross tolerance and naloxone reversal.

Prostaglandins, previously known to be involved in the mechanism of non-opiate analgesic action (aspirin-like), are now being implicated in opiate-induced analgesia, the development of tolerance and physical dependence. Morphine, and more recently the endogenous opiate peptides have been reported to inhibit basal as well as prostaglandin-induced stimulation of brain adenyl cyclase activity, resulting in a fall of c-AMP levels. Prolonged exposure to morphine then induces a compensatory increase in cyclase activity which in turn elevates the low c-AMP levels toward normal. This has been postulated as the mechanism of tolerance to opiate action. Dependence then develops at the cellular level for maintenance of normal levels of c-AMP. These effects of morphine can be reversed by the narcotic antagonist, naloxone. The sudden removal of morphine results in an overshoot in the synthesis of c-AMP, explaining biochemically the symptoms of opiate withdrawal. To support the hypothesis, increased levels of c-AMP have been found in abstinent rats. Thus, morphine may act centrally by blocking a normal function of prostaglandins in contrast to the peripheral inhibition of prostaglandin synthesis induced by aspirin and other non-narcotic analgesics.

Although the opiate receptor or receptors have not been clearly defined, many of their characteristics have been discovered. Certain lipids are known to act as ligands and to bind to opiates stereospecifically. To support the idea that the receptor is of a protein nature, proteolytic enzymes more selectively inhibit agonist binding than antagonist binding. The opiate receptor is believed to exist in two forms, sodium or no-sodium forms, which are normally in equi-

librium. The sodium form of the receptor has the capacity to bind to antagonists. Analgesic effects occur only when a drug binds to the agonist or no-sodium
form of the receptor. Thus, an antagonist by binding to the sodium form, will
shift the equilibrium toward the sodium form causing fewer free no-sodium receptors to be available for binding with an opiate agonist analgesic. Drugs with
mixed agonist-antagonist properties, such as pentazocine, have proportionate
intermediate sodium receptor affinities. Although it has not been reported,
a pure antagonist such as naloxone if administered chronically could theoretically induce a state of hyperalgesia since too few agonist receptors would be
free for enkephalin to act as the physiological modulator of pain perception.

Research continues on modifications of the opiate nucleus which might result
in compounds having the proper proportion of agonist-antagonist properties
within a single molecule so as to retain analgesic effectiveness with lower liability
for abuse, addiction, physical dependence and side effects. Several new compounds are in various stages of clinical investigation, such as buprenorphine,
butorphanol and oxilorphan. These are being studied either as predominate
analgesics (agonist $>$ antagonist actions) or antagonists (antagonist $>$ agonist
actions) for the treatment of narcotic addiction.

The use of anti-inflammatory analgesics for the treatment of non-arthritic
pain is on the increase. New drugs such as floctafenine, indoprofen and fen-
bufen are among those presently being evaluated for clinical analgesic efficacy
and safety. Ibuprofen, a newly introduced drug, has enjoyed wide patient-
physician acceptance. Research is still being conducted to improve the separation of therapeutic from toxic effects. A pharmacologically and chemically
unique compound, nefopam, has also undergone clinical trials.

Marijuana smoking has been shown to result in a significant increase in pain
threshold. Considerable chemical and pharmacological research is now being
conducted for the testing of synthetic cannabinoids as potential non-addicting
analgesics.

Laboratory animals are being used to evaluate the subjective effects induced
by various analgesic and psychotomimetic agents. Rats trained to discriminate
or recognize the subjective effects of morphine will also identify other active
opiates as being narcotic-like. The rats will not recognize inactive stereoisomers
of opiates or psychotomimetics as being morphine-like. Naloxone will block
the discriminative stimuli produced by morphine. The quality of subjective
effects induced by a drug is being used to estimate, preclinically, the relative
abuse liability for new drugs including analgesics.

Research in the analgesic area is presently leading to a basic understanding
of therapeutic effects as well as undesirable effects such as tolerance, addiction,
and physical dependence. In the future, various pharmacological manipulations
of drug-induced neurochemical changes should permit the attainment of greater
analgesic efficacy and safety.

Autoimmune Hemolytic Anemia

MICHAEL M. FRANK, M.D.
Clinical Immunology Section
Laboratory of Clinical Investigation
National Institute of Allergy and Infectious Diseases
Bethesda, Md.

Autoimmune hemolytic anemia, as the name indicates, is characterized by anemia due to lysis of a patient's own normal erythrocytes, either intravascularly or extravascularly, because of an immunopathologic reaction. Two major disease categories have been defined: (1) patients with antibodies which agglutinate their erythrocytes in the cold, but are not reactive with these cells at 37°C and (2) patients with antibodies which are reactive with red cells at 37°C. This classification generally defines patients whose disease differs both clinically and in terms of pathophysiological classification.

Patients with cold agglutinin disease usually have a cold reacting IgM antibody. Those patients with a disease of insidious onset, generally are elderly individuals who develop a mild hemolytic process associated at times with hepatomegaly, but only rarely with splenomegaly. This hemolytic process is generally mild, even though the patients red cells may be strongly positive with a Coombs reagent, unless the patient is exposed to a cold environmental temperature, after which brisk hemolysis may occur. These patients most commonly have a circulating antibody of the IgM class which reacts with a glycopeptide antigen on their own erythrocytes termed "I". This antibody is unusual in that it has highly restricted heterogeneity (monoclonal) and biochemically resembles the immunoglobulin of highly restricted heterogeneity found in patients with multiple myeloma. The I antigen on the erythrocyte membrane, with which the antibody reacts, is present on almost all normal adult erythrocytes but, interestingly, is not present on fetal erythrocytes or erythrocytes from umbilical cord blood. Thus, one can test a cold agglutinin with fetal and adult red cells as a first approximation of antigenic specificity. It has been shown that complement activation is required for these antibodies to mediate their immunopathologic effect. When the patient's erythrocytes are exposed to the antibody in the acral areas of the body such as finger tips, nose, etc., it reacts with the erythrocyte surface and initiates the complement activation sequence. When the cells return to core body temperature, the antibody may dissociate, but the complement activation sequence continues and proceeds to completion. Cold agglutinins differ from patient to patient in their thermal amplitude (in the temperature at which they will react with erythrocytes). Those which react at temperatures near 37°C tend to produce the most severe hemolysis. If relatively small numbers of complement molecules are activated at the cell surface, regulatory proteins in the circulation lead to cleavage of the components after their fixation to the erythrocyte. The erythrocyte is neither sequestered in the RES nor destroyed but remains Coombs positive. This cell may resist further complement activation because reactive erythrocyte membrane sites are blocked by the complement fragments.

Cold agglutinins are also found following a number of infectious illnesses including infectious mononucleosis and mycoplasma pneumonia. In these cases, the antibody does not have the same high degree of restricted heterogeneity, and has specificity for an antigen in the I series of antigens different from "I". This antigen termed "i" is present both on cord blood erythrocytes and adult cells. Thus, the antibody found in the postinfectious state can often be distinguished from that found in chronic cold agglutinin disease.

The second group of patients have an antibody which interacts with erythrocytes at 37°C. In general, these patients have a more severe hemolytic process often associated with splenomegaly. This illness can occur at any age and, in about half of the cases, careful patient evaluation will disclose an underlying illness with which the hemolytic anemia is associated. The nature of the underlying illness is variable but includes collagen diseases, especially systemic lupus erythematosus, and malignancies, especially those of the lymphoid series. The antibody responsible for this illness is usually of the IgG class, and often has specificity for the Rh group of antigens. Many of these patients also have complement on their erythrocytes in addition to their IgG autoantibody and, at times, when the number of erythrocyte bound IgG antibodies is low, only complement will be identified. IgG mediates erythrocyte clearance by virtue of its ability to interact with the IgG Fc receptor, thereby promoting phagocytosis. The complement in addition contributes to erythrocyte destruction. In these patients, extravascular hemolysis usually dominates the clinical picture, often occurring following sequestration of the sensitized erythrocytes by the spleen. Thus, splenectomy is one mode of therapy. At very high levels of cell sensitization, hepatic destruction may also occur.

These diseases differ not only in clinical presentation and pathophysiology but also in their response to therapy. Cold agglutinin mediated autoimmune hemolytic anemia often responds poorly to therapy, however, generally the patients do well if they can be maintained in warm environmental temperatures. Patients with warm antibody mediated disease often do respond well to therapy. These patients usually receive glucocorticoid as the first line of therapy. In patients who do not respond adequately to corticosteroids, or whose disease is not adequately controlled, a splenectomy is performed. This removes a possible site of antibody synthesis as well as a site of erythrocyte sequestration. Those patients poorly controlled on the above therapy often receive cytotoxic agents in addition. That group of patients with warm antibody mediated hemolytic anemia, with a primary associated illness, will usually improve when their primary illness is adequately treated and adequate treatment of the primary illness is the first goal of therapy.

Anesthesia

WEN-HSIEN WU, M.D., M.S.

Associate Professor and Director of Research
Department of Anesthesiology
New York University Medical Center
New York, New York

Chief, Anesthesiology Service
Veterans Administration Hospital
New York, New York

Anesthesiology is the art and science (of pain relief and life support) which embraces the scope of surgical anesthesia (general and regional), resuscitation, critical care medicine, fluid, electrolyte and transfusion therapy, management of pain, respiratory therapy and education. Development of modern anesthesia has been coupled with major advances in the art and technology of surgery, such as cardiac and pulmonary surgery, neurosurgery, reconstructive surgery and cancer surgery. Among the early anesthetics, ether and N_2O are still used widely today in different parts of the world. Introduction of cocaine and procaine at the turn of the century formed the bases of regional anesthesia, including spinal and epidural anesthesia and nerve blocks. Synthesis of new inhalation agents and short-acting barbiturates altered anesthesia practice of the 1930's to include the provision of unconsciousness, analgesia, muscle relaxation and inhibition of undesirable reflexes (general anesthesia). The introduction of neuromuscular blocking agents added a new dimension to provide pure muscle relaxation. Thus, clinical pharmacology of these drugs formed the basis of balanced anesthesia. Neuroleptanesthesia using dissociation and amnesia agents such as droperidol, diazepam and lorezopam expanded the scope of the balanced anesthesia, and added amnesia as an option to anesthesia. Development of hypothermia, extracoporeal circulation, and controlled hypotension using sympathetic blockade, positive airway pressure or sodium nitroprusside further enlarged the scope of anesthetic techniques. New inhalation hydrocarbon anesthetics, such as halothane, methoxyflurane, fluroxene, enflurane and isoflurane introduced within the last 15 years, continue to refine anesthesia management. The choice of inhalation or regional anesthesia is determined by many considerations such as patient age, prior anesthetic experience, preexisting disease(s), pregnancy, electrolyte and fluid balance, the proposed operation, surgeon's skill and approach and intraoperative position, and also by the anesthesiologist's skill and knowledge of pharmacology of anesthetics. Although there is more than one method to provide satisfactory anesthesia, there are specific contraindications to some anesthetic techniques and to certain drugs.

Higher anesthetic risk is associated with coronary disease, cardiac arrhythmias, fixed or reduced cardiac output due to embolic disease of the lung or to valvular diseases, and uncontrolled hypertension. Because all anesthetics tend to depress the myocardium and some produce myocardial irritability, proper selection of anesthetics and concentrations become essential. Critical organ perfusion pressure must be maintained by properly chosen fluid or vasopressor therapy.

Pulmonary diseases may be associated with central nervous, peripheral nervous or cardiac disease. Etiological diagnosis of the abnormalities should be made preoperatively. Bronchodilator therapy for bronchospasm, physiotherapy and antibiotic therapy for obstructive pulmonary diseases are frequently beneficial. Opiates and opioids should be avoided. Controlling ventilation with monitoring blood gases during anesthesia assures satisfactory gas exchange. Airway pressure should be within reasonable range to minimize the chance of rupturing blebs. Renal functions govern excretion of drugs, their metabolites and participate in electrolyte, acid-base and water balance. Drugs like phenobarbital and gallamine are mainly excreted by the kidney and should be avoided in renal diseases with markedly reduced excretory function. Hypertension, low sodium and calcium, high potassium, metabolic acidosis and therapeutic drugs for these conditions all increase the complexity of anesthetic management. Low perfusion state and massive blood transfusion may precipitate acute tubular necrosis. Vasodilators may be beneficial in management of hypertensive crisis.

Each endocrine disorder has its distinctive characteristic pathophysiology. Patients under hormone replacement or inhibition therapy present potential problems for anesthetic management. Carcinoma of the lung may be associated with myasthenic state or inappropriate vasopressin (ADH) secretion. In hyperthyroid state, antithyroid drugs and noncatecholamine-releasing anesthetics such as halothane or enflurane should be used. Pheochromocytoma requires proper use of α- and β-adrenergic blockade and vasodilators.

Drug interactions complicate anesthetic management. Many patients receive

multiple medications. Bioavailability of a drug depends on absorption, protein
binding locally and in plasma, receptor binding affinity, sites and rate of metabolism, route of excretion, tissue solubility and reabsorption. One drug may alter
the fate of another. Phenylephrine inactivates simultaneously infused penicillin.
Aspirin potentiates coumadin effect. Antacid decreases coumadin and digitalis
absorption from the digestive tract. Digitalis, diuretic, and antihypertensive
drugs are frequently used alone or in combination. They may produce hypovolemia, total body potassium depletion, altered neurotransmitter mechanism,
and exaggerate the depressant effect of anesthetics on the cardiovascular system.
Droperidol and other butyrophenone derivatives, e.g., haloperiodal, can produce
rigidity in patients with Parkinson's disease treated with high doses of levodopa
by antagonizing dopaminergic activity in basal ganglia. Monoamine oxidase
(MAO) inhibitors increase tissue norepinephrine, dopamine and 5-hydroxytrypt-
amine levels. They dangerously exaggerate the effect of catecholamines and
meperidine. Antibiotics (streptomycin, dihydrostreptomycin, neomycin,
kanamycin, gentamycin, colistin and polymixin B) cause mild neuromuscular
blockade. Hypoventilation or apnea can occur, if more than one of the above
drugs are given or given in the presence of residual neuromuscular blockade
from anesthesia. Alcohol inhibits hepatic microsomal enzyme activities and
slows down metabolic rates of barbiturates and some inhalation anesthetics.

General anesthesia can be accomplished by intravenous or inhalation agents.
Intravenous agents such as short-acting barbiturate (thiopental, methohexital)
or steroid derivative (Alphaxalone, Althesin) are used for short procedure, or
for rapid induction of anesthesia which is maintained by inhalation anesthetics.
Complete airway control during anesthesia is essential to avoid hypoventilation and aspiration. Commonly used inhalation anesthetics include gaseous
anesthetics (nitrous oxide, cyclopropane) and volatile liquids (diethyl ether,
halothane, methoxyflurane, trichloroethylene, enflurane and isoflurane). The
former are delivered by use of flow meters, and the latter by various types
of vaporizers.

Ether is the only anesthetic producing distinct signs of anesthetic depth.
Since the introduction of newer agents, the concept of minimal alveolar concentration (MAC) required to prevent movement in response to painful stimuli
was developed in an attempt to define anesthetic potency. MAC correlates
with the lipid solubility of anesthetics, and has been used to compare drug
effects. Despite the fact that MAC is influenced by many factors, such as age,
state of health, body temperature, concomitantly used central nervous depressants or metabolic state, it is useful in clinical anesthesia. Monitoring
expiratory anesthetic concentrations is useful clinically because they suggest
cerebral concentrations of the agent. Recent knowledge in biotransformation
shows that all inhalation anesthetics except nitrous oxide are metabolized to

various degrees. Hepatotoxic or nephrotoxic metabolites of halothane may be identified in animals. Liver microsomal enzyme induction by drug therapy, e.g., phenobarbital, may increase rate of biotransformation resulting in increased metabolites, and toxicity. Methoxyflurane is an example of this. Immunosuppressive effects of some inhalation agents have been reported. Its clinical significance is unsettled.

Influence of subanesthetic concentrations of anesthetics on cerebral functions can be detected by vigilance test and motor skill performance tasks. The interaction of stress and the above drug effects has not been established. Furthermore, the lowest concentration limits of the drug effects are still controversial.

Regional anesthesia including spinal, epidural anesthesia, nerve blocks and local infiltration is accomplished by using local anesthetics. Addition of epinephrine usually prolongs the duration of anesthesia. These types of anesthesia can be used for diagnostic purpose and to provide analgesia in the postoperative period or in patients with chronic or intractable pain.

Acupuncture has been used for surgical anesthesia successfully in selected patient populations in some parts of the world, but has not been completely accepted in the Western world due to lack of thorough understanding of it. However, it has been used in a limited fashion for surgical analgesia and in the management of chronic pain on a research basis.

Traditionally, anesthesiologists participate in cardiopulmonary resuscitation. Recently many of them are involved in the field of critical care medicine and respiratory therapy due to their special skills and knowledge in pathophysiology and pharmacology. Active participation in education programs is also an essential professional activity.

Inhalation Anesthetics

WEN-HSIEN WU, M.D., M.S.

Associate Professor and Director of Research
Department of Anesthesiology
New York University Medical Center
New York, New York

Chief, Anesthesiology Service
Veterans Administration Hospital
New York, New York

Inhalation anesthetics produce narcosis by altering cellular function in the central nervous system. Several theories have been advanced in an attempt to explain the phenomenon.

Colloid Theory: Claude Bernard (1875) proposed that aggregation of cellular colloids caused or was associated with anesthesia. This theory was not substantiated by subsequent workers.

Lipid Theory: Meyer (1899) and Overton (1901) proposed that lipid solubility of anesthetics was related to anesthesia. However, not all lipid-soluble substances are anesthetics.

Surface Tension: Traube (1904) and Lillie (1909) studied the relationship between potency and surface-tension-lowering of the nerve membrane. Clement and Wilson (1962) proposed that lowering surface tension might change the structure-related activity of the enzyme essential for oxidative phosphorylation and electron transport.

Cell Permeability Theory: Hober (1907) and Lillie (1909) proposed that anesthetics alter cerebral function by interfering with ionic flux through altered membrane permeability. Some anesthetics depress sodium influx, but cause-effect relationship has not been established.

Biochemical Theory: Clinically used halothane concentrations depress oxidation of glutamine and $NADH_2$ in brain mitochondria. However, whether this is the cause, producing anesthesia, or is the result of anesthesia is unclear.

Neurophysiologic Theories: Altered synaptic transmission, functional disorganization have been proposed to explain cerebral effects of anesthetics. Systematic explanation is not yet available.

Physical Theories: Linus Pauling suggested that anesthetics might form hydrated microcrystals in the brain and thereby interfere with brain conductance. To date, there has been no evidence of microcrystal formation.

Macromolecule Theory: Ueda and Kamaya (1969) proposed that a reversible change in the macromolecules of cell membranes might be a primary cause of narcosis. Buffington and Turndorf (1975) showed that nitrous oxide, cyclopropane and ethyl chloride reduce aqueous solubility of benzene, a nonpolar compound. They suggested a similar effect on non-polar components of membrane macromolecules would alter the energetics determining macromolecular conformation and thus affect the membrane property. Knowledge in this area is still scanty.

None of the above theories can explain anesthetic-induced narcosis completely. Wall (1967) suggested a multiple mechanistic theory. However, supporting data are lacking.

Inhalation anesthetics include two major groups, the gaseous (nitrous oxide, cyclopropane) and volatile liquids (diethyl ether, trichloroethylene halothane, methoxyflurane, enflurane and isoflurane). Diethyl ether produces distinct clinical signs to indicate anesthetic depth. Newer halogenated hydrocarbon anesthetics do not. Therefore, minimum alveolar concentrations (MAC) required to prevent movement in response to painful stimuli was developed by Eger and his associates in an attempt to form the basis for potency comparison.

Most inhalation anesthetics have limited biotransformation. Their uptake and elimination are achieved by ventilation through the lung. These aspects make the adjustment and control of anesthetic depth easier than agents administered rectally or intravenously. An agent such as methoxyflurane is metabolized up to 50% by hepatic microsomal enzymes. Its metabolites were associated with renal dysfunction. The relationship between metabolites of anesthetics and hepatotoxicity is still controversial. Inhalation anesthetics have different effects on cardiopulmonary, neuromuscular, hepatic and renal systems, teratogenicity, and analgesic potency.

Decreased abilities in vigilance and motor skill performance tasks have been observed in volunteers exposed to subanesthetic concentrations of halothane and N_2O. Increased incidence of abortion and lymphoma exists in anesthesia personnel. However, the cause-effect relationship is still unsettled. Stress and chronic exposure of anesthetic vapors in the operating room environment may also play a role in these phenomena. Additional information is still needed in this area.

GASEOUS ANESTHETICS

Nitrous Oxide (N_2O). It is a nonirritating, sweet-smelling colorless gas, non-explosive, but will support combustion even without oxygen.

It cannot produce unconsciousness uniformly when given in nonhypoxic concentrations, and is mainly used as an analgesic or carrier gas for other volatile anesthetics. It is a mild myocardial depressant.

Cyclopropane (C_6H_6). It is a fruity, sweet-smelling potent anesthetic gas. It is explosive in air (2.4-10.4%), in O_2 (2.5-60.0%) and in N_2O (3-30%).

Inhalation of cyclopropane produces analgesia (4%), light anesthesia (8%) and deep anesthesia (20-30%). Cyclopropane depresses respiration, mycardium (deep anesthesia), renal blood flow (up to 80%), hepatic perfusion and produces contractions of the gravid uterus. It inhibits the depressor neurones while stimulating the pressor neurones in the CNS. In light anesthesia it increases cardiac output with an increase of right ventricular stroke volume, stroke work and end-diastolic pressure as well as peripheral vasoconstriction due to stimulation of the sympathetic nervous system. It produces arrhythmias following administration of atropine or hypercapnea.

VOLATILE ANESTHETICS

Diethyl Ether ($(C_2H_5)_2O$). It is a colorless, pungent liquid and tends to decompose in air and light. Its vapor is flammable in air (1.9-48.0%) and in O_2(2.0-82.0%).

This is the only complete agent which gives distinct four clinical stages of anesthetic depth (alveolar concentrations 0.284-1.14% for Stage I, 1.14-2.27% for Stage II, 2.27-3.98% for Stage III, and 3.98-5.11% for Stage IV). It causes tachypnea in light anesthesia and depression in deep anesthesia. It stimulates adrenergic nerve activity and glycogenolysis, and blocks cholinergic activity. Its norepinephrine-releasing effect offsets partially the direct myocardial depression observed in the isolated heart-lung preparation. Therefore, caution should be observed in giving ether to patients with reduced or maximal catecholamine secretion, or altered adrenergic receptor functioning. Ether produces bronchial dilatation. It causes muscle relaxation by depressing the central nervous system, motor endplates, and post-junctional membranes. This results in a synergism with curarelike drugs. It also depresses renal blood flow.

Divinyl Ether ($(C_2H_3)_2O$). It is a colorless, nonirritating and sweetsmelling liquid. Its vapor is explosive both in air (1.7-27.0%) and in O_2(1.8-85.0%). A 4% vapor produces light anesthesia. It has similar cardiovascular effects as diethyl ether, produces bronchial dilatation and can cause hepatic necrosis after prolonged use.

Trichloroethylene (CCl_2CHCl). It is a colorless liquid with fruity odor. While being nonflammable, it decomposes in strong light or heat to form phosgene and hydrochloric acid; thus, it is usually not used with soda lime. It causes rapid and shallow respiration in concentrations above 3%. This may lead to hypoxia without respiratory assist. Various cardiac arrhythmias can occur especially when high inhaled concentrations are used. Higher than analgesic concentration (0.5% v/v) depresses uterine contraction.

Halothane (CF$_3$CHClBr). It is a colorless liquid with a sweet smell. It is nonflammable, nonreacting with soda lime.

It depresses respiration, myocardial contraction, heart rate, stroke volume, cardiac output and peripheral vascular resistance. The circulatory effects are due to norepinephrine receptor blockage in the central nervous system, the heart and peripheral tissue, and sensitization of cholinergic nerve endings. It increases cerbrospinal fluid pressure unless hypocapnea has preceded halothane administration. It reduces renal and hepatic blood flow and stimulates vasopressin release. It potentiates nondepolarizing muscle relaxants and antagonizes depolarizing relaxants.

The relationship between halothane and post-operative liver dysfunction remains unsettled. National Halothane Study (1966) concluded that halothane-related hepatic necrosis was rare. Mushin *et al.* (1971) suggested that it should not be reused within four weeks. However, McEwan (1976) could find no obvious relationship between the number of anesthetics and the disturbance of liver function tests.

Methoxyflurane (CHCl$_2$CF$_2$OCH$_3$). It is a colorless liquid with a fruity odor. Its vapor is nonflammable in anesthetic concentrations, and non-reacting with soda lime. It depresses respiration, circulation and the central nervous system to produce profound muscular relaxation. It does not stimulate catecholamine release. Prolonged deep anesthesia may be associated with renal failure which is believed to be caused by the metabolite inorganic fluoride.

Fluroxene (CF$_3$CH$_2$OC$_2$H$_3$). It is a colorless, light-sensitive liquid with a mild etherlike odor. It is flammable in concentrations above 4% (in O$_2$, N$_2$O/O$_2$ or air) and depresses respiration and circulation (deep anesthesia).

Enflurane (CFHClCF$_2$OCHF$_2$). Its basic physical characteristics resemble those of halothane. Its vapor is nonflammable. Anesthetic concentrations in man are 0.5–1.5%. Induction of and recovery from anesthesia are rapid. It has no analgesic effect before the patient loses consciousness. It increases respiratory rate, heart rate and cerebral blood flow, and decreases blood pressure. Cardiac output remains responsive to changes in CO$_2$ tension. Enflurane markedly potentiates the action of tubocurarine but not that of succinylcholine. This potentiated effect cannot be reversed by neostigmine. Electroencephalographic dysrrhythmia has been observed during and long after the use of enflurane. There is relatively little biotransformation of this agent, thus the risk of nephrotoxicity is minimal.

TABLE 1. Some Physical Properties of the Inhalation Anesthetics.

Agent	Formula	Boiling Point °C	Vapor Pressure 20°C	PARTITION COEFFICIENT Blood/Gas 37°C	PARTITION COEFFICIENT Oil/Gas 37°C	Concentration for 1st Plane S.A.* or MAC† Vols. %
Nitrous Oxide	N_2O	−89.0	680	0.468	1.4	120
Cyclopropane	C_3H_6	−33.0	78	0.415	11.2	13
Diethyl Ether	$(C_2H_5)_2O$	34.6	425	12.1	65	2.5
Trichloroethylene	C_2HCl_3	87.0	57	9.15	960	0.17
Halothane	$CF_3CHBrCl$	50.2	241	2.3	224	0.74
Methoxyflurane	$CHCl_2CFOCH_3$	104.6	23	13.0	825	0.20
Fluroxene	$CF_3CH_2OC_2H_3$	43.2	286	1.37	47.7	3.4
Enflurane	$CHFClCF_2OCHF_2$	56.5	180	1.91	98.5	1.68
Isoflurane	$CF_3CHClOCHF_2$	48.5	250	1.40	99.0	1.3

*S.A. = Surgical Anesthesia
†MAC = Minimal Alveolar Concentration

Isoflurane (Forane, $CF_2HOCHClCF_3$). The basic physical properties are similar to those of enflurane. It is a nonirritant and somewhat pungent vapor. The induction of and recovery from anesthesia is rapid due to its low blood/gas partition coefficient. It depresses tidal volume and minute ventilation more than most other agents. It produces hypotension in a dose-dependent fashion with little tendency to sensitize the myocardium to epinephrine and without vagal activation. Its papillary muscle depressant effect (cat with congestive heart failure) is greater than that produced by halothane. Because the refractory period of skeletal muscle is increased by isoflurane, the sustained response to a tetanic stimulation of higher frequencies is prevented. Thus, the required tubocurarine is markedly reduced. Succinylcholine effect is potentiated as well. Biotransformation is very little in the liver. This drug is still a drug under investigation in the U.S.A.

For a summary of the physical properties of the above compounds, see Table 1.

Regional Anesthesia

WEN-HSIEN WU, M.D., M.S.
Associate Professor and Director of Research
Department of Anesthesiology
New York University Medical Center
New York, New York

Chief, Anesthesiology Service
Veterans Administration Hospital
New York, New York

Regional anesthesia is abolition of painful impulses from any region or regions of the body by temporary interruption of the sensory nerve conduction. Motor function may or may not be involved, and the patient does not lose consciouness.

Nerve Conduction. A nerve fiber consists of a central exoplasm core, enclosed by the cell membrane. The cell membrane is believed to be composed of a bimolecular lipid pallisade, bounded inside and out by a monomolecular protein layer. Each fiber of a peripheral nerve is enclosed in neurilemma, from which it is separated by the myelin sheath, except at the nodes of Ranvier. The insulating myelin sheath is absent or nearly so in nonmyelinated nerves. Myelinated nerve fibers form bundles within the endoneurium. The perineurium surrounds a collection of bundles, and the epineurium encloses a whole nerve. Therefore, local anesthetic has to penetrate a substantial barrier to exert its action at the nerve cell membrane.

When a nerve is stimulated, partial depolarization of the membrane is accompanied by a release of calcium ions and leads to an increase in sodium permeability, resulting in massive depolarization. Thus, when the threshold for excitation potential is exceeded propagation of nerve impulse occurs. This phase is followed by an extracellular movement of potassium ions to establish the electrical neutrality. This is followed by a restoration phase in which sodium ions return to the extracellular space by the sodium pump and potassium ions reenter the cell membrane. In myelinated nerves, these changes take place only at the node of Ranvier.

Pharmacology of Local Anesthetics. Cell membrane stabilization is the basis of the local anesthetic action. Various theories have been proposed to explain the phenomenon. Firstly, the drug may bind the cell membrane and make the pore unavailable for electrolyte movement. Secondly, calcium binding to the cell membrane may be increased, thereby permeability to sodium ions is decreased. Thirdly, the drug may compete for the receptor sites for acetylcholine, the proposed transmitter for propagation of nerve impulses. Local anesthetics increase the threshold for excitation, slow the conduction and reduce the rate of rise of the action potential. The unionized (base) form of the local anesthetic molecule can penetrate tissue barriers. Therefore, the characteristics of local anesthetic action depend upon the pK. Thus, drug potency, lipid solubility, protein binding and rate of biotransformation are directly related to the acute toxicity.

Small, nonmyelinated or autonomic nerve fibers are blocked more readily than sensory fibers. They also exert its membrane stabilizing effect on cardiac tissues, and are used to treat cardiac arrhythmias by suppressing autorhythmicity in malfunctioning myocardium. Procaine and lidocaine have a quinidine-like effect to increase depolization threshold and reduce conduction. Cocaine causes cerebral stimulation. Other drugs can cause mild confusion, sedation and disorientation. Restlessness, paresthesia, tremor and twitching, and ultimately convulsion and unconsciousness can occur. Coma may be accompanied by apnea and cardiovascular collapse as a result of medullary depression. Whenever high spinal nerve blockade occurs, treatment should include airway control, oxygen supply, ventilatory and circulatory support, treating convulsion with intravenous thiopental Na or diazepam to reduce cerebral oxygen demand. Muscle relaxant may be necessary to reduce oxygen consumption.

Regional anesthesia is useful because its relative technical simplicity, decrease in autonomic and endocrine response to stress through interference with afferent impulse pathways and applicability in ambulatory patients and at high altitudes.

Some reasons for its limited usage are: 1) patient's hesitation in being conscious during operation, (2) impractically for some anatomical areas, (3) insufficient duration for the procedure, and (4) toxicity due to rapid absorption into the blood stream.

Regional anesthesia may not be preferred in young, hysterical or malingering-prone patient. Patients with hypertension, cardiac diseases, cardiovascular accidents and respiratory diseases undergoing major surgery may do better with inhalation anesthesia.

Regional anesthesia includes local infiltration, field block, nerve blocks (including spinal and peridural blocks). Commonly used drugs in regional anesthesia can be grouped according to chemical characteristics, as will be discussed below.

ESTER COMPOUNDS

Cocaine

It is a useful topical anesthetic for its vasoconstrictive effect. When applied to the nose, pharynx and tracheobronchial tree the dose should not exceed 200 mg in 1-10% solution. Its metabolites compete with acetylcholine for the receptors sites leading to central or peripheral synaptic transmission disturbance.

Procaine

It can be used in all forms of regional anesthesia except that topical activity is lacking. Doses for infiltration and epidural anesthesia should not exceed 1 g (0.5-2.0%) and 200 mg (5%) for spinal anesthesia.

Chloroprocaine

It is less toxic and has shorter action than procaine due to its rapid hydrolysis in plasma. It is not active when topically applied. Doses should not exceed 1 g.

Tetracaine

It is more potent, has longer action and higher toxicity than the above two. Dose limits are 25 mg (0.5%) for topical use in airways, 2 mg/Kg up to 200 mg (.25%) for infiltration or nerve block, 20 mg (1%) for spinal anesthesia and 125 mg (0.25%) for epidural block. Its action begins within 6-15 minutes and lasts 2.4-5.6 hours.

AMIDES

When an amide linkage is formed on an ester type of drug, it is more difficult to be metabolized resulting to greater toxicity tendency. Some show anti-arrhythmic property.

Dibucaine

It is a potent agent. Therapeutic-toxicity ratio is acceptable, because dilute concentrations are used. It is not commonly used topically except in ointment (0.2%) form, or other forms of regional anesthesia except spinal anesthesia, (10 mg, 0.25%).

Lidocaine and Mepivacaine

Both are heat stable synthetic base anilids, have similar quick onset, and toxicity (relatively low). They are metabolized slowly in plasma and in liver with a part excreted unchanged in urine.

Topically applied lidocaine is less effective than cocaine and does not have vasoconstrictive property. Topical use of mepivacaine has not been established. Lidocaine dose for topical use in 80 mg (2-4%). Dose limits for both drugs are 500 mg (0.5-2.0%) for infiltration and nerve block, 160 mg (4%) for spinal and 500 mg (1.0-2.0%) for epidural blocks. Addition of epinephrine (1:200,000) to lidocaine prolongs its action. Neuritis, paralysis and slough following infiltration, field block, and peripheral nerve block have been observed when injection of lidocaine in concentrations of 1.5% or greater is made in combination with epinephrine.

Bupivacaine

It is a synthetic anilid, more potent and markedly longer acting than the previous two drugs. When injected 84% of it is bound to protein. This may be responsible for its long action. Concentrations of 0.125-0.75% in doses not to exceed 200-500 mg are used. While 0.125% is adequate for epidural analgesia for vaginal delivery, 0.75% is needed for motor paralysis.

Prilocaine

It is a synthetic anilid and has a comparable potency to lidocaine. However, systemic toxicity is less due to its high protein-binding. Prilocaine without epinephrine lasts as long as lidocaine with the same dose. Its metabolite o-

toluidine produces methemoglobin resulting in cyanosis of mucosa and nail beds. This can be seen when the dose exceeds 600 mg. The treatment is intravenous methylene blue (1%, 1-2 mg/Kg) given over a 5 minute period.

Etidocaine

It is closely related to lidocaine chemically, has high lipid solubility and protein-binding. It retains the rapid onset of action of lidocaine and offers prolongation of action and profound motor blockade. Due to rapid redistribution to other tissues following absorption the blood levels are 40-50% lower than those of bupivacaine at all time intervals. It has been suggested that bupivacaine has a toxic potential in man 1.43-2.1 times greater than that of etidocaine. The absorption of etidocaine following caudal anesthesia is more rapid than following lumber epidural or brachial plexus blocks. The total dose administered and not the drug volume and concentration used appears to determine the plasma level. Addition of epinephrine to etidocaine does not prolong duration of epidural anesthesia and motor blockade, but does prolong duration of peripheral nerve blocks.

Botanicals

STACEY B. DAY, M.D., Ph.D., D.Sc.
Professor and Member
Sloan-Kettering Institute for Cancer Research
New York, New York

Botanicals were once an important part of the study of medicine. Herbs and herbals were fundamental to the healer's art, and *materia medica* played a vital role in the healing of the patient. The following sequence of illustrations exemplifies the sort of *life studies* that, beginning in the time of the early Monastery Gardens, became basic to medical training, first of priest-physician, and then later of the physician-doctor as he has become known today.

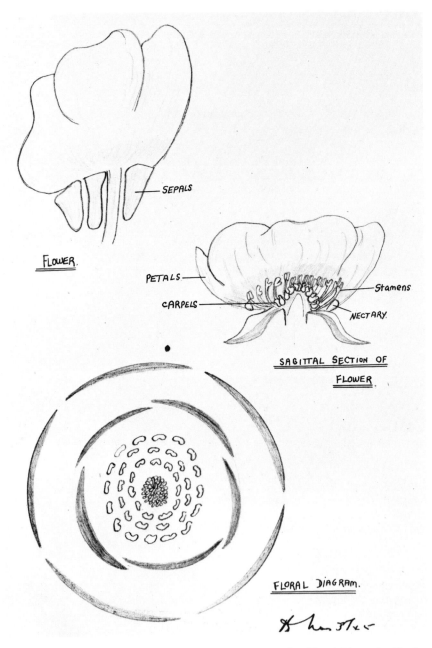

FIG. 1. Common Buttercup. Order Ranunculacae. $K_5C_5A_\infty G_\infty$ (Floral Diagram). (Study of Dr. Stacey B. Day.)

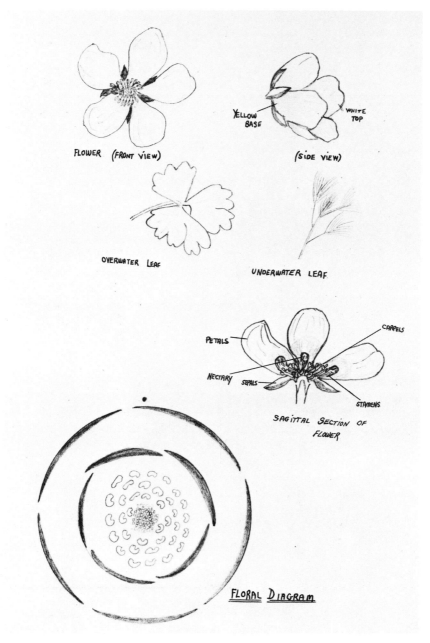

FIG. 2. Water Crowfoot. Order Ranunculacae. $K_5C_5A_\infty G_\infty$ (Floral Diagram). (Study of Dr. Stacey B. Day.)

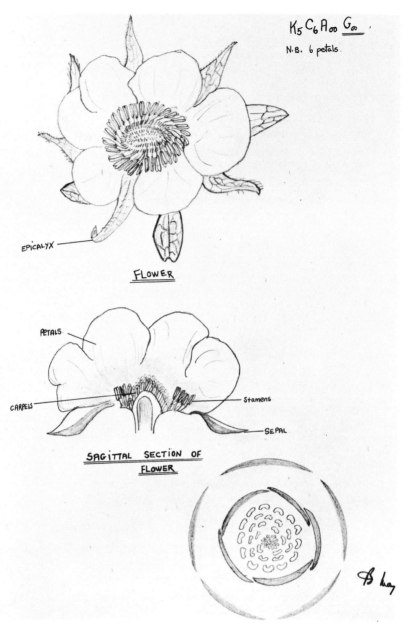

FIG. 3. Cultivated Strawberry. Order Rosaceae. $K_5C_6A_\infty G_\infty$ (Floral Diagram). (Study of Dr. Stacey B. Day.)

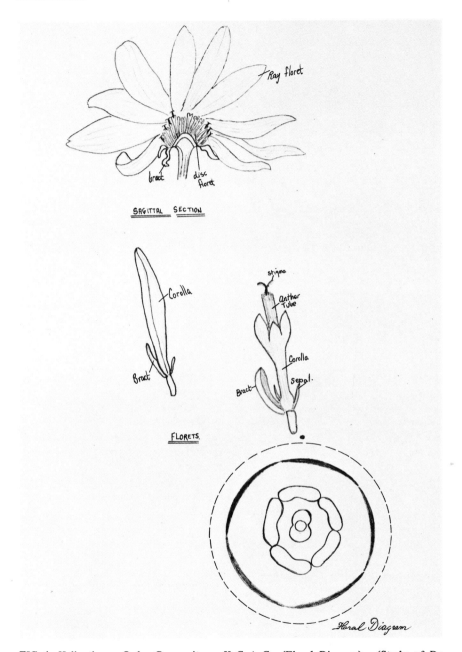

FIG. 4. Helianthus.   Order Compositae. $K_0C_5A_5G_2$ (Floral Diagram).   (Study of Dr. Stacey B. Day.)

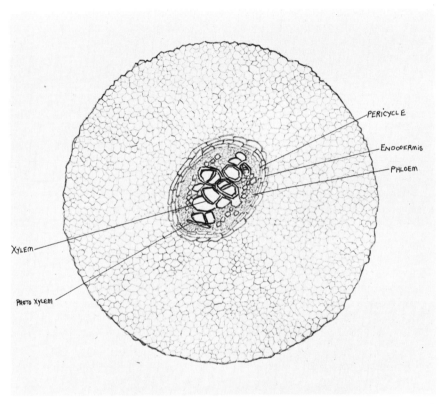

FIG. 5. Transverse Section of Root of Aspidium. (Microscope study of Dr. Stacey B. Day.)

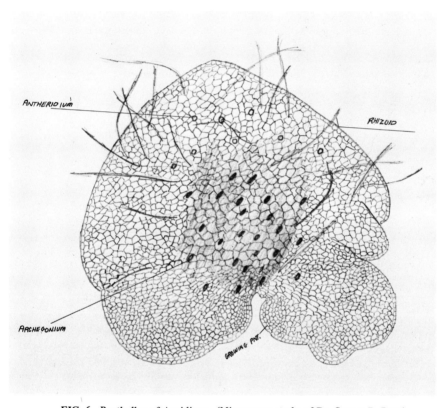

FIG. 6. Prothallus of Aspidium. (Microscope study of Dr. Stacey B. Day.)

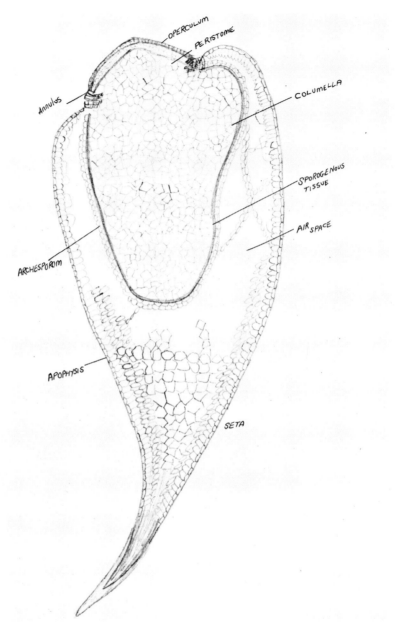

FIG. 7. Capsule of Funaria (Bryophyta). (Microscope study of Dr. Stacey B. Day.)

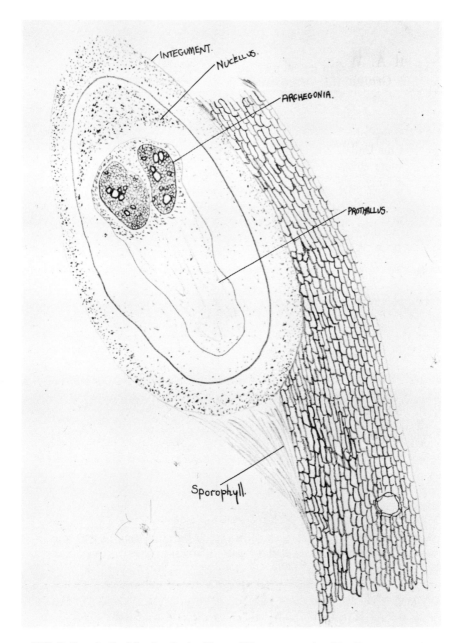

FIG. 8. Longitudinal Section Ovule. Pinus. (Microscope study of Dr. Stacey B. Day.)

Cancer

JOSEPH A. WILBER, M.D.
Atlanta, Georgia

Cancer is one of man's most dreaded diseases. It is the second leading cause of death in the United States and is on the increase. Cancer can occur at any age. It kills more children than any other *disease*. In teenagers and young adults cancer is fairly rare, but it begins to increase in frequency at age 35, especially in women, and is rapidly becoming a leading cause of death.

Cancer is a disorderly, purposeless growth of cells regardless of the consequences to the rest of the body. Normally growth and repair of injury is a regulated process. Cells by some unknown mechanism know when to stop dividing and multiplying when proper size and function have been achieved. In cancer something interferes with the mysterious controls that give messages to cells regarding growth. In cancer abnormal cells grow wildly without stopping, building up masses of tissue called tumors. "Benign" tumors are composed of more normal appearing cells and while they may get large, they do not invade neighboring organs or parts of the body. They can be removed and usually will not come back. "Malignant tumors" (cancers) not only invade and destroy the tissues near them but microscopic particles of cells break off the original mass and are carried by the blood stream to distant parts of the body where they can grow causing multiple malignant tumors throughout the body. These distant new growths are known as *metastases*.

The causes of cancer are not exactly known. Much research is going on and it is evident that genetic as well as environmental factors are involved. Like many of the diseases of modern civilization it is likely that cancer is a disease with multiple causal factors. Certain substances, particularly chemicals that are used in industry, are clearly strong causes of cancer. Radiation in the form of X-rays and radioactive materials also are proven causes of cancer. Viruses are strongly suspected as playing a role in certain forms of cancer. An example of an environmental cause related to one's lifestyle is cigarette smoking. Lung cancer has increased drastically over the past 50 years and this is closely related to the increase in use of cigarettes. Evidence is now overwhelming that cigarette smoking is the principle cause of lung cancer in men and women. Similar lifestyle factors have been found in epidemiological studies of other forms of cancer. For example cancer of the cervix occurs predominately in women who begin their sexual activity at an early age and who practice poor hygiene and have multiple sexual partners. A relationship has been established between cancer of the cervix and a form of herpes virus that in another form produces the common "fever blister" or "cold sore" in the region of the lips.

Approximately one-third of the various kinds of internal cancer, particularly breast cancer, can be cured if discovered early. Unfortunately cancer is another "silent killer" like high blood pressure and atherosclerosis and gives little or no warning of its presence in the early stages. Worse yet, some types of cancer such as leukemia arise simultaneously in many different parts of the body or spread at such a rapid rate that even if detected early, the patient cannot be cured.

Our hope lies in prevention. Yet, with a few exceptions, we do not know methods of prevention in most types of cancer. One common type of cancer, that associated with smoking, can be prevented. Two common cancers, breast and cervical, if found early, can be cured. Survival rates for breast and cervical cancer have increased since the 1940's. This is undoubtedly due to early detection and treatment programs. Unfortunately cigarette smoking and lung cancer continue to increase. Reducing or eliminating cigarette smoking, which is a self-imposed risk, would do more than any other factor to reduce cancer deaths. Moreover, the reduction in smoking would also reduce the incidence of other lung diseases that are increasing rapidly such as emphysema and chronic bronchitis. Coronary heart disease, cancer of the bladder, cancer of the esophagus and cancer of the oral cavity also are more common in cigarette smokers.

The public should insist on stronger legislation and funds for programs to decrease cigarette smoking. Public health programs should be increased for early detection of breast and cervical cancer. The Pap smear is a simple painless method of gently scraping the cervix and by special stains of the cells obtained determining the presence of early changes that show the beginning of cancer. Similarly, if all women were taught to examine their breasts at regular intervals, they can become familiar with their normal texture and then frequently will be able to detect early cancer which can be cured.

All adults should know the seven danger signals of cancer. They are as follows:

1. Unusual bleeding or discharge from any part of the body.
2. A lump or thickening in the breast or elsewhere.
3. A sore that does not heal.
4. A change in normal bowel or bladder habits.
5. Continuous hoarseness or cough.
6. Persistent indigestion or difficulty in swallowing.
7. A change in a wart or mole.

In Georgia there are public health programs making it possible for nearly every woman in the state to obtain regular cervical Pap smears conveniently and at little or no cost. In cooperation with the American Cancer Society, Georgia Division, the Public Health Service offers free Pap smear screening clinics and also programs teaching women self breast examination. The Georgia Heart Association, the Georgia Lung Association, and the Cancer Society all actively promote anti-cigarette smoking educational programs aimed at preventing smoking as well as helping the already addicted individual stop smoking. A method of conveniently detecting hidden blood in stool specimens (an early sign of intestinal cancer) has been developed and it is hoped that screening programs aimed at those over age 50 who have a high risk of colon and rectal cancer will be developed for early detection of this second leading cause of cancer deaths in Georgia. The feasibility of this type of screening is not proven as yet but seems promising.

Expanded cancer registries are needed to detect and locate clusters of cancer in certain socioeconomic or occupational groups. In this way new causal factors for cancer will be learned and we may prevent many of the cancers related to environment and lifestyle. Several states and many large hospitals have "cancer registeries." These are collections of data about all of the cancers detected and treated in the population served. They are used to keep track of individual patients and make sure they are seen regularly to determine recurrence and outcome, and also to search for causal factors such as exposure to a chemical in a particular type of industry that would give us clues as to prevention. There are considerable geographic variations in the incidence of cancer. For example, cancer of the colon and the rectum is very rare in those countries that eat diets containing high amounts of roughage from certain vegetables and grains. Whereas in countries like the United States that eat a much softer and refined diet with large amounts of meat, cancer of the colon and rectum is the second major cause of cancer death.

Treatment for cancer is also making progress. There are three major modalities of treatment; surgery, radiation, and drug treatment (chemotherapy). In recent years the simultaneous use of all three types of therapy in patients with cancer, even before metastases are found, is showing promising results. Some patients with Hodgkin's Disease and some bone cancers are now surviving five years where very few or none did prior to this form of combined therapy. Combined or "triple" therapy in children with leukemia is now producing five year "cures" in a small percentage whereas previously none would have ever survived this long.

These treatments are often very debilitating, time consuming, and expensive, and frequently only one in ten respond. The possible financial costs to society, are staggering if these methods would be applied to all patients with cancer. Therefore, we need community public health programs to prevent cancer in those cases where known causal factors are amenable to changes in lifestyle. Finally, screening and early detection programs for those cancers where treatment is curative should be available to all as part of a low cost regular health checkup.

What I Think of Cancer

LUCIEN ISRAEL, M.D.
University of Paris
Professor and Head
Medical Oncology Service
Centre Hospitalier Universitaire
Bobigny, France

The point of view expressed below is that of a physician who finds himself at the cross-roads of clinical medicine, epidemiology, cell kinetics, immunology and therapeutics.

What we call cancer is basically a disorder of the system which enables the normal cells that are part of a multicellular organism to recognize signals controlling their division rate, function and mobility and to obey these signals. We should keep in mind that, viewed from this angle, cancer is an event which is likely to occur, at least in an occult state, in most complex multicellular organisms. What I want to do is to discuss briefly why this event is a permanent "temptation" for all cell lines, why it is capable of being triggered by multiple causes and why it should cease to be viewed as an accidental phenomenon.

Cancer is a permanent "temptation" because, as long as the host is alive, the cancer cell is privileged from the standpoint of nutrition, synthesis growth and mobility as compared to normal cells. Let us not forget that cancer cells are also capable of protecting themselves against the host's defence mechanisms, either by secreting substances which neutralize these mechanisms or, as our own work has contributed in showing, by inducing secretion of such substances by the host himself. Thus, from a short-term Darwinian standpoint, as soon as cancer cells develop they are potentially the "fittest to survive."

Regarding the multiplicity of causes, it is well known that such diverse factors as various radiations, chemical carcinogens, exogenous viruses, viruses contained in the genome, and even chromic traumatic or thermal injuries have been accused of participating in the chain of mutation leading to cancerization. This means that the mechanisms of signal recognition and obediance are highly complex and fragile ones, capable of being altered by a wide variety of events.

These two facts help us understand why cancer must be regarded as a natural and common phenomenon. The multiplicity of possible causes, the fragility of the intracellular repression mechanisms, as well as the "benefit" derived from cancerization for a given cell line may explain the high incidence of transformation and why its incidence increases with time, and with the normal turnover of the various tissues—since the probability of one cell being "caught" itself increases with these factors.

It is well known however that higher organisms have a specialized system of immunosurveillance which recognizes and rejects those cells that have undergone antigenic changes and lost their membrane contact inhibition properties. Thus clinical cases represent only the successful cases in which the rejecting system is permanently or temporarily deficient for congenital or acquired reasons. Owing to longer life expectancy, to the deterioration of the immunosurveillance system induced by environmental factors, and to the increasing statistical chances of cellular mutation also induced by these factors, it is to be anticipated that both the incidence of clinical—that is successful—cancer will increase and that mean age of onset will decrease in the near future.

The present therapeutic tendency, which is to kill cancer cells by refining the selectivity of anticancer agents and by enhancing immunological defence mechanisms, could thus prove to be largely inadequate. "Curing" the cancer cells in the complex organism to which they belong, rather than killing them, may become a plausible alternative, as our understanding of the mechanisms underlying cell transformation evolves.

Energy State of a Cancerous Cell

OKAN GUREL
IBM
Westchester, New York

Cancer is defined as a dynamic state of an individual cell. The state of a cancerous cell is theorized by the existence of a biodynamic field in and around the cell. The characteristic behavior of this field is its instability (Instability Principle). Cancerous activities and mitotic activities are shown to be dual as corresponding to the negative gradient and the positive gradient portions of the energy space, respectively (Duality Principle). It is therefore concluded that any system (a cell) which goes through mitotic activity, i.e., all cells by definition, is *susceptible* to going through a cancerous transformation. Another significant conclusion based on the characteristics of the energy space of cancerous state is its *irreversibility*. Mitotic state can be initiated with a diminishingly increasing rate, can reach its peak and finally the process can return to its equilibrium state (dormant state of cell, G_0). However, cancerous state, once initiated increases irreversibly. It is therefore concluded that a cure of cancerous state as returning to the equilibrium state can only be achieved by absorption of excess energy. Thus the theory has obvious implications in terms of therapy. Since the cancerous state is represented by a higher energy level and radio therapy is fundamentally energy introducing, a cancerous cell may not be "cured" by this technique. Chemotherapy may be effective in the case of a nonaccelerated cancerous state (with saddle point type instability) where chemicals may enter the unstable biodynamic field around the cell via stable manifolds, and be effective. In terms of the system dynamics this is, of course, a difficult task. One of the most striking results of the theory is that since a cell is "susceptible" but "not immune" to going through a cancerous transformation, immunotherapy of a cancerous state in the sense of immunotherapy as applied to diseases attacking the immune system of the body is not applicable. Instead, immunotherapeutic agents, particularly organisms, such as BCG, C. parvum may well be demonstrated to absorb energy from a cancerous state thus, be effective in that fashion, rather than restoring "immunity to cancer."

Geographic Pathology of Cancer and Computer Cartography of Cancer Data

STACEY B. DAY, M.D., Ph.D., D.Sc.*
Sloan-Kettering Institute for Cancer Research
New York, New York

Strong emphasis on the prevention of cancer and on determining its environmental causes is a direction that has not really been as vigorously pursued as it ought. Improving advances in geographically oriented programs (software) and computer graphic machines and devices (hardware) in the health care area, plus the realization that most health statistics currently being generated represent geographic data, makes the possibility at hand for new control of cancer information and environmental ties by means of computer display and visualization of cancer data.

For some time a few interested pathologists, notably Howard Hopps of Missouri and J. N. P. Davies of Albany, New York, have stressed what should be more widely known, that perhaps 80% of human cancer is attributable to environmental factors. This does not mean that the environmental factor(s) in question is the total cause, rather it means that it is a principal contributing imperative. Day, developing strategies for communication and information exchange of cancer data, convened a group to explore the great potential advantages of graphic display of spatial data representing cancer statistics. Some aspects of computer modelling and cancer data as explored by this group in devising potentially useful methodologies are discussed here.

An important factor in these studies has been the experience of J. N. P. Davies, who long interested in the cartography of cancer, had developed, in Uganda, mapping strategies of the distribution of the Burkitt Lymphoma which brought to light its climatic and geographic limitations. Included in such approaches are the following important points:

1. The mapping of total cancers adds little to knowledge. Advances must come from studying individual types of cancer. In this research maps are of particular value because in showing spatial coincidence, particularly contour maps, they show rates of change and are unlikely to be influenced

---

*Computer generated cartographic expressions of cancer data by courtesy of Professor Eric Teicholz, Harvard University.

by such matters as political boundaries that may bear little relevance to disease (Hopps).

2. Usefulness of cartographic methods comes from associating specific cancers with other specific factors, e.g., water supply, natural ecological distribution of area, foods, industrial exposure, blood groups, etc. (Davies).

3. Much usefulness should come from studies of wind patterns in relation to industrial pollution, as seen in asbestos exposure around mines. Similar data would be valuable for vinyl chloride. Wind patterns, in fact, may be as important as sea currents, water drainage and contamination. Tragic emphasis on this point is illustrated by the recent *Cloud of Seveso* incident in 1976 (Italy). Dioxin, chemically tetrachlorodibenzodioxin, is formed as an impurity in the manufacture of trichlorophenol. According to Garattini this material is "one of the most dangerous chemicals ever detected" and one about which very little is known. In an industrial accident, Dioxin exploded out of a reactor in Seveso, a small town 12 miles north of Milan, in a huge white cloud of droplets, *spread by the wind* over an area of thousands of acres. Both imminent and potential long range injuries became manifest via farm animals, crops and fruit consumption, water drainage, as well as long term genetic potential, for Dioxin is not disposed of by the body but gradually builds up with doses over time.

4. It is important today to map *yesterday's* carcinogenic pollutants. Davies has suggested that maps of schools, kindergartens, clinics and other places where people spend much time are important, and in such cases as pediatric cancers, may be critically relevant to the disease.

5. The value of geographic studies lies within the interdisciplinary background of its nature—such scientific fields as industrial medicine, meteorology, toxicology, genetics, all are important in this approach. I believe also that sociological and cultural habits are important. Such habits, customs and societal ways of people include what they eat, where they spend their time, what sort of environments they share—all reflect possibilities that might be important in induction of cancer, and are easily studied by cartographic technics. Further, sociological studies and cultural anthropologic interpretation might yield clues to *obstacles* to cancer surveillance, social barriers to seeking treatment, social "exposure" to disease provoking situations, food cults, migration of peoples and innumerable other perspectives that go to make cancer a pluricausal disease, which, while not all perhaps acting simultaneously, must nevertheless be assessed separately, and together. Cartographic studies offer great possibilities in these approaches.

COMPUTERS IN CANCER DATA EVALUATION

As surveillance for cancer data widens, and organization and management of disease information becomes structured on geographic and cartographic strategies, computer methodologies, especially those able to present visual graphic display of cancer statistics and epidemiological profiles, afford possibilities for enhanced research in the cancer field. In the present early phases initial problems are being solved in these methodologies. Specifically our group has concerned itself with exploring study disciplines and structuring the data set. These issues include evaluating problems of pattern recognition including machine understanding of symbols, generation of data base from recognized patterns, modeling strategies and application of graphic (cartographic) displays of the filed cancer data. Analysis of cancer data by the following routes has been suggested.

Data Collection: (After Gurel).

Data Level	Study Method	Study Discipline
Molecular Data	Chemical Structure Chemical Kinetics	Chemistry Biochemistry Mathematics
Cellular Data	Cellular Structure Cellular Dynamics (metastasis included)	Cytology Pathology Mathematics Pattern recognition
Organ Data	Geometry Dynamics	Pathology Chronobiology Pattern recognition
Environmental Data Atmospheric Nutritional Artificial	Geometry	Chemistry Microbiology Virology Epidemiology Cartography Statistics Pharmacology Statistics Chemistry Radiation Statistics Cartography

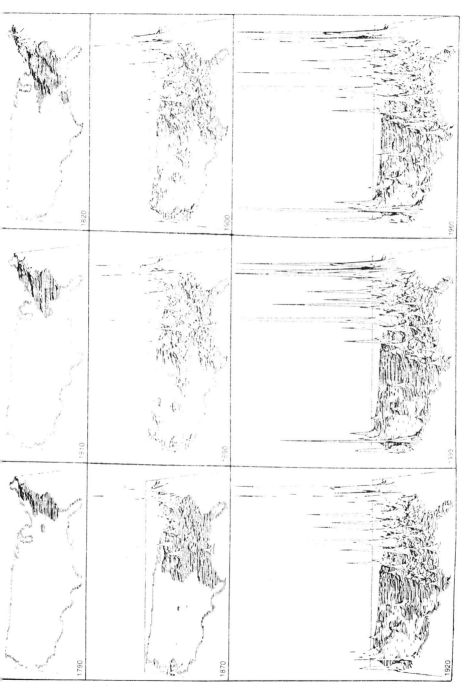

FIG. 1. Automated map display of spatial data permits visualisation of changes over time. (Reproduced by courtesy of Professor Eric Teicholz.)

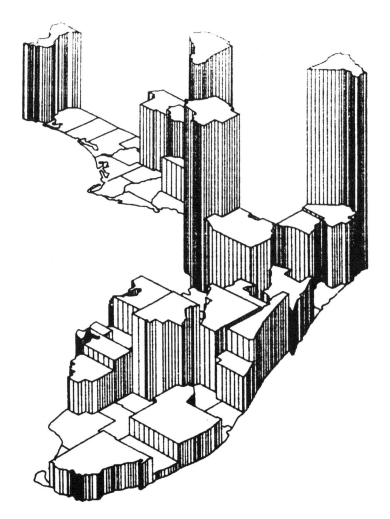

A Prism Map by   XXXXXXXXXXXX on   1-AUG-77 11:03:18
FLORIDA ELF   1.0030MAY77 CHAIN   FLORIDA COUNTIES STERECGR ON TAMPA
WHITE MALES LIP CANCER RATES

FIG. 2. Rates for hip cancer in a 100,000 population of white males. Graphic display of spatial data representing cancer statistics developed by the Laboratory for Computer Graphics and Spatial Analysis, Harvard University. Data file covers period 1950–1969 and was supplied by Dr. Thomas Mason, Environmental Epidemiology Branch, National Cancer Institute, National Institute of Health. (Presented by courtesy of Professor Eric Teicholz, Harvard University.)

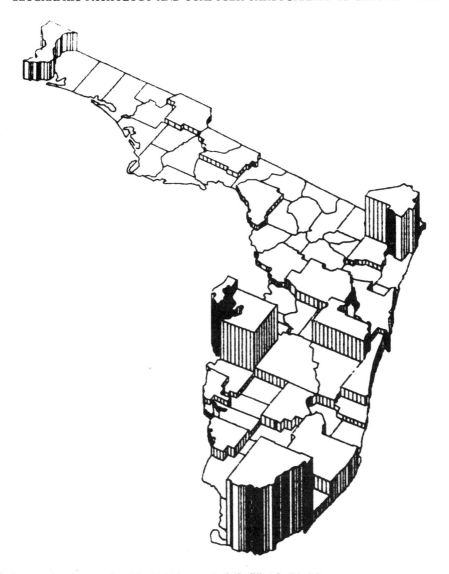

A Prism Map by XXXXXXXXXXXX on 1-AUG-77 12:03:53
FLORIDA ELF   1.0030MAY77 CHAIN   FLORIDA COUNTIES STEREOGR ON TAMPA
COUNT OF WHITE MALE DEATHS FROM LIP CANC

FIG. 3. Counts of hip cancer from white males (Florida) represented as spatial data in the form of a map. File data courtesy Dr. Thomas Mason and generation of map by the laboratory for Computer Graphics and Spatial Analysis, Harvard University. (Reproduced by courtesy of Professor Eric Teicholz.)

Data Usage:

Data Level	Prevention	Detection	Therapy
Molecular Data	–	X	Cheme- Immuno-
Cellular Data	–	X	Immunotherapy
Organ Data	–	X	Surgery Radio
Environmental Data	X	–	–

Data Processing

Data Collected: Input
Data Analyzed: CUP
Data Presented: OUTPUT

	Hardware	Software
Input-Output	1/10 terminals	Conversational, Interactive system
Data Base	Disk, tape, etc.	Storage: DB management Data communication
Analysis	CPU	Algorithmic programs

Teicholz has recently conducted a pioneer demonstration project to illustrate one of several possible applications for computer generated maps to health statistics. At the laboratory for Computer Graphics and Spatial Analysis (Harvard University), he has manipulated and displayed a statistical file reporting deaths from cancer for each of 3300 counties in the United States. The data file presents a death count and rate per one hundred thousand population for thirty-five different types of cancer supplied by the Environmental Epidemiology Branch of the National Cancer Institute. Figures 2, 3, and 4 present such maps displaying spatial and geographic cancer statistics for selected criteria in Florida.

Such collaborative interdisciplinary studies developed in association among multidisciplinary expertises, many not recognized or traditionally thought of as being in the cancer field, will, in our view offer much better promise for understanding of cancer, and contribute via superior epidemiologic and biologic interpretations to the ultimate control of cancer based on prevention of the diseases rather than on cure.

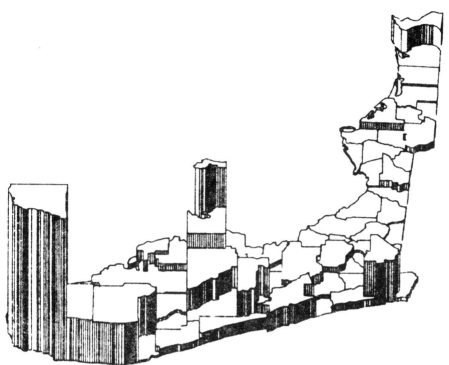

A Prints Map by X/XXXX/XXXXXX on 27-SEP-77 16:41:40
FLORIDA FOR   1.0030MAY77 CHAIN   FLORIDA COUNTIES STEREOGR ON TAMPA
VARIABLE 22 COUNTS

FIG. 4. Counts of white female deaths from breast cancer. File data courtesy of Dr. Thomas Mason; generation of map by the Laboratory for Computer Graphics, Harvard University. (Reproduced by courtesy of Professor Eric Teicholz.)

References

1. Summary and Remarks as made at Seminar Conference on *Computer Modelling and Cancer Data*, April 24th, 1977, New York.

   (Stacey B. Day, M.D., New York, Howard C. Hopps, M.D., Columbia, Missouri, Okan Gurel, Ph.D., IBM, White Plains, New York, S. S. Hyder, University of Montreal, Quebec, Eric Teicholz, Ph.D., Harvard University, Cambridge, Massachusetts, Thomas Lincoln, M.D., Rand Corporation, Santa Monica, California, and J. N. P. Davies, Albany, New York).

2. *Environment and Cancer.* 24th Annual Symposium on Fundamental Cancer Research 1971. Williams and Wilkins (Baltimore). See John Higginson, pages 69–89, *The Role of Geographic Pathology in Environmental Carcinogenesis.*

3. Hopps, Howard, et al. *Computerized Mapping of Disease and Environmental Data.* Army Research Office Publication, 1968.

4. Teicholz, Eric and Hanson, Mark. *Computer Display of Cancer Data.* In press. *Biosciences Communications* 1978. Also see publications of the Laboratory for Computer Graphics and Spatial Analysis, Harvard Graduate School of Design.

5. File Data used by Professor Eric Teicholz from Environmental Epidemiology Branch, National Cancer Institute, National Institute of Health, courtesy of Dr. Thomas Mason.

Overview: Cancer Metastasis

Isaiah J. Fidler, D.V.M., Ph.D.
Head, Biology of Metastasis
Cancer Biology Program
Frederick Cancer Research Center
Frederick, Maryland

The spread of neoplastic cells from a primary site to and growth in distant organs is defined as metastasis. The major cause of death from cancer, no matter what the modality of therapy, is due to the disseminated and uncontrolled growth of tumor cells (metastases) in various organs in the body. Therefore, until cancer can be prevented, the goal of the oncologist should be to prevent and/or treat metastases. Most malignant neoplasms will, in time, spread to distant organs. The number, size, and distribution of metastases depend on both host and tumor cell factors. Tumor cells can spread by three major routes: (a) by direct extension or transplantation, (b) through the lymphatic system, and (c) through the bloodstream.

The development of tumor metastasis involves several sequential and highly selective steps: (1) invasion of cells from the primary tumor into the surrounding tissue, with penetration of blood and/or lymph vessels; (2) detachment or release of single or multiple tumor cell emboli into the circulation; (3) arrest of the circulating emboli in small vascular beds of organs; (4) tumor cell invasion of the wall of the arresting vessel, infiltration into adjacent tissue, and multiplication; and (5) growth of vascularized host stroma. The subsequent growth of the arrested tumor emboli leads to the formation of multiple tumor colonies. Here, too, the processes of invasion, embolization, arrest and cell multiplication can take place once again to yield other metastases.

Tumor cell invasion. The first step in the pathogenesis of metastasis is the invasion of normal host tissues by malignant cells. Three different mechanisms of tumor invasion have been proposed. The first suggests that as the malignant primary tumor mass grows in size, it exerts mechanical pressure on surrounding normal tissues. A second theory is based on the observations that highly invasive tumors possess cell-released or cell surface-bound enzymes that degrade the normal tissue matrix, and as the tumor expands into the regions of the disrupted normal cell matrix, the host cells may be damaged mechanically or enzymatically. A third proposal states that because of their greater ameboid motility, malignant cells simply infiltrate the more static normal tissues.

Spread by direct extension. Tumors that invade body cavities could release cells or fragments that would travel freely in the cavity to seed distant serosal and/or muscosal surfaces and develop into new growths. Primary or metastatic cancers of the ovary often shed many cells into the peritoneal cavity, where they subsequently grow to cover all peritoneal surfaces. Also, primary tumors of the central nervous system, although highly invasive, rarely metastasize to organs outside the nervous system. Their mode of spread appears to be by direct extension or via the cerebrospinal fluid.

Spread via the lymphatic system. The initial spread of tumor in the lymphatic system takes place by tumor embolism in lymph vessels. Tumor emboli can be trapped in the first lymph node encountered on their route; alternatively, they can traverse lymph nodes or even by-pass them to form distant nodal metastases. Originally, it was thought that the only lymphatic-venous communications occurred at the venular angles in the neck; however, several studies established that the vascular and lymphatic systems are intimately connected at various levels. Tumor cells can readily pass from blood to lymphatic channels and back again, indicating that the two systems are probably inseparable in the pathogenesis of the disease. Lymph nodes in the area of a primary neoplasm often are enlarged and clinically palpable. Histologically, the increase in their size can be due to active growth of tumor cells or to hyperplasia of follicles accompanied by proliferation of reticulum cells and sinus endothelium. Often, lymph node sinuses are filled with histiocytes. Lymph node hyperplasia and sinus histiocytosis may indicate a host reaction against a growing tumor and a favorable prognosis.

Spread via the bloodstream. Malignant tumor cells usually also invade blood vessels. The thin-walled veins, like the lymphatic channels, offer little resistance to penetration, and thus provide the most common pathway for blood-borne tumor emboli. In contrast, the arterioles and arteries, whose walls contain elastic and collagen fibers, are rarely invaded by tumor cells. Once primary malignant cell invasion has occurred, the next step appears to be cell detachment, when individual tumor cells or tumor emboli separate from the primary tumor. After infiltrating the vessels, tumor cells can be carried away passively in the bloodstream, or remain localized and proliferate at the site of invasion. Frequently, a thrombus can form around the actively growing tumor cells. Embolization of tumor cells is probably a continuous process. Multiple tumor cell emboli may be released rapidly due to a sudden change in venous pressure, such as may occur during a cough. Diagnostic procedures, surgical trauma, or general manipulation of primary invasive neoplasms can also cause a sudden increase in the number of tumor cells circulating in the blood of patients. However, tumor cell presence in the blood does not indicate that distant metastases will form. The overwhelming majority of circulating cancer cells die quickly in the blood and only a very few survive to eventually form secondary tumors. The hostile circulatory environment and trauma of transcapillary passage probably account for the considerable tumor cell death.

Prior to and during circulatory transport, tumor cells can undergo a variety of cellular interactions, including aggregation with other tumor cells and host cells such as platelets and lymphocytes. In addition, arrested tumor cells may interact with noncirculating host cells such as endothelial cells. Moreover, some tumor cells are thromboplastic and elicit fibrin formation either during their circulation or soon after their arrest in capillary beds. If blood-borne tumor cells are aggregated into large emboli, their success in forming tumors following arrest in the microcirculation is increased. Thus, purely mechanical factors such as embolic size and deformability (as well as capillary diameter) are important to implantation. The rates at which tumor cells pass through capillary beds are not related to individual cell size, but are related to their deformability. After their arrest, some tumor cells are surrounded by a fibrin matrix which may aid their survival. Secondary invasion of the capillary endothelium (extravasation) and underlying basement membrane then occurs, which may depend on the release of tumor and perhaps also host (polymorphonuclear cells) degradative enzymes. Cell surface-associated and secreted proteases are known to be higher in tumor than surrounding normal tissues. When malignant cells reach an extravascular environment, they usually continue to proliferate until growth is slowed due to nutrient and hormonal depletion; but in malignant lesions, vascularization of the micrometastases is probably stimulated by a tumor angiogenesis factor, a glycoprotein that stimulates endothelial cell movement and division.

Metastases are found more frequently in some organs, such as the lungs and liver in both man and animals and the bone marrow of human patients. Metastases are rarely found in spleen, skeletal muscle, cartilage and others. The major reason may be that most of the venous blood drains into the lung and liver; whereas in muscles, spleen and thyroid, tumor cells cannot extravasate through arteriole walls. Collectively the evidence suggests that both mechanical and local "soil" factors determine whether metastases develop after tumor emboli are arrested in an organ.

Dormancy of metastatic cells. Metastatic growths may sometimes appear many years after a primary tumor was excised, suggesting that tumor cells can survive for long periods. It has been postulated that dormant cells are in a nonproliferating state (G_0). Alternatively, tumor cell division may be balanced by tumor cell destruction by host defenses. Indeed, in several experimental systems localized trauma to an organ or immunosuppression of hosts that exhibited no evidence of metastases led to the formation of clinically visible tumor colonies, suggesting that the balance had been shifted in favor of tumor cell division.

Summary. The determinant, then, for the outcome of metastasis appears to be an interplay between the properties of the host and the unique characteristics of malignant tumor cells themselves. The development of metastasis is the result of several highly selective processes. Very few cells residing within a primary neoplasm may be capable of invasion, spread and survival in a hostile environment. Host factors greatly influence the outcome of the phenomenon. Indeed, the majority of tumor cells probably succumb to the tremendous selective pressures (including host immune system) exerted by the host. However, such pressures may in fact permit the cloning of tumor cells with unique properties that are associated with their resistance. The phenomenon of metastatis is thus affected by a multitude of factors. The main factor could be the malignant characteristics of cancer cells, including loss of adhesiveness, increased motility, secretion of proteolytic enzymes, and cell surface characteristics that allow for survival (clumping) in the circulation and preferential arrest and subsequent growth in organs. The probability that patients with solid primary tumor have disseminated disease is great, and therefore, systemic treatment would be recommended. Manipulations of both host defense mechanisms and tumor cell properties could determine the ideal therapeutic approaches to the elimination of micrometastases.

Cancer Rehabilitation

A. TÜRKYILMAZ ÖZEL, M.D., M.S.*
Associate Professor
FREDERIC J. KOTTKE, M.D., Ph.D.
Professor and Head
Department of Physical Medicine and Rehabilitation
University of Minnesota Medical School
Minneapolis, Minnesota

The survival rate of patients with cancer has been increased by recent progress in the early detection of cancer through better diagnostic methods and education of the public, and the development and refinement of the new treatment modalities. Longer survival can sometimes be achieved at the expense of resec-

*Professor Özel died suddenly of a massive coronary occlusion in February 1978. He will be much missed.

tion of an important internal organ or amputation of a body part, or can be accompanied by severe involvement of central and peripheral nervous systems. Therefore, the survival of more patients with physical disabilities may be expected as a result of development in cancer treatment. Equally as important as survival of the patient is the quality of life measured by intellectual, emotional, and physical satisfaction. Each of us must receive gratifications from living in order for life to have any meaning. Therefore, severe physical disability with the consequent psychological, social and economical problems which result, requires comprehensive management in order to achieve the optimal level of satisfactory function for each patient.

Cancer of the face and mouth frequently results in facial disfigurement as well as functional deficits in communication, saliva control, mastication and swallowing. Facial disfigurement has a psychosocial impact on the patient. Intensive counseling before and after the surgery, and when indicated and possible, maxillofacial prostheses or reconstruction plastic surgery are required. Patients undergoing hemiglossectomy or other resection procedures in the maxillofacial area can relearn to speak, masticate, swallow and control their saliva with the help of a speech therapist. Similarly, laryngectomy creates a severe communication problem. Development of esophageal speech is preferred to an artificial larynx in communication.

Radical neck dissection results in a dropped painful shoulder and a rotated scapula secondary to a paralysis of accessory nerve. Prevention of overstretching of trapezius muscle and strengthening exercises to levator scapula and rhomboids are essential in the management of this condition. Nerve grafts to replace the removed part of the accessory nerve can also be used.

Severe lymphedema follows radical mastectomy in more than 10% of the cases. Post-mastectomy lymphedema can be treated by elevation of the extremity, manual or pneumatic massage, optimum amount of isometric muscle contraction of the involved side, external elastic support and diuretics. An appropriate prosthesis provides excellent cosmesis after mastectomy.

When amputation of an extremity is the necessary treatment of cancer, the goal of rehabilitation will be to fit a prosthesis which will meet the patient's needs. Consideration must be given to age, occupation, general health, strength, coordination and ability to maintain balance in an erect position. Physical therapy prepares the stump for fitting by bandaging and exercises, provides pre-prosthetic and prosthetic training and aids the adjustment of the patient to his disability.

Involvement of the central or peripheral nervous system may result in upper or lower motor neuron type of paralysis or paresis. Also, either sensory deficit, pain or both are frequent. Exercises for the maintenance of normal mobility, improvement of muscle strength and coordination, use of braces, corsets and splints to relieve pain, to support weakened muscles or unstable joints, to prevent contractures and to improve function by substituting for a weak muscle are important steps in the rehabilitation process.

Pain that cannot be controlled by medication, heat therapy or a peripheral nerve stimulator may require a neurosurgical approach. A chairback brace with a sternal pad may be used to diminish pain and to improve the spinal stability following compression fractures of the vertebrae.

Several medications and treatment modalities used in cancer such as chemotherapeutic agents, steroids, and radio therapy, may also cause additional neuropathies and myopathies that need further rehabilitative efforts.

Cancer of the bladder, rectum or colon may result in ileal loop, colostomy, or ileostomy procedures. Relief of the emotional trauma is frequently much more difficult than to train the patient for the management of the stoma and its appliance.

In any case, cancer, besides its life threatening and disabling physical problems, creates many psychosocial problems. The adjustment of the patient and his family to the disability requires the collective efforts of the various rehabilitation disciplines. A patient younger than the retirement age may not be able to return to his previous job and would need vocational guidance and training. Without the support and guidance of the rehabilitation team many cancer patients do not make successful psychosocial adjustments.

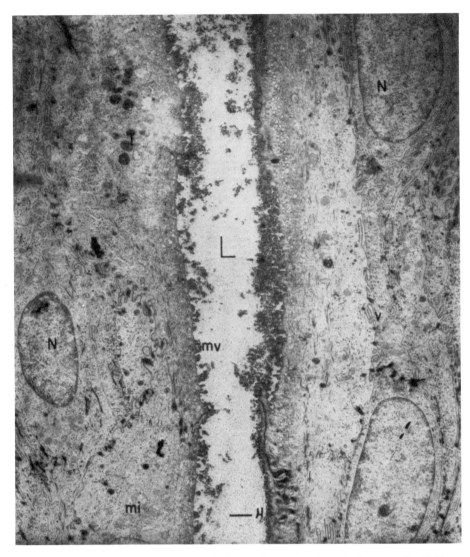

This is a most exceptional longitudinal electromicrograph through the normally highly coiled epidermal sweat duct unit and rarely seen in this plane of section. (L-lumen: note the degree of specialization of the cells bordering the lumen: mv-micro vilae: N-nucleus: mi-mitochondria). Note the degree of membrane specialization of the cells bordering the lumen. (Study of Professor Alvin Zelickson, University of Minnesota.)

The Cell

AARON BENDICH, Ph.D.
Professor and Member
Sloan-Kettering Cancer Research Institute
New York, New York

Cells are the smallest units possessing all the characteristics of life. These fundamental units repair or replicate themselves or their components and are often capable of prolonged existence. Viruses are incapable of reproduction without the intervention of a cell and are thus regarded as nonliving. There are approximately 2 million species of organisms, some of which consist of more than a hundred trillion cells behaving differently or in concert. Despite this diversity, all the cells on Earth are remarkably similar in several basic aspects. All contain an active, functional membrane (the plasma membrane) which surrounds and interacts with internal cellular components, and responds to various external and internal stimuli. All contain genetic determinants inherited from the parent cells which dictate and control their characteristic appearance and function. The genetic information is arranged in chromosomes which are either contained in a distinct membrane-bound nucleus separated from the cytoplasm [eukaryotes (animals and plants)] or free within the cytoplasm [prokaryotes (bacteria and blue-green algae)]. The presence or absence of a separate nucleus thus distinguishes these classes of organisms even though the nuclear membrane disappears briefly during the reproduction of eukaryote cells.

Extrachromosomal genetic determinants called plasmids exist within most bacteria. They are replicated independently of the host chromosomes and can be transferred to neighboring cells and thus endow the recipients with altered heritable behavioral properties. Other cell-made nonliving extrachromosomal entities consisting of tiny collections of genes either encased in membranelike enclosures (viruses) or free (viroids) can infect cells and add to their genetic activities. Within the cytoplasm of the eukaryote cell (but absent in prokaryotes) are several hundred indispensable bacterialike membrane-encased mitochondria which function as power plants, using nutrients to provide the energy required for all cellular activities. Although the mitochondria are not capable of independent existence outside the cell, they contain genetic determinants of their own, distinct from those in the nucleus of the cell in which they reside; thus, eukaryotes have a dual genetic system. In addition, plant cells contain other semi-autonomous self-replicating organelles, i.e., chloroplasts, which absorb solar radiation for photosynthesis.

All cells contain thousands of cytoplasmic ribosomes which function in the synthesis of proteins. Other general anatomical features of the eukaryote cytoplasm include the undulating membranous endoplasmic reticulum which functions as the site for the synthesis, assembly, and transport of proteins; the Golgi complex which consists of vesicles and membranes active in membrane assembly and in the processing and secretion of cell products; and lysosomes which are membrane-enclosed sacs containing destructive enzymes which can digest either internal components during fasting or cell death or unwanted ingested or invading materials. There is an extensive network of internal membranes in eukaryote cells which encloses and organizes the cytoplasm into multiple compartments permitting the coordinated processes needed for normal cell function and motility. These in turn, especially in animal cells, are influenced by an endoskeletal framework of contractile filaments similar to those comprising the muscle fibers. Bundles of filaments are involved in chromosome movement during mitosis and in cell division, in cell-cell interactions, in the maintenance of cell morphology and in cell-surface membrane phenomena. Various cellular components are thus made and function in tiny microenvironments. They are not randomly distributed throughout the cell but are specially organized and are frequently found in very high, and shifting, local concentrations. At predetermined times, depending on environmental conditions and cell type, these components and especially the genetic determinants, are concentrated, replicated, and realigned in the processes of cell division and differentiation. It is remarkable that this almost always occurs so smoothly, even in an unfriendly environment. All cells including the smallest organisms show these features.

The simplest, a cell belonging to a genus of bacteria called mycoplasma, is 0.1 μ in diameter (0.1 μ, or 0.1 μm, is 1/254,000 inch), which is equivalent in length to 1,000 hydrogen atoms laid end-to-end. These cells contain as genetic determinants some 100 genes which are regarded as representing the barest minimum essential for life. Most bacteria (such as E. coli) are about 100 times larger in size and more complex, and their genetic information content is correspondingly higher; mammalian cells are approximately 1,000 times larger and complex than E. Coli and contain roughly a million genes. Although constituted differently, the very similar components and functions in all cells suggest that they may have arisen from common progenitors.

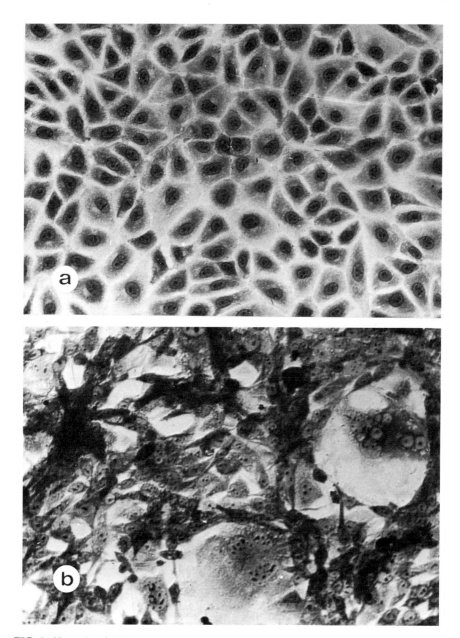

FIG. 1. Normal and Malignant Liver Cells. Cells were removed from the liver of a normal rat and allowed to grow in nutrient fluid in petri dishes. A culture of these cells was developed that continued to multiply and grow for over two years without alteration. If it were

(continued on p. 288)

possible to collect in one place all the cells that would have grown during this time, the weight of the cells would have exceeded that of the Earth! To manage such a situation, the cell harvest of 1–2 week's growth is ordinarily "split" and a small part again grown in a dish, the remainder used up for various experiments. The cells grow in a regular cobblestone fashion in a monolayer one cell thick as shown in a). However, the regularity of their growth and appearance changes drastically [see b)] after the additon of a small, measured amount of a carcinogen (i.e., cancer-producer) to the fluids in the culture dish. The carcinogen used was methylazoxymethanol acetate, a relatively simple chemical closely related to cycasin, a very potent carcinogen found in cycads, palm-like trees whose nuts, containing this poison, were used in the past as a food source. These cells grow in an haphazard fashion, and move past or pile on top of one another; some of the cells contain large nuclei, others several nuclei. Unlike the cells in a), those in b) form cancers when inoculated into test animals.

The cells shown were fixed and stained to make them more visible; magnification: 370X. (Study of Professor A. Bendich.)

FIG. 2. Autogradiograph of Chinese Hamster Cancer Cells. ♂ Chinese hamster cells transformed to malignancy by the carcinogen benzpyrene which occurs in cigarette and other smoke. Chinese hamster cells were grown in a nutrient fluid in a cell-culturing dish and exposed to a cancer-inducing chemical, benzpyrene, which occurs in cigarette and other smoke. The cells were transformed from normal to the malignant state by this treatment as evidenced by the ability they acquired to form a cancer in a hamster. Cells taken from this cancer were grown on a microscope slide in the presence of tritium-containing thymidine which specifically enters and labels only the DNA (i.e., the genes). This is revealed by the technique of autoradiography (see legend to Fig. 4 for details). The above autoradiograph shows that the DNA is located exclusively in the chromosomes (upper cell). There is an abnormally large number of chromosomes (40 instead of 22). The upper cell is in a mitotic phase (metaphase) and has not yet separated from the lower cell (interphase) in which the label is dispersed over the nucleus (N); in this latter phase of division, the chromosomes themselves are so dispersed and spread out as to make identification impossible. (Magnification 2,140X.) (Study of Professor A. Bendich.)

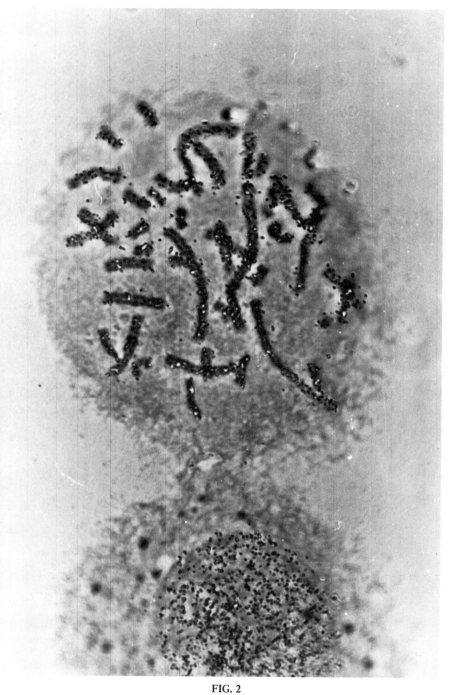

FIG. 2

FIG. 3. Chinese Hamster Chromosomes. ♂ Chinese hamster chromosomes. Chinese hamsters are useful laboratory animals for biological research since their body cells contain 22 easily recognizable chromosomes, a convenient number for study (the human has 46). The chromosomes are the structures in which the genes are contained and arranged in a specific manner. The figure shows the chromosomes of a normal cell in an intermediate stage of multiplication known as metaphase. Multiplication will be completed when each chromosome is divided in half longitudinally so that each offspring cell will contain the same amount of genes. The chromosomes above are from a cell of a male Chinese hamster. It contains one X and Y chromosome while the female has 2 X chromosomes and no Y; these determine the sex. Exposure of cells to ionizing or ultraviolet radiation, or to mutation- or cancer-causing chemicals can lead to abnormalities in the number and appearance of the chromosomes. Chromosomes from individuals with specific genetic abnormalities have tell-tale markers which help in the identification of the disease. (Magnification: 2,720X.) (Study of Professor A Bendich.)

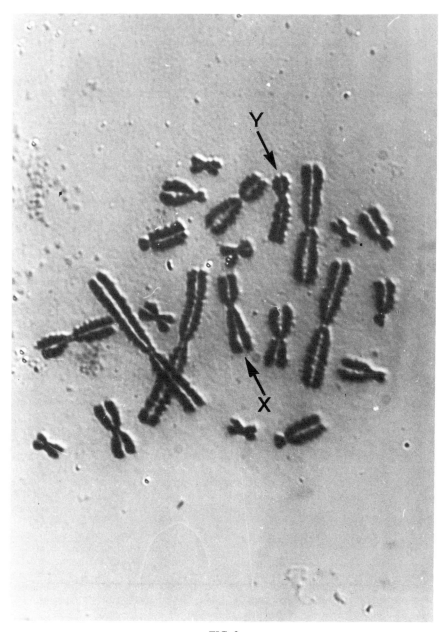

FIG. 3

FIG. 4. Chinese Hamster Cells plus Mouse Sperm. ♂ Chinese hamster cells plus mouse sperm the ger.es (DNA) of which had been labeled with a radioactive tag to help study the path and ate of the sperm. Experiments were carried out to learn whether sperm can "fertilize" a cell other than an egg (i.e. somatic or body cell). A colony of Chinese hamster bone marrow cells was grown on a microscope slide in a culture dish; living mouse sperm were then added. The sperm were from a mouse that had previously been injected with a radioactive hydrogen tag (^{3}H, or tritium) for specific labeling of the DNA (i.e., the genes) to facilitate study of their path and fate. The specific radioactive label employed was ^{3}H-thymidine; this enters only the DNA, during the growth of a cell. Some of the cells were penetrated by the sperm. After 3 days of co-culturing, the specimens were kept in position by adding a fixative and dried and then coated with a very thin film of photographic emulsion (in the dark!) while the cells were still attached to the microscope slide. After an appropriate time of exposure to the photographic film, the entire specimen was developed as is done in ordinary photography. The tritium label is detected by the presence, in the developed photographic film, of grains of silver, shown above by the small black dots carrying a white cap. Each such dot originates in the overlaying film from the irradiation due to the radioactive decay of a single tritium atom associated with the DNA that had been labeled. To localize the position of the silver grains with respect to the cellular structures beneath, the entire slide with adhering cells and film was stained to reveal the cellular structures, and photographed. The figure shows many silver grains just over the sperm head(s) and over the nucleus of some of the cells. In two of the cells, the sperm head appears to have penetrated to the nucleus and some of its radioactive label has entered into the nucleus. Study of these cells in culture over longer periods of time revealed that some of the multiplying cells had acquired attributes of the mouse fetus and showed abnormalities such as are seen when Chinese hamster cells are treated with cancer-causing chemicals. The above technique employing radioactively-tagged substances to locate specific components of cells is known as autoradiography. (Magnification 1,500X.) (Study of Professor A. Bendich and Professor E. Borenfreund.)

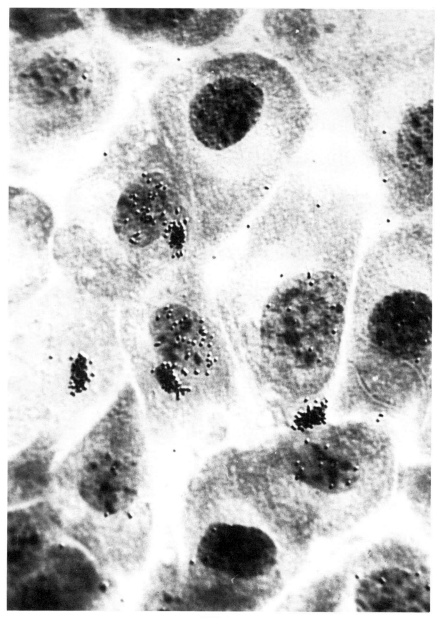

FIG. 4

Cell Metabolism

M. EARL BALIS, Ph.D.
Professor and Member
Sloan-Kettering Institute for Cancer Research
New York, New York

Cell metabolic reactions can arbitrarily be divided into two classes. The first are those that occur during the life of all cells and the second those that are found only in more highly specialized cells and which are largely responsible for the behavior and function of individual organs. We shall limit this article to the former class. Most, if not all, chemical transformations that occur in cells are catalysed by enzymes, organic catalysts that affect the rate at which reactions occur. Thus, the rate at which intracellular chemical transformations occur depends upon the amount and activity of various enzymes.

Complicated regulatory systems exist that control the synthesis and degradation of these enzymes. Some are so short lived that their life time is measured in minutes; others last for many days or weeks. The natural assumption has been made that extremely short-lived enzymes regulate sensitive systems that must respond to sudden stresses. The long-lived would seem valuable to present a constant baseline for fundamental processes on which the cells viability depends.

Survival of a cell requires, ideally, that the cell respond to its environment to take advantage of benefits that may occur and to meet adverse challenges. To a considerable extent this is done by changes in the rates of many metabolic reactions. If a necessary substance is made available from the outside, the cell has mechanisms to shut down its own production of that nutrient. This can be done by stopping the synthesis of a given enzyme, a process called enzyme repression. Obviously the speed with which this can take place is a function of the normal rate of renewal of that enzyme. Alternatively an existing enzyme can be prevented from functioning. This often occurs by the attachment of some of the product to the enzyme at a place referred to as the active site, a mechanism called end-product inhibition. Alternatively a more distant product often serves as a regulatory substance by binding to a place on the enzyme molecule other than the active site. Changing the activity of an enzyme in this manner is called allosteric regulation.

A complex grid of regulatory system exists that in many way resembles the feedback systems used to maintain stability in electronic apparati. These controls are of value to the organism in buffering necessary metabolic reactions against untoward changes in the environment and in taking advantage of beneficial shifts in the external milieu.

In a multicellular organism the metabolism of each cell must not only function in the manner that benefits that cell *per se* but also as it interacts with other cells for the good of the entire organism. Invasive uncontrolled metabolism might, in fact, be viewed as one major facet of cancer. The balance among cells is achieved by a complex set of signals that regulate both adjacent and distant cells. These signals can involve chemical, electrical or physical transmission of information. In all these cases the ultimate expression is a change in certain metabolic activities.

The end results of the metabolic activities of a cell are the conversion of exogenous molecules into new molecules needed for growth and/or maintainance of the *status quo* and the degradation and oxidation of compounds with the concurrent release of energy in a usable form. For example, as mentioned above, many enzymes have very limited life spans. Replacement of these proteins requires both new structural molecules and amino acids, some of which can be taken in from the extracellular medium and others that the cell must synthesize itself. Synthesis of these amino acids and the polymerization of them into proteins requires energy. Not only is energy required but it must be in highly specialized forms.

The availability of a constant source of energy in the proper form is a *sine qua non* of cell survival and growth. All forms of life on the planet transport most of their energy in a chemical form known as "high-energy phosphate bonds." The core about which almost all metabolic reactions revolve is this high energy phosphate. Energy flows into and out of the high energy pool as reactions use or produce available energy. Many sequences of reactions, such as the conversion of sugar to carbon dioxide and water are major sources of this energy. This is the same overall reaction that occurs in the burning of sugar. In the cell however, there is a multi-step sequence of chemical changes that results in the production of high energy phosphate, carbon dioxide, and water. In this long sequence there are metabolic reactions that produce and some that use energy, and some that involve no change in the high energy pool but a rearrangement of molecular configurations.

One of the most striking aspects of the immensely complex network of interdependent metabolic reactions known to occur are certain similarities among widely differing species. Rat, monkey, and man use the same systems to degrade glucose, synthesize nucleic acids and polymerize amino acids into proteins. The enzymes that catalyse these reactions are distinguishable by physical, chemical or antigenic properties, but the reactions themselves are the same. This has permitted biochemists to learn much about the metabolism of man by the study of reactions in experimental animals in situations that could not be duplicated in man. One must, of course, always be aware that past records do not insure that every reaction will be the same in all species and, in fact, exceptions have already been noted in several reactions.

For many years a group of diseases has been known that can be called metabolic defects. Patients who are unable to carry out certain normal biochemical reactions can suffer severe and often bizarre symptoms. Many of these diseases are fatal. In most cases these aberrations are due to abnormal kinds or amounts of enzymes. It is not possible to cure these diseases by inducing the production of missing enzymes although some believe this to be theoretically feasible in the future. Knowledge of the nature of the metabolic reactions involved has permitted better therapy of these patients by altering diets and the use of drugs that can partially compensate for the inherent defect. In addition knowledge of the specific error has permitted the identification of carriers of those metabolic defects that are heritable in nature. This has led to genetic counseling of prospective parents. Diagnosis of feti at risk is also possible by culturing amniotic cells from pregnant women.

Junctional Communication between Animal Cells

FRED SHERIDAN, Ph.D.
Department of Zoology
University of Minnesota
Minneapolis, Minnesota

Within the past decade an important mechanism for coordinating the activities of cell populations in multicellular animals has received considerable attention. This mechanism relies on specialized membrane contacts that allow small molecules to pass directly from cell to cell. Most tissues are, therefore, functional syncytia with respect to small ions, nutrients, and metabolites. Consequently, the biological unit for many processes is not the cell, but a group of cells.

The sites of cell-to-cell transfer is believed to be "gap junctions." As characterized by electron microscopy, these junctions are composed of two aggregates of complementary subunits embedded in each cell membrane. The subunits, which are seen as intramembranous particles in freeze-fracture replicas, extend into the extracellular space where they interact, linking the two cells together. The subunits contain protein and lipid, but no carbohydrate, and they lack detectable enzyme activity. X-ray analysis confirms the suggestion from freeze-fracture that each pair of subunits has a hydrophilic central channel of some 1–2 nm. connecting the cell cytoplasms.

Physiological studies have shown these junctions to be permeable to small inorganic ions as well as to tracer molecules up to about 1000 daltons in size. Of particular functional significance is the evidence that nucleotides and other phosphorylated cell metabolites pass through the junctions as well.

Gap junctions occur in lower animals, such as sponges, and in higher animals such as man. The general structural and permeability patterns for gap junctions have undergone little evolutionary variation except in the arthropods, where their structure and susceptibility to chemical and physical disruption are somewhat different.

Although seen as static structures in the electron microscope, gap junctions are in reality quite dynamic. They can form within minutes after cells come into contact and can similarly be disrupted as rapidly in some cases. Formation probably involves self-assembly of precursors in the membrane, but can be under hormonal control.

By virtue of their permeability to inorganic ions, gap junctions couple the electrical activity of certain excitable cells, notably cardiac and smooth muscle. Nerve cells are only infrequently joined by gap junctions, but where they occur the junctions serve as electrical synapses and often synchronize the electrical activities of groups of neurons.

In nonexcitable tissues, the junctions presumably have nonelectrical functions. They almost certainly permit sharing of metabolite and nutrient pools, averaging out the effects of differences in transporting ability and in enzyme activities. Since many hormones and some nerve transmitters act by changing intracellular levels of small molecules, the junctions may serve to coordinate cell responses in the event of unequal innervation, hormone receptor density, or vascularization. Although these functions are reasonable, and follow nearly axiomatically from the properties and distribution of the junctions, none has been clearly proved.

The discovery of the junctions in embryos prompted the suggestion that the junctions were somehow involved in the regulation of cell growth and differentiation. It was argued further that abnormalities in these cellular processes might occur when the junctions were absent or defective. Certain early experiments on cancer cells seemed to support this idea by indicating a lack of junctions. However, subsequent work has shown that most cancer cells have gap junctions although they may be decreased in number and size.

The possibility that cancer cells have deficient junctions has diverted attention from the equally reasonable possibility that deficiencies in gap junctions might contribute to other types of pathology. An important area of future research will undoubtedly be to look for alterations in gap junctions in other diseases. Some possibilities are atherosclerosis and psoriasis, which have components of hyperplasia, and cystic fibrosis, which involves abnormally regulated cellular transport. Through such studies, we might not only learn important facts about the etiology of the diseases, but perhaps find the elusive clues to the general functional importance of gap junctions and junctional communication.

Aging and Death at the Cellular Level

AARON BENDICH, Ph.D.
Professor and Member
Memorial Sloan-Kettering Cancer Center
New York, New York

The conversion of mammalian cells from a condition of normalcy to that of malignancy is a phenomenon which renders the cell immortal. Proliferation of normal cells in tissues is orderly and self-limiting, whereas that of cancer cells is usually hapazard and poorly controlled leading to ever increasing numbers of cells. These features are also observed when both kinds of cells are grown under laboratory conditions. Thus, investigation of cells in culture can proceed without the encumbering complexities of the entire individual from whom the cells were taken. Differences in the way cells of normal and cancerous tissues are maintained or grow have guided investigators interested in the genesis, diagnosis, treatment and prevention of cancer. Their findings and ideas also have bearing on problems of senescence and death.

When grown on the surface of a small petri dish in an appropriate growth medium, nonmalignant cells move about and multiply forming a layer one cell thick until the entire surface is covered. Growth and mobility are then inhibited when the cells are in maximum contact with one another, even in the presence of adequate nutrients in an optimal environment. If the confluent layer is disrupted, and a portion of the cells placed into fresh culture dishes, the cycle of growth can be repeated. An analogous resurgence of *in vivo* growth in the whole mammal is seen during wound healing and in tissue regeneration, both of which are self-limiting. There is a strict limit to the number of times this can be done with truly normal cells since these show a phenomenon akin to senescence and die out eventually, unless a rare spontaneous malignant (i.e., oncogenic) transformation occurs. It is not clear why the latter change takes place, even when scrupulous care in cell culturing techniques is exercised; it would appear that the senescence results from an intrinsic attribute of the cellular genetic apparatus. Viewed in this light, it can be surmised that the normal senescence of an individual results from the intrinsic aging of his cells. Indeed, cells from young individuals survive longer in culture than do those from older ones.

However, the deliberate application of a variety of physical as well as chemical factors, to which we are all frequently exposed, can induce cancerous changes (i.e., immortality) in such cells, in culture and in our bodies. Indeed, most human cancers appear to be induced by such factors in our environment; these lead the cells along the multi-step pathways to immortality provided they can survive the toxic action exerted by these agents along the way.

The change of a normal cell into a cancer cell appears to have a genetic basis since the ensuing properties of unrestrained growth are passed on to offspring cells indefinitely. Whether this involves an actual change (i.e., a classical mutation) in the cellular genetic determinants (DNA) is a matter of much dispute. Although most carcinogens are mutagenic, there are sufficient examples of the reversibility of the malignant state to raise serious questions regarding mutagenesis as a full explanation. The reversal or loss of malignant expression does not produce truly normal cells but rather those that still retain an indeterminate lifespan. Immortalization of cells which lack oncogenic potential has also been achieved, albeit rarely, by patient long-term culturing of cells taken from normal tissues. Such established cells, however, are more readily transformed to malignancy by various chemical agents than are the cells before culturing.

The various features of the in vitro growth and senescence of cells are best considered in the context of a network of autoregulatory mechanisms (i.e., the genetic determinants) the operation of which can be manipulated by external factors. It can also be suggested that the only truly immortal component of the cell is its DNA since its functions can still be demonstrated even after extraction from dead cells.

(It is not suggested here that the oncogenic pathway should be regarded as the most desirable way to achieve immortality.)

Experimental Approaches to Cellular Engineering

ROBERT A. GOOD, Ph.D., M.D.
President and Director
Sloan-Kettering Institute for Cancer Research
New York, New York

Cellular Engineering is a procedure by which active healthy cells from a normal individual are transplanted into an individual lacking those cells. Since 1968 we have shown this procedure to be successful in permanently restoring the immune function of patients with various forms of primary immunodeficiency and aplastic anemia by transplanting cells from bone marrow, fetal liver, and thymus.

Cellular engineering has great promise in correcting immune deficiencies or abnormalities of cellular development which are involved in producing various diseases of blood development, including leukemia, chronic infections, enzymatic and metabolic deficiencies, autoimmune diseases, such as rheumatoid arthritis and other age-associated disorders. The main obstacle for a more general use of Cellular Engineering against primary immunodeficiencies and against the above, more common diseases, is the difficulty of finding sufficient numbers of suitably matched donors of cells and the problem of graft versus host disease in certain instances following transplantation.

Both of these problems will be addressed in future research. We have recently found that by using new information about the major histocompatibility genes, we can seek matched donors for cellular engineering from distantly related individuals and even unrelated persons from the general population. This will expand the number of potential matched donors. However, the critical issue is graft-versus-host (GvHD) which occurs because certain of the donor's immune cells recognize the recipient's tissues as foreign and attack them. The ideal solution would be to eliminate from the donor cells those cells which produce this sometimes fatal GvHD so that we need no longer worry about selecting matched individuals as donors.

Our experimental work in mice clearly indicates that unrelated, deliberately mismatched donor cells can be treated to remove the GvHD-causing cells and these then can be used to reconstitute recipient animals without causing any GvHD at all. The major aim of proposed research is to more generally prove the validity of these techniques in animals and expand their use to man, first using matched individuals and then progressing to completely mismatched persons. When this stage is reached, the way will be clear to use Cellular Engineering against many of the major chronic debilitating diseases of mankind.

PRECIS

Cellular engineering via transplantation of bone marrow or other sources of immune stem cells is a most attractive new therapeutic approach for the treatment of immunodeficiency diseases[1], aplastic anemia[2] and leukemias[3]. The possibility of extending cellular engineering to the treatment of other abnormalities of blood development, chronic infections, autoimmune diseases, certain metabolic and enzymatic deficiencies, and even diseases associated with aging, has been raised by recent clinical and experimental observations. We discuss the means we intend to use to develop the promise of cellular engineering in the broadest terms possible in order that it may eventually be applied to the treatment of this wide range of persistent human diseases.

BACKGROUND

Among the earliest experimental attempts to use cellular engineering were those of Thomas et al.[4,5] in the late fifties who used allogeneic bone marrow to immunologically and hematopoietically reconstitute inbred beagle dogs which had been lethally irradiated. Their few successfully transplanted animals survived for long periods in good health. They were limited to only a handful of successes due to complications following transplantations which today still frequently stand as obstacles to the wider exploitation of this technique. Failure of graft take, graft rejection, fatal graft-versus-host (GvH) reactions, and infection were all encountered in these first experiments in cellular engineering.

At about the same time, our own work with endocrine organ transplantation[6] and thymus transplantation to reconstitute neonatally thymectomized mice[7] indicated that GvH could be avoided if the donor and recipient were matched at the major histocompatibility region, the H2 locus on chromosome 17[8,9]. This finding permitted further studies[10-13] on immune reconstitution which provided sufficient information about the scheme of immune development and primary immunodeficiency to plan clinical immune reconstitution of human patients born without any immune system at all, as in severe combined immunodeficiency (SCID).

Rapid concomitant development of knowledge about the HLA histocompatibility system on chromosome 6 in man, and recognition of its parallel to the H2 region of the mouse, encouraged us, after an initial failure due to GvH, to attempt hematopoietic tissue transplantation as therapy for SCID. In August

1968 we used bone marrow from a sibling matched at the major histocompatibility locus but mismatched at the HLA-A to treat a child with SCID[15,16]. In spite of a GvH-produced aplastic anemia necessitating a second marrow transplant, this first attempt was successful and the child, presently nine years old, remains healthy. He was cured of two fatal diseases without recurrence of SCID of the aplastic anemia produced by the first transplant.

Simultaneously, Bach et al.[17] successfully treated a child with fatal Wiscott-Aldrich immunodeficiency using bone marrow from an HLA Mixed Lymphocyte Culture (MLC)-matched sibling donor. These clinical successes indicated that bone marrow transplantation can be used to correct fully certain inborn errors of metabolism and acquired diseases associated with inadequacy of bone marrow function. Since these first successes, about 40 children with SCID around the world have been cured by bone marrow transplantation.

The early and continued[18] studies of Thomas et al. indicating that bone marrow transplantation can be used to rescue fatally irradiated dogs, prompted Mathe et al.[19], Thomas et al.[20] and us to begin treating aplastic anemia and leukemia with lethal irradiation and bone marrow transplants from matched sibling donors. At present, 50% cure of otherwise incurable aplastics and 20–30% prolonged remissions or even cure of patients with end-stage leukemia can be achieved using bone marrow transplantation following high dose cyclophosphamide or lethal irradiation together with vigorous chemotherapy.

OBSTACLES TO BROADER USE OF CELLULAR ENGINEERING

It is quite clear that GvH reactions are the prime obstacle to the success of cellular engineering and the possibility of their occurrence must be dealt with at all stages of the procedure. Treatment of SCID using marrow from a matched sibling donor, although often producing some GvH disease, has never produced *fatal* GvH. On the other hand, 20–30% of patients transplanted after irradiation, cyclophosphamide and/or other cytotoxic drugs suffer lethal GvH reactions. The reasons for this are not completely understood but may have to do with the possibility that those tissues damaged by irradiation and/or cytotoxic drugs are the same ones assaulted by GvH reactions perhaps due to mismatching at other histocompatibility determinants aside from the MHC.

Further obstacles also can complicate any bone marrow transplantation or other cellular engineering procedure. Even if a matched sibling donor is available, treatment of SCID occasionally fails, treatment of aplastic anemia often fails, and treatment of leukemia usually fails, as intercurrent viral infections, fungal or bacterial infections, graft rejection, or recurrence of drug-resistant leukemia prove to be obstacles in addition but frequently related to GvH, which are too insurmountable for current management. However, new approaches to these obstacles, particularly GvH reactions, are being developed concurrently in laboratory and clinic.

APPROACHES TO BROADEN THE USE OF CELLULAR ENGINEERING

There are a number of approaches that are in various stages of development which offer promise in overcoming GvH reactions and associated complications following bone marrow transplantation or other forms of cellular engineering. It seems logical to separate these approaches into techniques currently under investigation which have acquired sufficient experimental basis to be used in the clinic and those which are at various stages of experimental development. It is clear, of course, that this separation is somewhat artificial since the constant interaction between clinic and laboratory requires and insures that all progress is made simultaneously.

Clinical

The deterrent to clinical progress in cellular engineering is that only a small proportion of potential recipients have a matched sibling donor available. Due to mendelian segregation of the genes operational at the MHC, the probability that one sibling will be matched with another at this locus is one in four[9,14]. The problem of scarcity of matched siblings for patients with SCID is being resolved by seeking among more distantly related persons and even among the general population, those adequately matched donors at the MLC locus. This has been made possible by recent advances in histocompatibility research by Dupont et al. of this Center[21,22] which permit extremely accurate matching of donor and recipient, the donor now capable of being selected by computer. These critical experimental advances have made possible for the first time the recent immune reconstitution of two children with SCID, one with marrow donated by an uncle and the other with marrow donated by a completely unrelated person[21,22,23]. Thus, using advanced histocompatibility testing along with laminar flow isolation and decontaminating for bacteria and fungi, wider use of donors can be made for the treatment of SCID, and the chances of GvH further reduced. Following our lead this approach has successfully been used to treat several diseases including chronic granulomatous diseases of childhood, SCID, and aplastic anemia using marrow from matched sibling donors and members of the general population.

This approach is one way to avoid GvH but a more generally applicable approach would be preferable and this is where we hope to place a great deal of emphasis. Liver from mouse fetuses of 16-18 days gestation (depending upon strain), neonatal spleen from mice, or spleen from neonatally thymectomized 15-day-old donors have been used in mice to achieve regularly successful hematologic reconstitution following lethal irradiation. This procedure has resulted in long-lived chimeras which have the ability to reject third party skin graft but are tolerant of grafts from either donor or recipient. The use of fetal tissues containing no post-thymic T lymphocytes which cause GvH but rich in hematopoietic stem cells, permits hematopoietic and immunologic reconstitution without any GvH at all[24,25], and this has been successfully applied in humans with immunodeficiency disease. Fetal thymus transplantation has been used to fully or partly correct the T lymphocyte deficiency of patients with the 2nd and 3rd pharyngeal pouch syndrome of DiGeorge. These achievements were based upon experiments using only an epithelial anlagen from syngeneic or allogeneic donors to reconstitute the normal T cell population of neonatally thymectomized or nude (nu/nu, athymic) mice without encountering GvH. In addition, full immunologic reconstitution of SCID patients has been achieved using fetal thymus as a supplement to fetal liver transplantation[29,30]. These findings and principles have now been successfully applied in patients with SCID[26,27,28,29] and we are attempting to apply this approach to aplastic anemia. Further progress exploiting these principles against these and a host of other diseases (see below) must await developments in the laboratory.

Experimental

Our use of fetal liver to treat certain cases of primary immunodeficiency successfully without producing any evidence whatever of graft-versus-host disease serves to illustrate and confirm the practical applicability of the principle of using a stem cell source lacking post thymic elements (T cells) as a means for immunologic reconstitution while avoiding GvH disease. However, the absolute number of fetal liver cells is limited and it is now felt that little can be accomplished from the use of fetal liver transplantation to treat aplastic anemia, leukemia, or hematopoietic failure from irradiation or drugs because of the substantial number of transplanted cells required. Technological developments, particularly the augmentation of hematopoietic stem cells by culturing, could alter this prospect by producing a sufficient and perhaps constantly available supply of cells.

Progress in another direction based upon similar principles, may hold more immediate possibilities for general application. Muller-Ruchholtz et al. have prepared heteroantisera in rabbits that can, in combination with complement, be used as a reagent to eliminate all lymphocytes from rat bone marrow while apparently leaving essential stem cells intact[31]. They were able to achieve successful bone marrow transplantation of lethally irradiated rats by treating these bone marrow samples and then transplanting them across major histocompatibility barriers without GvH disease being observed. We have recently shown that properly absorbed antilymphocyte and antithymocyte sera can remove all T lymphocytes, and probably all lymphocytes from marrow without harming committed stem cells identified and quantitated by soft agar culture[32].

Our overall objective, then, is to determine whether sibling marrow, matched at the major histocompatibility complex and freed of T lymphocytes, or even of all lymphocytes, can be used to treat primary immunodeficiencies, aplastic anemia, leukemias, and therapy-induced marrow failure without encountering as much or any GvH and without decreasing the number of successful takes in comparison to the cellular engineering techniques currently in use. If this approach works as we predict it will based upon experimental work, we will proceed to the next stage in which incompletely matched donor marrows will be treated with antilymphocyte antisera and used against these diseases. However, we will not consider ourselves completely successful until we reach the ultimate stage: the successful transplantation of bone marrow from a mismatched donor. Only when any healthy person can be made, through the proposed techniques, to be a suitable donor for any patient suffering from these diseases, will we have achieved the practical realization of the "universal donor" concept.

In concert with broadening the technique of cellular engineering by crossing increasingly greater histocompatibility barriers, we must also assess whether the immunologically reconstituted chimeras are indeed functioning normally to address viruses and other persisting infections, as well as delayed hypersensitivity and other manifestations of cellular immunity. Clinical and laboratory observations of currently successful transplants indicate that normal immunological functions do indeed exist and that in human and animal recipients, viruses and bacteria, both intra and extracellularly, can be effectively addressed when the hematopoietic and immunologic systems are foreign to the host, even in mismatched systems. However, the possibility exists[33] that cooperative interactions between lymphoid populations required in the defense against certain viruses may require histocompatibility between the host and the foreign lymphoid cells. We are hopeful and the experiments in animals suggest, that these interactions may be mediated by a mechanism whose action can be independent of the major histocompatibility complex. These are questions we hope to answer as we proceed with the main general objectives already discussed.

OUTLOOK

Cellular engineering has already permitted effective treatment and frequently cure of diseases for which no other effective treatment has yet been developed.

This has been possible because of the direct application of experimentally demonstrated immunological principles and techniques to the solution of challenges and experiments of nature presented in the clinic. The expansion of this general technique to enable its applicability to a still wider range of the more common and persistant afflictions of mankind must once more depend upon this clinical-experimental interaction and cross-fertilization.

The remaining problems and the potentially promising means for their solution which have been described here clearly have implications for the entire future of medicine. Not only do these approaches broaden the use of cellular engineering promise to put a wider range of diseases within our preventative and curative grasp, but they are already leading to a far greater understanding of lymphoid and hematopoietic differentiation and development. The wider knowledge thus gained will, as it is doing already, provide access to even newer and far more simple and direct approaches to disease, such as macromolecular engineering which is based on strengthening hormonal mediation of immune and somatic interactions.

References

1. Good, R. A. and Bach, F. H. Bone marrow and thymus transplants; cellular engineering to correct primary immunodeficiency. *Clin. Immunobiol.* 2: 63–114 (1974).
2. Storb, R., Thomas E. D., Weiden, P. L., Buckner, C. D., Clift, R. A., Fefer, A., Fernando, L. P., Giblett, E. R., Goodell, B. W., Johnson, F. L., Lerner, K. G., Neiman, P. E. and Sanders, J. E. Aplastic anemia treated by allogeneic bone marrow transplantation: a report on 49 new cases from Seattle. *Blood* 48: 814–841 (1976).
3. Thomas, E. D. Bone marrow transplantation. *Clin. Immunobiol.* 2: 1–32 (1974).
4. Thomas, E. D., Ashley, C. A., Lochte, H. L., Jr., Jaretzki, A., III, Sahler, O. D. and Ferrebee, J. W. Homografts of bone marrow in dogs after lethal total-body radiation. *Blood* 14: 720–736 (1959).
5. Thomas, E. D. and Ferrebee, J. W. Prolonged storage of marrow and its use in the treatment of radiation injury. *Transfusion* 2: 115–117 (1962).
6. Good, R. A., Martinez, C. and Gabrielsen, A. E. Progress toward transplantation of tissue in man. *Adv. Pediatr.* 13: 93–127 (1964).
7. Ynis, E. J. Hilgard, H. R., Martinez, C. and Good, R. A. Studies on immunologic reconstitution of thymectomized mice. *J. Exp. Med.* 121: 607–632 (1965).
8. Gorer, P. A. The detection of antigenic differences in mouse erythrocytes by the employment of immune sera. *Br. J. Exp. Biol.* 17: 42 (1936).
9. Klein, J. *Biology of the Mouse Histocompatibility-2 Complex.* Springer-Verlag, New York, 1975.
10. Cooper, M. D., Peterson, R. D. A. and Good, R. A. Delineation of the thymic and bursal lymphoid systems in the chicken. *Nature* 205: 143–146, 1965.
11. Cooper, M. D., Peterson, R. D. A., South, M. A. and Good, R. A. The functions of the thymus system and the bursa system in the chicken. *J. Exp. Med.* 123-75-102, 1966.
12. Peterson, R. D. A., Cooper, M. D. and Good, R. A. The pathogenesis of immunologic deficiency diseases. *Am. J. Med.* 38: 579–604 (1965).

13. Good, R. A. and Bergsma, D., (Eds). *Immunologic deficiency diseases in man; proceedings*. The National Foundation Press, New York, 1968. (Birth defects; Original article series, Vol. IV, No. 1, 1968).
14. Schreffler, D. C. and David, C. S. The H-2 major histocompatibility complex and the I immune response region: genetic variation, function, and organization. *Adv. Immunol.* **20**: 125–195 (1975).
15. Gatti, R. A., Meuwissen, H. J., Allen, H. D., Hong, R. and Good, R. A. Immunologic reconstitution of sex-linked lymphopenic immunologic deficiency. *Lancet* **2**: 1366–1369 (1968).
16. Good, R. A. Immunologic reconstitution: the achievement and its meaning. In *Immunobiology* (R. A. Good and D. W. Fisher, eds). Stamford, Conn.: Sinauer Associates, Inc., 1971, pp. 230–238.
17. Bach, F. H., Albertini, R. J., Joo, P., Anderson, J. L. Y. and Bortin, M. M. Bone marrow transplantation in a patient with the Wiskott-Aldrich symdrome. *Lancet* **1**: 1364 (1968).
18. Storb, R., Weiden, P. L., Graham, T. C. and Thomas, E. D. Studies of marrow transplantation in dogs. *Transplant Proc.* **8**: 545–549 (1976).
19. Mathe, G. and Schwarzenberg, L. Treatment of bone marrow aplasia by mismatched bone marrow transplantation after conditioning with antilymphocyte globulin—long term results. *Transplant Proc.* **8**: 595–602 (1976).
20. Thomas, E. D., Buckner, C. D., Cheever, M. A., Clift, R. A., Einstein, A. B., Fefer, A., Nieman, P. E., Sanders, J., Strob, R. and Weiden, P. L. Marrow transplantation for leukemia and aplastic anemia. *Transplant Proc.* **8**: 603–605 (1976).
21. Dupont, B., Anderson V., Ernst, P., Faber, V., Good, R. A., Hansen, G. S., Henriksen, K., Jensen, K., Juhl, F., Killman, S. A., Koch, C., Muller-Berat, N., Park, B. H., Svejgaard, A., Thomsen, M. and Wiik, A. Immunologic reconstitution in severe combined immunodeficiency with HLA-incompatible bone marrow graft: donor selection by mixed lymphocyte culture. *Transplant Proc.* **5**: 905–908 (1973).
22. Koch, C., Henriksen, K., Juhl, F., Wiik, A., Faber, V., Anderson V., Dupont, B., Hansen, G. S., Svejgaard, A., Thomsen, M., Ernst, P., Killman, S. A., Good, R. A., Jensen, K. and Muller-Berat, N. Bone marrow transplantation from an HLA-nonidentical but mixed-lymphocyte-culture identical donor. *Lancet* **1**: 1146–1149 (1973).
23. O'Reilly, R. J., Dupont, B., Pahwa, S., Grimes, E., Pahwa, R., Schwartz, S., Smithwick, E. M., Svejgaard, A., Jersild, C., Hansen, J. and Good, R. A. Reconstitution of child with severe combined immunodeficiency (SCID) and transplant-induced aplasia with marrow from an unrelated, HLA and ABO nonidentical, MLC-compatible donor. *Clin. Res.* **24**: 482A (1976).
24. Tulunay, O., Good, R. A. and Yunis, E. J. Protection of lethally irradiated mice with allogeneic fetal liver cells: Influence of irradiation dose on immunologic reconstitution. *P N A S* **72**: 4100–4104 (1975).
25. Yunis, E. J., Fernandes, G., Smith J. and Good, R. A. Long survival and immunological reconstitution following transplantation with syngeneic or allogeneic fetal liver and neonatal spleen cells. *Transplant Proc.* **8**: 521–525 (1976).
26. Keightley, R. G., Lawton, A. R., Cooper, M. D. and Yunis, E. J. Successful fetal liver transplantation in a child with severe combined immunodeficiency disease. *Lancet* **2**: 850, 1975.
27. Buckley, R. H., Whisnant, J. K., Schiff, R. I., Gibertsen, R. B., Huang, A. T. and Platt, M. S. Correction of severe combined immunodeficiency by fetal liver cells. *New Engl. J. Med.* **294**: 1076–1081 (1976).

28a. Cleveland, W. W., Fogel, B., Brown, W. T. et al. Foetal thymic transplant in a case of DiGeorge's syndrome. *Lancet* **2**: 1211–1214 (1968).

28b. August, C. S., Rosen, F. S., Filler, R. M., Janeway, C. A., Markowski, B. and Kay, H. E. M. Implantation of a foetal thymus, restoring immunological competence in a patient with thymic aplasia (DiGeorge's syndrome). *Lancet* **2**: 1210–1211 (1968).

29. O'Reilly, R. J., Pahwa, R., Pahwa, S., Smithwick, E., Schwartz, S., Grimes, E. and Good, R. A. Successful immunologic reconstitution with split chimerism following fetal liver and thymus transplantation in SCID. In *International Cooperative Group for Bone Marrow Transplantation in Man*, Third Workshop, Tarrytown, New York, 1976.

30. O'Reilly, R. J. Unpublished observations.

31. Muller-Ruchholtz, W., Wottge, H. U. and Muller-Hermelink, H. K. Bone marrow transplantation in rats across strong histocompatibility barriers by selective elimination of lymphoid cells in donor marrow. *Transplant Proc.* 8537–8541, 1976.

32. Ascensao, J., Pahwa, R., Kagan, W., Hansen, J., Moore, M. and Good, R. A. Aplastic anemia: evidence for an immunological mechanism. *Lancet* **1**: 669–671 (1976).

33. Zinkernagel, R. M. and Doherty, P. C. Restriction of in vitro T cell-mediated cytotoxicity in lymphocyte choriomenegitis within a syngeneic or semiallogeneic system. *Nature* **248**: 701 (1974).

34. Blaese, M. Personal communication.

Cholinergic System

H. LADINSKY
*Istituto di Ricerche Farmacologiche Mario Negri
Milan, Italy*

The three functional components of the cholinergic system are acetylcholine, choline acetyltransferase and acetylcholinesterase. Choline acetyltransferase is distributed throughout the neuron and is transported to the terminals by fast and intermediate axonal flows where it is concentrated for acetylcholine synthesis. The substrate acetylcoenzyme A is mainly synthesized in mitochondria. The other substrate, choline, probably cannot be synthesized by neuronal tissue and is taken up into the terminals by a low affinity ($K_m = 100$ μM) and a high affinity ($K_m = 2$ μM) choline uptake system. The high affinity choline uptake system is a specific component of cholinergic nerve terminals, is Na^+ dependent, is blocked by a low concentration of hemicholinium, provides choline for acetylcholine synthesis and may play a strategic role in the regulation of acetylcholine synthesis. No high affinity uptake system for acetylcholine exists.

Acetylcholine nicotinic receptors, from the electric organ of certain fish, and muscarinic receptors, from rat brain, have been obtained in a high degree of purity. The nicotinic receptor is a true intrinsic membrane protein while the muscarinic receptor is an exomembrane protein.

In brain, acetylcholine is ubiquitous but is not distributed homogeneously. The acetylcholine content in some 40 rat brain nuclei varies almost 20-fold from $0.07-1.27$ nmol/mg protein. Highest levels are in the nucleus interpeduncularis, dorsal raphé, caudate and nucleus accumbens and the lowest levels are in the substantia nigra, olfactory tubercle, cerebellum and white matter. The regional distribution of acetylcholine correlates well with the distributions of choline acetyltransferase activity, muscarinic receptor density, and the K_m for high affinity choline uptake, but not with acetylcholinesterase activity which is associated not only with cholinergic neurons but also with glia and aminergic neurons.

Several cholinergic tracts have been localized by the technique of producing electrolytic lesions in specific nuclei combined with regional biochemical determinations of choline acetyltransferase activity and acetylcholine concentration. The tracts are: septo-hippocampal; habenulo-interpeduncularis; parafascicular nucleus-caudate and short axon cholinergic interneurons in the caudate. Histochemical and immunofluorescence techniques for demonstrating choline acetyltransferase are being perfected and have yielded initial results on the localization of cholinergic tracts.

The regulations of cholinergic activity by other neuronal systems in the brain is stimulating widespread interest. Advances in this area were made possible by several significant factors:

1. the recent development of sensitive, specific, and rapid chemical and radioenzymatic methods for measuring acetylcholine;
2. the identification of aminergic tracts arising in midbrain nuclei;
3. identification of putative neurotransmitters in high concentration in selective brain nuclei; and
4. the availability of powerful, specific receptor agonists and antagonists. The nigroneostriatal dopaminergic neurons inhibit striatal cholinergic interneurons and is the only neuronal interaction presently firmly established. Evidence for a gabergic (inhibitory)-dopaminergic (inhibitory)-cholinergic link terminating in the striatum and a serotoninergic (inhibitory)-cholinergic link terminating in the striatum exists.

References

1. Hanin, I. *Choline and Acetylcholine: Handbook of Chemical Assay Methods.* New York: Raven Press, 1974.
2. Waser, P. G. *Cholinergic Mechanisms.* New York: Raven Press, 1975.

Chronic Obstructive Pulmonary Disease

RAYMOND F. CORPE, M.D.
Director, Tuberculosis Control Division of Physical Health
Dept. of Human Resources
Rome, Georgia

The term Chronic Obstructive Pulmonary Disease (COPD) embraces various clinical syndromes characterized by dyspnea with or without cough and sputum production. The major physiological lesion is airway obstruction, intermittent as in asthma, or continuous as in chronic bronchitis and emphysema, which often occur together and may be difficult to separate clinically. Chronic bronchitis is characterized by excessive bronchial mucus secretion, cough and sputum production for at least three months every year for two successive years in patients free of other disease that could cause these symptoms. A history of prolonged smoking, cough and expectoration with wheezes, coarse rales or rhonchi on auscultation suggest chronic bronchitis. Emphysema is permanent anatomical enlargement of airspaces distal to terminal bronchioles with destructive changes. Dyspnea, inappropriate for level of exertion, decreased breath sounds and constant wheezing suggest emphysema.

Cigarette smoking is an important risk factor for COPD. Other factors include air pollution, pulmonary infections, pipe and cigar smoking, dusty occupations, increasing age, allergic reactions and genetic predispositions such as asthma.

COPD begins early in life although symptoms may not be manifest before the fourth decade, and disability may not appear until after the sixth decade. The earliest lesion may be small airway disease and is the sole potentially reversible change. Obstructive changes have been noted in teenagers dying of accidental causes, school children exposed to heavy air pollution, asymptomatic young adults with protease deficiencies and with a history of smoking.

The early diagnosis of COPD is important because of high mortality, morbidity and economic costs. The ability to breathe adequately declines more rapidly in COPD patients than in normal subjects. An estimated 15,000,000 Americans have some form of COPD. Chronic bronchitis and emphysema caused about 30,000 American deaths in 1969 and the death rate doubles every five years. It is now 25 per 100,000 population compared to 50 per 100,000 for lung cancer. Based on age, sex, and smoking history, many COPD patients are prime candidates for heart disease. The lessened ability to breathe properly and exchange gases may lead to a chronic low blood oxygen content with resulting irregularities of the heart beat. There may be increased amounts of carbon dioxide in the blood stream at times severe enough to lead to coma. COPD may directly cause high blood pressure in the lung circulation and heart failure, respiratory failure, and death.

Since the potential for curing COPD patients is dismally low, preventive measures offer a better investment in time and money. These include treatment of infections, reducing or omitting exposure to general and occupational air pollution, and altering the citizen's life style to exclude cigarette smoking by starting with vigorous elementary educational programs on smoking. Benefit to the COPD patient can be achieved by removing him from smoking and pollution.

The prompt diagnosis and treatment of infections, systematic approaches to home and ambulatory respiratory care, patient and family education with regard to effective cough conditioning exercises, breathing exercises and oxygen therapy are necessary parts of an adequate therapeutic program.

The Complement System

Noorbibi K. Day, Ph.D.
Sloan-Kettering Institute for Cancer Research
New York, New York

It is now clear that the complement system (a system of nine main components, C1-C9) is a fundamental biological mechanism of major importance to the body economy. It is involved in many disease processes since it stands at the crossroads of immunological defences, nonspecific inflammation, small vessel reactivity and blood coagulation. Activation of the C system results in several important biological functions including enhancement of phagocytosis, cell lysis, bactericidal activity, chemotactic factors and anaphylatoxins. Although much information is available on these C-dependent biological functions, a whole new concept is developing of the importance of the C system in the specific immune response in relation to cell to cell cooperation, antigen focussing and secondary signals.

COMPLEMENT ACTIVATION

Two major pathways of C activation are now recognized, i.e., the classical pathway and the alternative pathway. The essential difference between the two pathways is that in the former the earlier components C1, C4 and C2 react with antigen-antibody complexes, which in turn activate C3-C9. The type of immunoglobulin varies in its ability to activate C1, e.g., IgM is more effective than IgG and IgA antibodies lack the ability to interact with C1. In the alternative pathway the participation of the earlier components C1, C4 and C2 is not detectable, but activation occurs via C3 and C3b itself along with at least four or five additional interacting proteins termed factor D, factor B, properdin (P), initiating factor (IF) and/or nephritic factor (NF). Other additional pathways to the activation of the terminal C components that involve the proteins of the properdin system are also known to occur, one involving the interaction of C1 with a still undefined properdin-system factor and another involving a new enzyme termed properdin convertase.

It should, however, be noted that there are several other pathways of C activation which are exclusive of both main pathways. Some of these include the direct activation of C1s by enzymes and the nonenzymatic activation of C1 involving C1q by certain polyanions and/or polyanion-polycation complexes, certain viruses and C-reactive protein (CRP).

Direct activation of C components can also occur at the cell membrane. Silicic acid, tannic acid, carbowax 4000 and polyethylene glycol induce the attachment of C components to unsensitized erythrocytes. C-dependent lysis of rabbit erythrocytes in human serum by an antibody-independent properdin system and platelet serum interactions resulting in the assembly and attachment of the attack mechanism in autologous sera have also been described.

RELEASE OF BIOLOGICAL PRODUCTS DURING C ACTIVATION AND INTERRELATIONSHIP OF THE C SYSTEM WITH OTHER BIOLOGICAL SYSTEMS

In both major pathways of C activation release of important biological peptides occurs during the molecular interaction of later components, i.e., (a) chemotactic factor for polymorpholeucocytes consisting of two low molecular weight fragments, C3a and C5a, and a macromolecular complex made of C5, C6 and C7; (b) the anaphylatoxins which are low molecular weight fragments derived from C3a (molecular weight 7000) and which cause contractions of smooth muscle. The two anaphylatoxins are biologically and chemically distinct.

Complement mediated reactions requiring interactions of the earlier C components are: (1) release of a complement derived kinin thought to be a split product of C2, with vasoactive properties, distinguishable from bradykinin and isolated from patients with hereditary angioneurotic edema; (2) complement mediated histamine release which is distinct from anaphylatoxin; (3) immune adherence, the combination of antigen, either soluble or particulate, with a complement-fixing antibody with complement adhering to primate erythrocytes or nonprimate platelets. Immune adherence requires the presence of C3 in the immune aggregate. Before C3 is fixed, C1, C4 and C2 must react with the immune aggregate. Nelson has proposed that the phagocytosis promoting activity of C3 is due to activation of the immune adherence mechanism.

The biological phenomenon brought about by C activation is interrelated with the clotting system, the kinin system and the fibrinolytic mechanism. All three mechanisms can be initiated by a variety of biological substances of Hageman factor, a component of the coagulation system.

ACQUIRED AND INHERITED DEFICIENCIES OF C

The importance of the complement system to host defense and in the development of tissue injury stems from the study of both congenital and acquired C deficiencies in man and in experimental animals. The biological functions of C in effector mechanisms of inflammation, i.e., enhancement of phagocytosis, anaphylatoxins, and chemotactic factor relate directly to the pattern of clinical diseases produced in acquired states of C deficiency. The pathogenesis of glomerular damage seen in many types of renal disease is a good example.

On the other hand, the study of the congenital C deficiencies has broadened our understanding of the role of C in disease states by examining the lack of effector mechanism(s) and control proteins. The absence of these control proteins or inhibitors in the C system has been clearly associated with specific clinical patterns of disease. Absence of C1 esterase inhibitor is seen in hereditary angioneurotic edema and has been reported to be associated with renal disease. Recently Michael Frank of the National Institutes of Health in Bethesda, Md. and his associates have demonstrated that Danazol, an androgen derivative, effectively prevents attacks in hereditary angioedema and acts to correct the underlying biochemical abnormalities, i.e., C1 esterase inhibitor levels and levels of the fourth component of complement.

Dr. Henry Kunkel of Rockefeller University, Dr. Robert A. Good of Sloan-Kettering Institute, his associates and I were the first to show that C2 deficiency is associated with collagen, vascular, and renal disease. With the discovery of more C2 deficiencies and other complement deficiencies, quite a few of which we have studied, it became increasingly clear that isolated C deficiencies are associated with serious diseases. The lack of earlier C components (C1s, C1r, C4) and later components (C5 and C7) in renal diseases has now been recognized. Of interest is the high incidence of infections with gram negative bacteria in patients with deficiencies of C6, C7 and C8. In another family studied in our laboratory C8 deficiency was associated with xeroderma pigmentosum. One patient with C6 deficiency and one with C7 deficiency are apparently healthy. Deficiencies of C3 are of particular importance because C3 is situated at the pivot of the classical and alternative pathways. Four cases of C3 deficiency and one case of C3b inactivator deficiency have now been described. One C3 deficient individual has been studied extensively in our laboratory.

HEREDITARY C DEFICIENCY AND ASSOCIATION WITH THE MAJOR HISTOCOMPATIBILITY SYSTEM (MHC)

C2 deficiency is the most frequently observed deficiency of C components in man. The relatively frequent occurrence of the $C2^\circ$ genes has facilitated studies on their genetic association with HLA which were described for the first time by Dr. Shu Man Fu and his associates of Dr. Henry Kunkel's laboratories. In their C2 deficient family the $C2^\circ$ genes segregated in close linkage with a particular haplotype, the C2 deficient patients being HLA-A10, B18 homozygous. These observations were confirmed by us in a large pedigree (94 members). Only one of the $C2^\circ$ genes was associated with an HLA-A10, B18 haplotype while the other was carried by a rare HLA-A2, B4A2* haplotype. Of additional interest in this study was the observation of a recombinant between the $C2^\circ$ and HLA-A2, B4A2*. Several other $C2^\circ$ deficient pedigrees associated with HLA have since been reported. It was subsequently shown that the C2-carrying haplotypes included the HLA-DW2 determinant and that there was a strong disequilibrium between A10, B18, DW2 and $C2^\circ$.

To date close linkage of Factor B, C4 and C2 with HLA has been described. Other C components are not closely linked. There is a good likelihood therefore that the genetic association between the C components that activate C3 and MHC has a distinct evolutionary significance. Studies on the relation between the activation of C3, the MHC and the control of the immune responses and the linkage of MHC and C genes during phylogeny are indicated. Although there is little information on C deficiencies associated with cancer, we now have studied a C2-deficient patient with Hodgkin's disease and one with osteogenic sarcoma. From these studies it is clear that isolated deficiencies of the complement system in man are frequently associated with disease and regularly reflect functional deficiences, decreased concentrations or even complete absences of individual components of the C system. This was predicted from our phylogenetic studies where we showed the presence of factor B-like components in invertebrates, e.g., in the starfish. A system like the complement system which comprises a cascade of interactions, proteins, proenzymes and enzymes could not have been maintained against the pressures of genetic drift after several hundred million years unless the system as a whole and each of its major components had survival advantage. What had not been anticipated was that some of the primary deficiencies of the C system would be closely linked to the major histocompatibility complex. This extraordinary associations prompts much speculation, but its true meaning will be revealed only by further analysis.

STUDIES OF C IN CANCER

Previous studies in man have given some indication of C activation in certain cancer patients. Deposition of earlier C components—C1, C4 and C2—with circulating 7sIgM has been reported in two patients with Iymphosarcoma. C3 was normal, C9 was elevated, and both patients had reduced serum levels of C1 1NH (C1 esterase inhibitor). One of these patients had episodes of angioedema associated with the disease which were clinically indistinguishable from those seen in patients with hereditary disease. Similar findings have been reported by us. Elevated levels of a component of the alternative pathway were found in various forms of neoplasia. Recently, abnormalities of C and its components in patients with leukemia, Hodgkin's disease and sarcoma have been reported.

STUDIES OF FIXED IMMUNE COMPLEXES IN PATIENTS WITH CANCER

A few studies have noted the presence of fixed immune complexes (Ig and C) in the glomeruli of patients with leukemia, lymphoma, Hodgkin's disease, colon carcinoma, lung carcinoma, African Burkitt's lymphoma and other malignancies. Recently Dr. M. Weksler of Cornell University Medical College in collaboration with our laboratories showed the presence of Ig and C in a patient with melanoma associated with decreased levels of complement.

STUDIES OF CIRCULATING IMMUNE COMPLEXES IN PATIENTS WITH CANCER

We have now screened 9000 sera from patients with various forms of malignancies and have observed hypocomplementemia in a large number. This finding is of potential significance in the understanding of the immune response to tumor antigens. We showed that circulating immune complexes were present in over 50% of 500 cancer patients we studied. The amount of immune complexes was widely scattered and did not appear to be specific for a particular malignancy.

COMPLEMENT DEFICIENCIES ASSOCIATED WITH CANCER IN ANIMALS

Complement deficiency in the AKR mouse is associated with leukemia. We have shown that inbred Bio 424 with a deficiency of C6 have a high incidence of tumors of the adrenals. Studies by some investigators have indicated that complement, not specific antibody, may be the limiting factor in determining the effectiveness of immune reaction to tumor in the mouse. Progressive growth of leukemic cells in AKR mice does not appear to be due to a lack of specific cytotoxic antibody but rather to a lack of C5. Dr. L. Old of Sloan-Kettering Institute has observed that H-2 incompatible ascite sarcomas can grow in highly immunized mice (in this case, from a $C5^+$ strain). Even though the tumor cells were shown to have attached cytotoxic alloantibody, injection of guinea pig serum as a complement source brought about rapid rejection of these far-advanced ascite tumors. Thus C deficiency, whether relative or absolute, can be viewed as another mechanism whereby tumor cells escape the consequences of their antigenicity. Appropriately, reduced levels of complement have been found in sera of tumor bearing mice in our laboratories and in leukemic cats.

Suggested Reading

1. Day, N. K. and Good, R. A., editors. *Biological Amplification Systems in Immunity*, Plenum Publishing Co., New York, 1977.
2. Muller-Eberhard, H. J. Complement, *Ann. Rev. Biochem.* 44: 697–724 (1975).
3. Teshima, H., Wanebo, H., Pinsky, C. and Day, N. K. Circulating immune complexes detected by ^{125}I-C1q deviation test in sera of cancer patients. *J. Clinical Invest.* 59: 1134–1142 (1977).

Coronary Heart Disease and Stroke

JOSEPH A. WILBER, M.D.
Atlanta, Georgia

These two major "killers and cripplers" are discussed together in this section because they are both the result of a poorly understood single disease called "atherosclerosis." This disease affects all the arteries of the body, but unfortunately the arteries to two vital organs, the heart muscle (coronary arteries) and the brain (cerebral and carotid arteries), are the ones most frequently affected with disastrous results in the form of heart attack and stroke.

Atherosclerosis is often also called arteriosclerosis or "hardening of the arteries." It was once thought to be a natural aging process, but in the past 50 years as the epidemic of heart attacks and strokes developed in the affluent industrialized countries of the world and as younger and younger adults died or became disabled, it was realized that this was no natural process but rather a disease somehow related to modern society. Coronary heart disease is the leading cause of death in the U.S. and stroke is the third major cause after cancer. Coronary heart disease causes one third of all deaths under age 65.

Atherosclerosis is a degenerative process involving the inner layers of the large and medium sized arteries. Slowly and insidiously over many years on the inner surface of arteries there is the gradual formation of raised irregular yellowish masses composed of fatty, gritty material containing cholesterol and scar tissue. These masses or lumps protrude out and gradually occlude the lumen of the artery and a blood clot (thrombosis) may form on the rough surface with sudden shutting off of the blood supply downstream. This results in death of the heart or brain tissues supplied by that artery.

There is no simple diagnostic test to determine what is going on inside the arteries, and all too often the patient and his doctor first become aware of the disease in the form of sudden death or a devastating heart attack or paralyzing stroke. In some cases there may be warning signs, chest pain on exertion (angina pectoris) or transient loss of vision or paralysis, little strokes (transient cerebral ischemic attacks). If these warning signs are recognized early, surgical procedures removing the block or bypassing it may prevent the heart attack or stroke. Atherosclerosis is not as yet reversible by any medical treatment, and since it is very difficult to diagnose this disease early and since death or permanent disability can occur in a few minutes, it is clear that our major hope in reducing the toll lies in prevention.

The disease process usually begins in childhood. In the Korean and Vietnam Wars the hearts of American soldiers killed in action were examined. In approximately 30% of these young men, their arteries revealed atherosclerosis of varying severity. We do not know what causes atherosclerosis, but long-term prospective epidemiologic studies, such as the Framingham Study, where large populations of apparently healthy men and women were examined regularly and followed over a 25-year period have revealed "risk factors" or predictors that are statistically associated with an increased risk of stroke or heart attack. The strongest modifiable predictors of coronary heart disease and stroke risk are high blood pressure (hypertension), high cholesterol levels in the blood, cigarette smoking and diabetes. Age and genetic inheritance are also strong factors, but we can do nothing about them.

Emotional stress, lack of exercise and obesity appear to be weaker predictors of risk and we are less certain of their importance. All investigators are convinced that there are important "missing risk factors" that either contribute to the disease or protect against it.

Since some people without known risk factors such as cholesterol die or are disabled by heart attack and stroke, and conversely since a few people with high blood pressure, high cholesterol and who smoke cigarettes live to a ripe old age, it is likely that we still have a gap in our knowledge of the multiple causes of the disease atherosclerosis and that there are as yet unknown risk factors or protective factors that nullify the known risk factors.

As previously mentioned, diabetes is also a strong risk factor in heart attack. Control of this disease is discussed in another section of this volume.

PREVENTION

Controlled scientific studies have proved that stopping of smoking and lowering of high blood pressure *immediately* reduce the incidence (risk) of fatal heart attacks and strokes. Blood cholesterol levels can be lowered by changing our diet, reducing fat and cholesterol intake. However, as yet the difficult and very expensive study that is necessary to completely prove the value of lowering cholesterol by changing the diet of several thousand volunteers, beginning in childhood and continuing to age 50, has not been attempted and probably will never be done. Thus the beneficial and preventive effects of a low cholesterol-low fat diet are unproven, but it is prudent, safe and reasonable to reduce this *proven* predictor of increased risk—high blood pressure.

No cigarettes, a low blood pressure (below 140 systolic and 90 diastolic) and a low cholesterol (under 200 mgm %) can be achieved by lifestyle modification and/or medications, and will definitely greatly decrease one's risk of premature death.

Control of body weight, regular sustained exercise such as jogging or bicycling and if possible, reduction of anger, frustration and fear will make life more enjoyable probably for a longer time and also probably will prevent heart attack and stroke.

The Adult Health Section of the Georgia Department of Human Resources in 1975 began a program to reduce the mortality and morbidity from heart disease and stroke by offering free periodic health screening to all adults in Georgia to identify those at high risk and give them the opportunity to decrease their risk by controlling elevated blood pressure, reducing their blood cholesterol and quitting or decreasing smoking. Public health risk factor reduction clinics ("well-adult clinics") have been established in some counties, and at present about

8,000 individuals are having their blood pressure and other risk factors reduced in these clinics. Our goals are:

1. By 1982 to reduce the stroke and heart attack death rates in adults under 65 years of age by 25%.
2. By 1982 to reduce the stroke and heart attack death rates for all ages by 10%.
3. By 1982 to screen 500,000 people for hypertension and other modifiable risk factors.
4. By 1982 to assist private physicians in medical management of hypertension and other risk factors in 50,000 people by routine management, health and nutrition education and lifetime follow up.

WHAT THE INDIVIDUAL CAN DO NOW

Parents

1. Ask your pediatrician to routinely check your children's blood pressure and blood cholesterol levels.
2. Show good example by not smoking.
3. Provide the family with a low fat-low cholesterol weight control diet.
4. Learn cardiopulmonary resuscitation (emergency first aid for sudden death).
5. Ask your schools to teach health lifestyles and methods of preventing the common diseases of our society.

Adults

1. Do not smoke.
2. Get your blood pressure checked regularly.
3. Know whether your cholesterol, blood fats, and blood sugar are high through regular checkups.
4. Exercise at least three days per week; particularly sustained heart and lung strengthening exercises like jogging, bicycling or swimming laps. If you are over 40, check with your doctor before beginning an exercise program.
5. Know and be aware of the warning symptoms of heart attack and stroke and see your physician immediately if they occur.
6. Influence your local, state and federal government to pass laws that help prevent smoking, make low cost healthy food available to all, that provide convenient access to more physical exercise, and that provide low cost health checkups and treatment for those at high risk.

Cyclic Nucleotides

THOMAS H. HUDSON
JOHN R. SHEPPARD
University of Minnesota Medical School
Minneapolis, Minnesota

The concept that hormones mediate their cellular effects through "second messengers" such as cyclic nucleotides was postulated by Sutherland and Rall twenty years ago. Since then, some of the molecular components of this biological communications system have been identified but the complete picture is far from clear. Two naturally occurring cyclic nucleotides exist: cyclic adenosine 3',5' monophosphate (cyclic AMP) and cyclic guanosine 3',5' monophosphate (cyclic GMP). While cyclic AMP has been shown to participate in regulating the activity of a wide variety of biological functions, the biological significance of cycle GMP remains to be established. An intriguing possibility suggested by N.D. Goldberg is that the two cyclic nucleotides act antagonistically in a "Yin-Yang" fashion to regulate cellular functions.

The widespread occurence of cyclic AMP in both prokaryotic and eukaryotic cells indicates the biological importance of this simple molecule. In the prokaryotes shown to possess cyclic AMP (where it is common but not ubiquitous) cyclic AMP serves as a stress signal and directly regulates gene expression. The bacterial cyclic AMP level rises in response to a depleted glucose supply and induces the synthesis of messenger RNAs that code for enzymes which provide an alternate mode of carbohydrate metabolism, e.g., lactose utilization (the lac operon) is induced by cyclic AMP. Thus cyclic AMP reverses the catabolic repression that is maintained by a sufficient glucose supply. The mechanism through which this occurs involves a catabolite receptor protein (CRP) which binds the cyclic AMP. This CRP-cyclic AMP complex binds directly to the DNA and opens up the double helix thereby allowing the RNA polymerase to attach and commence transcription.

In the cellular slime mold, *dictyostelium discoideum*, cyclic AMP serves an an extracellular chemotactic agent that directs the movement of single cells to generate an aggregate and to ultimately form the fruiting body. Thus, in these cells, cyclic AMP is a diffusable signal for differentiation and is again synthesized in response to a reduced energy supply. The secreted cyclic AMP is sensed by the specific receptors on the membranes of individual cells. The single cells respond by changing their shape through a rearrangement of the membrane-associated contractile proteins. This causes cell movement and increased cellular adhesiveness resulting in stalk formation and a mature fruiting body.

The mechanism by which cyclic AMP elicits physiological effects in higher cells requires the participation of protein kinase enzymes and is described in another essay in this volume on "Protein Phosphorylation."

The enzyme responsible for the synthesis of cyclic AMP is adenylate cyclase. It is unusual because of its bifunctional properties; in addition to catalyzing cyclic AMP production from ATP, it is sensitive to a variety of biological signals foremost of which are the polypeptides and catecholamine hormones.

The molecular receptors for these hormones are distinct from the catalytic components of the enzyme; they are separate gene products. Thus adenylate cyclase is really an enzymatic complex of separate receptors, transducing, and catalytic subunits. The receptor concept and some current thoughts about biological communication are discussed in "Hormone Receptors," also in this volume.

The mechanism of cyclic nucleotide degradation involves hydrolysis of the $3',5'$ cyclic phosphodiester bond by a phosphodiesterase enzyme. Several different isozymes have been identified, each exhibiting different kinetic characteristics and substrate affinitives. In addition to degrading the cyclic nucleotide to the $5'$ nucleotide, elimination of this potent regulatory molecule also occurs via export from the cell. In this context the cyclic nucleotide may function as a primary messenger (as it does in the cellular slime mold). Brain cells which are sensitive to catecholamine hormones have the capacity to increase their cyclic AMP level by 1000-fold. Much of this cyclic AMP is secreted or transported out of the cell and may serve to modulate neurotransmission.

While it is apparent that cyclic nucleotides are important regulatory molecules such as divalent cations (e.g., calcium), fatty acids, and the phospholipids also participate in regulating cellular physiology. Determining how these varied molecules act in concert to control biological function is a major challenge for medical researchers today.

Diabetes Mellitus

JOHN K. DAVIDSON, M.D., Ph.D.
Atlanta, Georgia

Of the approximately ten million Americans who have diabetes, about four million are unaware that they have the disease. Most of those with undiagnosed diabetes are among the approximately nine and one-half million who developed the disease after age 20 years. In older individuals, the disease develops subtly, and may be present for many years before it causes symptoms that prompt a visit to the doctor. Symptoms include frequent urination, thirst with frequent water-drinking, excess eating with craving of sweets, itching, blurring of vision, infection, numbness or pain in the arms and legs, and unexplained weight loss. Of the approximately one-half million who developed diabetes before age 20 years, many developed it almost explosively, being well one day and in diabetic coma a few days later. This group of patients will need expert medical management, including insulin administration. An abnormal amount of glucose in the urine is usually found, and a diagnosis of diabetes is established if an abnormal amount of glucose is found in the blood fasting or after a drink containing a large amount of glucose has been swallowed (glucose tolerance test).

The relative importance of heredity and environment in causing diabetes is not fully understood. Although it has been known for over three hundred years that diabetes is much more common in some families than in others, the mechanism of genetic transmission remains unknown. Only half of those whose disease is diagnosed before age 20 years have a relative who is known to have the disease at that time. It is estimated that about 35% of Americans have at least one blood relative who is known to have the disease. Environmental factors probably play an important causal role in diabetes: (1) excess caloric intake and inadequate exercise produce obesity and every 20% excess weight above ideal body weight doubles one's chance of developing the disease; and (2) it is strongly suspected that certain viral infections may destroy the cells in the pancreas that produce insulin with diabetes developing as a result.

The endocrine and metabolic aspects of diabetes are now understood rather well. The beta cells of the islets of Langerhans of the pancreas synthesize and secrete insulin, a hormone which promotes uptake of glucose and other calorie-containing nutrients. The beta cells in normal adults at ideal body weight secrete about one unit per hour while the individual is fasting and secrete an additional three units after each of three meals for a total of about 33 units per day. Thus in normal individuals enough insulin is secreted to keep the glucose level in blood normal during fasting; after food intake, the glucose level rises modestly and additional insulin is secreted which brings the glucose level back to normal within three hours. Patients at or below ideal body weight who develop diabetes are unable to produce enough insulin to keep the blood glucose level normal. Most obese individuals who develop diabetes are producing more than 33 units of insulin per day; some are producing as much as 120 units per day but the blood glucose remains abnormally high.

If the pancreas is producing large amounts of insulin, why does the blood glucose remain high? Some recent very important research has answered this question, at least in part. Insulin does not initiate its effects on chemical activities in the cells of the body until it combines with a specific receptor site on the cell surface.

In overweight patients with diabetes, the insulin receptor sites are reduced in number, sometimes to as little as 25% of normal. When weight is reduced to normal, the number of insulin receptors increases to normal, and the blood glucose returns to normal. Thus diabetes is at least two distinct disease entities, one due to an absolute deficiency of insulin due to an insufficient number of insulin-producing cells in the pancreas, the other due to a deficiency of insulin receptors on insulin-responsive cells.

In diabetes, there is not only an abnormally high level of glucose in the blood, there is also an abnormally high level of other sugars, fats, sugar-protein compounds, and fat-protein compounds. These chemicals accumulate in and damage the tissues, especially the nerves, capillaries, and arteries. These events cause the chronic complications of diabetes from hardened arteries such as heart attack, stroke, and gangrene of the feet, from leaky capillaries with thick basement membranes such as kidney failure and hemorrhage in the eye with blindness, and damaged nerves with pain, paralysis, and marked decrease in sensation.

Before the discovery of insulin in 1921, the most common cause of death was diabetic coma. The most common cause of death in patients with diabetes currently is from heart attacks; strokes and gangrene are also much more common in diabetic than in nondiabetic patients. Diabetes is the most common cause of new cases of blindness, and it is estimated that by 1980 it will be the most common cause of kidney failure. High blood pressure and obesity are frequently associated with diabetes.

Diabetes is much less common in countries that have little food with only a few overweight individuals than it is in countries that encourage excessive food intake and have many overweight individuals. Prevalence of diabetes always decreases during famine. Seventy million Americans are now classified as obese in that they are 20% or more above ideal body weight. Since the prevalence of diabetes doubles for every 20% incremental increase in excess body weight, an individual whose ideal body weight is 120 lbs, but whose actual body weight is

240 lbs would be 32 times as likely to have diabetes as would have been the case had ideal body weight been maintained. Ideal body weight for adults may be calculated as follows: (1) Males, medium frame, allow 106 pounds for first 5 feet of height, add 6 pounds for each additional inch, subtract 10% for small frame, add 10% for large frame; (2) Females, medium frame, allow 100 pounds for first 5 feet of height add 5 pounds for each additional inch, subtract 10% for small frame, add 10% for large frame.

The sociocultural and family milieu are very important determinants of the lifestyle that leads to malnutrition, especially as it relates to excessive caloric intake and obesity. Patients can be taught to purchase foods that are appropriate. Food labeling is helpful in this respect. Although it is likely to take some time, it seems quite feasible to mount an intensive education campaign to teach the general public good nutrition principles.

Personnel health services should concentrate on early diabetes detection before the patient becomes symptomatic. Once detected those who are overweight can be reduced with the blood glucose level returning to normal in many. Those at or below ideal body weight, pregnant diabetics, nearly all who develop the disease before age 20 years, and all in diabetic acidosis need insulin therapy. Patients should be educated preferably by a physician-nurse-dietician team concerning meal planning, exercise, foot care, urine testing, and the importance of continued follow-up. Insulin-dependent diabetics must be medically managed and also must be taught insulin administration techniques, hypoglycemia prevention, and appropriate action which must be taken when diabetes is out of control.

DNA

Liebe F. Cavalieri, Ph.D.
Member, Sloan-Kettering Institute For Cancer Research
New York, New York

DNA (deoxyribonucleic acid) occurs in all living cells as well as in many bacterial and animal viruses. DNA contains the genetic information which determines the fundamental characteristics of the cell or virus. In higher cells the DNA is associated with protein and is confined to the nucleus while in bacteria the DNA occurs free and is not confined to a given region. Cells which are about to divide have doubled their DNA so that the daughter cells each have their complement of DNA which is identical. The life cycle of a virus includes the replication of many identical copies of the virus arising from a single, infected cell. Cells and viruses therefore contain heretible characteristics which are due solely to DNA.

DNA is a large molecule (a macromolecule) composed of a very large number of smaller monomeric molecules called nucleotides. In *E. coli* the DNA contains about 10 million nucleotides; in higher (eukaryotic) cells there is about 500-1000 times more DNA and a correspondingly larger number of nucleotides. The DNA of the genome (the entire cellular genetic content) occurs in the form of a double-helix which is composed of two spirals wound around each other. The spirals are intertwined so that they cannot be pulled apart by a lateral displacement. Each spiral, often called a strand of DNA, contains half of the total amount of DNA; the nucleotides are therefore evenly divided between the two strands.

There are four types of nucleotides: deoxyadenylic acid, thymidylic acid, deoxyguanylic acid, and deoxycytidylic acid. The nucleotides are composed of a purine or pyrimidine base attached to a deoxyribose phosphate residue. There are four bases—adenine, thymine, guanine and cytosine—which comprise these four nucleotides.

The nucleotides can occur in any order along a DNA strand; the order or sequence constitutes the genetic instructions contained within any given length of DNA. This will be discussed below. A central feature of the double helical structure for DNA is its complementarity. Regardless of the order of nucleotides on one strand they will be complemented by the appropriate nucleotides on the other strand. Thus adenine will always pair with thymine and guanine will always pair with cytosine. This is an invariant. A second feature of this structure is that the helix is right-handed. A third feature is that the sugar-phosphate backbone has a $3'$-$5'$ linkage; the hydroxyl group at the 3 position of the deoxyribose (called the $3'$-OH) is esterified to phosphate which in turn is esterified to the $5'$-OH of the succeeding nucleotide. A fourth feature of the helical structure is that it exhibits a steriochemical polarity: the $3'$-$5'$ linkages run in opposite directions in the two strands (anti-parallel).

DNA replication, once thought to involve one or two enzymes, has turned out to be quite complex. At least a dozen proteins are involved in bacterial replication. There are at least three kinds of DNA polymerases, in addition to unwinding, and winding proteins, cofactors and ligase (to seal together adjacent phosphate and OH-groups). DNA synthesis commences on a small fragment of RNA called an Okazaki primer which is probably about 10 nucleotides long. DNA synthesis continues until a fragment of about 1000 nucleotides is made; synthesis then stops. Then a new RNA primer is encountered and serves as a primer for another DNA fragment, and so on. After the RNA primers are digested away, the fragments are joined into a large DNA molecule by sealing together the small ones with ligase. The newly synthesized DNA, which is now in the form of a double helix, contains the old parental strand and the new one. During synthesis the original two parental strands unwind and serve as templates for the two new daughter strands. Thus if a parental strand contains an adenine at a particular site the daughter strand will contain a thymine at that site. Synthesis therefore occurs in a complementary fashion. The other parental strand contained a thymine at that same site; therefore its daughter strand contains an adenine. In this manner the two daughter helices are seen to be identical to the parental helix.

Since there are four bases in DNA and each occurs a large number of times (n) then there are 4^n possible permutations. Thus if a gene has a molecular weight of 1 million it contains 1500 base-pairs and there would be 4^{1500} possibilities—a very large number.

Since there are 20 amino acids and 4 bases it is clear that more than one base is required to determine an amino acid. Genetic data show that 3 bases are required to determine one amino acid in a protein. Furthermore the sequence is read in a linear fashion. It sometimes happens that there is a base change at one site, say an adenine for a guanine; this gives rise to a mutation which changes a characteristic of the gene product. An altered enzyme may be the result. An average gene of 900 base pairs would give 300 reading units and this would correspond to a protein of about 30,000 molecular weight. Note that the calculation considers a base *pair* since only one of the two plays a role in determining the amino acid.

RNA

LIEBE F. CAVALIERI, Ph.D.
Member, Sloan-Kettering Institute for Cancer Research
New York, New York

RNA is very similar to DNA in its chemical constitution. There are two differences: uracil replaces thymine; ribose replaces deoxyribose. The latter is most important in terms of chemical stability and reactivity.

There are various types of RNA in bacterial cells. These include two classes of ribosomal RNA (rRNA); a variety of messenger RNAs (mRNA); transfer

RNA (tRNA); and 5S RNA. In mammalian cells in addition to these there is an important class called heterogeneous nuclear RNA (hnRNA). The two ribosomal RNAs are structural components of ribosomes which are essential matrices required for *in vivo* protein synthesis. The ribosomal RNAs occur in the two structural units of the ribosome.

There are a large number of messenger RNAs in a cell, each different from the other. This is so because the mRNAs represent the immediate instructions for the synthesis of proteins of which there are probably 3000–4000 per bacterial cell. These RNAs are transcribed from the various regions of DNA. Each may contain genetic information for one or several proteins. The RNA is templated from only one of the two strands of DNA. RNAs are therefore linear molecules and since only one DNA strand is copied these RNAs are single-stranded; their complementary strands have no reason to exist.

Transfer RNA (tRNA) is a low molecular weight RNA which is shaped like a clover leaf. It is essentially single stranded but because there are many complementary nucleotides, the molecule has a great deal of secondary structure, i.e., hydrogen bonded base pairs. Transfer RNA is essential for protein synthesis. Together with mRNA and ribosomes it mediates the faithful translation of information from the mRNA. It does this by virtue of its anticodon contained in one of the loops. The anticodon "reads" the codon from the mRNA. The reading occurs while the tRNA and mRNA are situated at precise sites on the ribosome. There are two such sites. One contains the amino acid which has just been covalently joined to the rest of the peptide chain and the other site contains a tRNA molecule which is about to release an amino acid previously joined to it. When it is released it is simultaneously covalently attached to the amino acid adjacent to it on the second site of the ribosome. This process continues until all of the mRNA has been translated, yielding the gene product whose information was originally obtained from DNA. There is a large class of tRNAs which are required for their translation process. Sometimes the same tRNA permits the insertion of two types of amino acid; sometimes two transfer RNAs are capable of inserting the same amino acid.

All RNAs are synthesized on DNA as the template. Different regions of the DNA are used for different types of RNA. Ribosomal and transfer RNAs are structural elements. Messenger RNA belongs to a different class since it contains the instructions for the amino acid sequence of the proteins. If mRNA contains information for more than one protein, it is called a polycistronic messenger.

RNA synthesis is accomplished with RNA polymerase. The monomers used for its synthesis are the ribonucleoside-5'-triphosphates. The basic mechanism for RNA synthesis is similar to that for DNA synthesis—base-pairing with one strand of the DNA acting as the template. When one side is the template the complementary strand is not read. However not all genes are read from the same strand; both may be used, but at different locations along the DNA. RNA synthesis begins at particular sites and proceeds in a $5' \rightarrow 3'$ direction.

RNA polymerase contains 5 different polypeptide chains; these subunits are not covalently attached to each other. One of the subunits allows for the recognition of the initial site. The second step occurs a short distance where the polymerase binds more firmly. Synthesis then starts but recognition sequences themselves are not transcribed. RNA synthesis proceeds until a stop signal is encountered.

Recombinant DNA

LIEBE F. CAVALIERI, Ph.D.
Member, Sloan-Kettering Institute for Cancer Research
New York, New York

Recombinant DNA is made by recombining DNA fragments from any source into a single DNA molecule. This is made possible by the fact that certain enzymes (restriction enzymes) recognize specific sequences in the DNA and cleave them in a staggered manner; i.e., the cleavage points do not contain flush strands of DNA. DNA from two sources will adhere to each other at these sites because of the complementarity of the bases; the ends are often referred to as "sticky ends." When ligase is added to the reaction mixture in a test tube the ends become covalently attached and a permanent hybrid DNA has been produced. There are many types of restriction enzymes, each carrying a recognition site for the specific sequence in the DNA. Thus many types of recombinants can be produced provided the same enzyme is used to create the sticky ends.

To be useful from a biochemical standpoint the recombinant DNA must be amplified many times. To achieve this the small amount of recombinant DNA is absorbed into $CaCl_2$-treated *E. coli*. After a number of generations the bacteria are harvested and the recombinant DNA isolated.

A common vector for carrying out this research are the bacterial plasmids. These are small circular DNA molecules (MW $\approx 3 \times 10^6$) which grow autonomously in *E. coli*. They are easily isolated from supernatants and are therefore quite convenient. The recombinant DNA is also a circular plasmid of somewhat larger size than the original molecule.

As noted before any types of DNA can be used. A major scientific aim of this research is to study the chromosomes of eukaryotic cells. The aims of such studies include the mapping of the mammalian genome as well as the study of initiation and termination sequences for transcription.

Experimental Surgery

STACEY B. DAY, M.D., Ph.D., D.Sc.
Sloan-Kettering Institute for Cancer Research
New York, New York

In a general sense, as commonly understood, the validity of scientific knowledge for the surgeon has rested on certain technical manipulative skills, rather than on an understanding of the philosophical background dealing with the order and dependence of the problems that he is dealing with in a surgical respect. Thus, experimental surgery has been seen in terms of methods and operative technics as used in clinical situations rather than as a discipline circumscribing thought and philosophical analysis underlying the physiology of the surgery used, as well as the rationale comprehending the basis of the propositions that one is faced with in the patient oriented clinical situation. In reality then service and education, as they underlie the basic surgical skills, should not be left to empirical study alone. It is clear today, that experimental surgery includes aspects of all the basic sciences, and is, generally speaking, not a matter of exploring operative technics. In my view, *experimental surgery* is to be understood as the *pursuit of those researches that best establish the physiological principles underlying surgical practice.*

It should not escape notice that perhaps the most outstanding of the "experimental surgeons" in the historic past, have been primarily biologists who have seldom been concerned with clinical or operative surgery at all.

Background Reading

1. *Collected Writings of the Abbé Lazzaro Spallanzani, 1729-1799.* (Including Dissertation relative to the Natural History of Animals and Vegetables. Translated from the Italian, London, J. Murray, 1784.)

2. Bernard, Claude. *An Introduction to Experimental Medicine.*
3. Hunter, John.   Including *Observations on Certain Parts of the Animal Oeconomy,* *London, 1786.*
4. Pavlov, I. P.   *Conditioned Reflexes. An Investigation of the Physiologic Activity of the* *Cerebral Cortex.* (Translated and edited by G. V. Anrep). Dover Publications.

The Surgical Treatment of Ischemic Heart Disease

STACEY B. DAY, M.D., Ph.D., D.Sc.
Sloan-Kettering Institute for Cancer Research
New York, New York

I. BACKGROUND

Surgery, more than any other branch of the physician's learning, is influenced by physiology and medicine. This was particularly true for the advances made in cardiac surgery in the early 1950's. As early as 1923 Cutler and Levine had reported the first successful operation for mitral stenosis, but it was not until the early fifties that the confluence of several interdisciplinary and clinical innovations made possible real advances in cardiac surgery. Foremost among these developments were the organization of vastly improved blood banks, new and broader spectrum antibiotics, physiologically sound and technically improved anesthesia methodologies, as well as inauguration of new experimental possibilities for studying both normal and abnormal changes in the heart. At the same time *technologies* for heart surgery began to be introduced; hypothermia, hyperbaric pressure chambers, and extracorporeal pump oxygenators based on physiologic principles such as the azygos (low flow) principle for maintenance of the circulation. Much of this new progress was attributable to vigorous intellectual perspectives advanced in experimental surgical studies. It might be said here that such "experimental surgery" is not so much a matter of exploring surgical technics, but is preferably understood as *studies investigating the physiological principles of surgical practice*—a far different thing. This improved knowledge of the physiology of the heart, included biochemical and biophysical understanding of cardiac function, as well as insight into the contractile and electrical behavior of the cardiac muscle.

Suddenly, so it seemed, there was hope for cure of functional heart disease. In succession diseases of the mitral valve, congenital heart malformations, aortic and tricuspid disease yielded to surgical intervention. But with ischemic heart disease surgical interest became almost a stigma for discredit. It was almost axiomatic, in the early days of surgery for occlusive coronary disease, that no patient came to the surgeon until he was almost abandoned by the physician. No surgeon can perform miracles; and a myocardium that is destroyed beyond repair could hardly be expected to be made new by "revascularization".

Meanwhile deaths from coronary heart disease steadily mounted. As recently as 1973 there were 684,000 reported deaths in the United States alone. Increasingly younger age groups seemed afflicted (under age 35), and it was inevitable that surgeons would seek to devise operative approaches that could conceivably restore normal coronary blood flow to an ischemic myocardium. This was a supremely challenging surgical physiologic problem. If it could be solved, some 4,000,000 beneficiaries in the United States alone would be heirs to surgical relief!

II. THE ANATOMIC BASIS OF THE PROBLEM

Comprehensive understanding of coronary artery disease rests on knowledge of the anatomicopathologic basis of the disease and on knowledge of the nature of the coronary artery distribution in normal and in diseased hearts. Differences in coronary artery patterns may be indicative not only of possible evolutionary significance, but may also define possible functional and therefore clinically "different" aspects of the disease. Moreover ethnic and cultural features related to differing populations have been implicated with differing anatomy: the Bantu of South Africa and the mountain dwellers of the high Andes seem to have features of coronary artery anatomy that are unique to these groups.

The coronary arteries are effectively end arteries, and while their pattern in man may differ from heart to heart, in relative length of right and left main vessels, number of collateral and intercoronary connections, as well as size of such vessels, three general classifications for hearts are encountered:

GROUP I. This is the most frequent group encountered in human hearts and comprises those distributions in which the right coronary artery is predominant. In this group the right coronary artery extends beyond the posterior ventricular septum and supplies a portion of the left ventricle. The posterior descending coronary artery arises as a branch of the right coronary artery. Forty-eight percent of all human hearts have been observed to fall within this classification. (Schlessinger.)

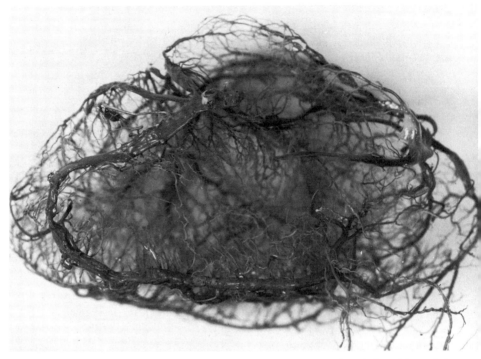

FIG. 1. Unusual and excellent birds-eye-view of the entire length and disposition of the main circumflex branch of the left coronary artery. Observe the way in which it almost completely circumscribes the heart. Corrosion Cast. (Study of S. B. Day.)

GROUP II. In the second or "balanced" group, the right coronary artery supplies the right ventricle and the posterior septum, but does not supply the left ventricle beyond the posterior septum. Thirty-four percent of human hearts fall into this description and are alleged to suffer least from the effects of an acute coronary disaster.

GROUP III. In these hearts there are varying degrees of left coronary preponderance in which the left circumflex coronary artery supplies part or all of the posterior septum and supplies part of the right ventricle as well. Eighteen percent of human hearts have been estimated to fall within this group. Recovery rates following occlusion in this group are reported to be poor.

INTERCORONARY ANASTOMOSES

The word "anastomosis" is probably derived from the greek suggesting "a coming together through a mouth." This usage appears in the description of the *Body of Man* written by Helkiah Crooke in 1615. In 1669 Richard Lower clearly expressed the concept in his statement that "the vessels which carry blood to the heart come together again and here and there communicate." Thebesius, thirty nine years after Lower, demonstrated by careful dissection the presence of anastomoses between the right and left coronary arteries. Later, Vieussens presented an original description of the course of the coronary arteries in man, in which accompanying illustrations depicted collateral channels.

Controversy over the function of interarterial coronary channels in the human heart, and their role in pathological circumstances, has always been perplexing. Anatomists such as Von Haller and Ruysch on one hand could clearly see collateral vessels. On the other hand pathologists observed necrosis of the myocardium following acute ligation of the coronary arteries. On this evidence such prominent investigators as Henle, Hyrtle, Cohnheim and Von Schultheiss-Reichberg denied their existence. These conflicting views dragged on for years; Neelson, Koester, Rojecki, Bianchi, Jamin and Merkel adding opinions. In 1896 Dragneff endeavored to evaluate the whole problem. So did Piquand in 1910, and later Spalteholz and Gross. The basis for so deeply conflicting views was undoubtedly due to the fact, that while most certainly the collateral vessels were anatomically present, *time* was required to establish *effective* coronary flow which would protect the myocardium. If death occurred before these channels became functionally useful, one can see that there might well be grounds for conflicting viewpoints. Principal credit must go to the work of Schlesinger, Blumgart, and Zoll for in large part resolving many of the earlier differences.

One can clearly summarise the situation, in the understanding that interarterial intercoronary anastomoses are unquestionably end arteries physiologically, in that they are not of sufficient size and number to be *functionally* significant in immediately preventing myocardial damage after sudden coronary artery occlusion. However, while not of immediate functional significance, they do serve as a potential vascular framework to limit the size of infarct, and to promote later revascularisation via functional emergence of new developing channels. Demonstrations by the writer have unquestionably shown these vessels to exist in a large spectrum of human and animal hearts, and to be readily provoked in experimental studies.

EXTRACARDIAC ANASTOMOSES

The precise role of these vessels is still ill defined. Such channels were first mentioned by Von Haller. Langer described anastomoses with vessels in the mediastinum, the parietal pericardium, the diaphragm and the hila of the lungs. Woodruff and Wearn described anastomoses with the vasa vasorum of the aorta; and Hudson, Moritz, and Wearn found extensive communication between the auricular branches of the coronary arteries and vessels in the pericardial fat, with the pericardiacophrenic arteries, the internal mammary arteries, and with the anterior mediastinal, bronchial, superior phrenics, and intercostal and esophageal branches of the aorta.

OTHER ARTERIES CONTRIBUTORY TO THE SURGICAL ANATOMY

So-called supernumerary coronary arteries have been noted since early days. Quain in 1849 observed that "the evidence of three coronary arteries is not a very rare occurrence, the third being smaller and arising close by one of the others." This vessel has frequently been sighted and variously named, (Merckel, Bianchi, Piquand, and Monroe Schlessinger). Such a vessel is probably now best known as the Conus Artery, or Third Coronary Artery.

The Septal Artery may be important to the coronary circulation both in man and animals. A unique anterior septal artery is normal to the circulation of dogs and rabbits. In man the septum is supplied by rich brushworks of vessels arising from the anterior and posterior descending coronary arteries.

III. THE PHYSIOLOGIC BASIS OF THE PROBLEM

Operations for coronary artery disease have generally been based on two physiologic principles:

1. Redistribution of blood available to the heart via interconnecting interarterial intercoronary anastomoses and other juxtaposed vessels. Much effort was spent in the fifties to encourage development and/or to increase such networks. However precise knowledge of the factors which control the rate and extent of coronary arterial collateral development were inadequately explained, and no operation described could assure production of such functioning vessels. Objective evidence was presented to support the importance of anoxia in the formation of such new vessels. Other theories postulated increased differential pressures, metabolites, and reflex nerve action as being potential mechanisms for the opening of pre-existing but non functioning interarterial intercoronary vessels.

2. The restoration of blood flow to previous ischemic areas of myocardium following by passing of stenotic or occluded segments of coronary artery by coronary by-pass operations. Physiologically speaking, these approaches seem most sound, but they must take into account numerous qualifying

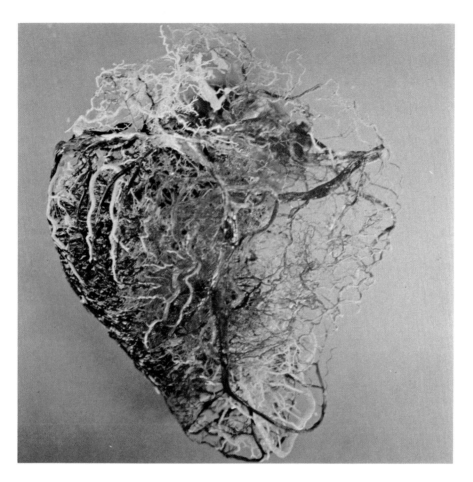

FIG. 2. Corrosion cast of Human Heart to demonstrate coronary arteries. Balanced coronary pattern. (Study of S. B. Day).

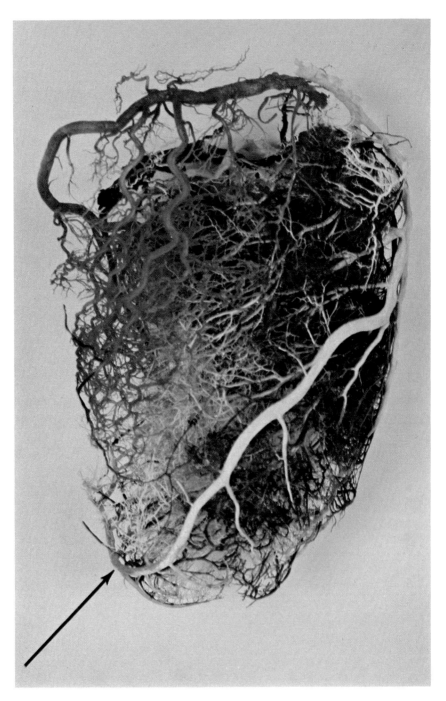

FIG. 3. Corrosion cast of Human Heart to demonstrate the coronary artery vasculature. *Left Coronary Artery Preponderant Pattern.* Observe presence of intercoronary anastomosis at apex of heart—red dye from right coronary artery and yellow dye from left coronary artery. (Study of S. B. Day).

parameters including the extent and spread of disease, the state of the ischemic myocardium, hemodynamic and functional assessment of the myocardium, as well as, most important of all, the general condition of the patient, with particular respect to coincident vascular disease (renal, cerebrovascular, peripheral circulatory, diabetes etc.) *The extraordinary developments in technology of the last years and the great investments in heart surgical units*, plus such improvements as coronary cineangiography, led to an explosive increase in coronary by-pass procedures during the last decade. Generally speaking the most favored procedure has been use of a saphenous vein graft interposed between the ascending aorta and the occluded artery, thus permitting blood to flow from the aorta to the coronary artery beyond the obstruction—the so called aortocoronary by-pass. A modification of the old Vineberg procedure occasionally used is an end to end anastomosis of the internal mamary artery with the coronary artery beyond the point of obstruction. In the United States, at least, these two interventions are those most widely described as being currently acceptable for the surgical relief of ischemic heart disease.

Review of Surgical Efforts

The operative treatment of angina pectoris was proposed originally by François Frank in 1899 and practiced first by Jonnesco in the form of a neurosurgical approach in 1916. This was a purely symptomatic endeavor not intended to cure. No suggestion of revascularizing the myocardium was made. On the contrary, the avowed object of interrupting the cervico-sympathetic pathway to the heart was to ameliorate the doleful pain of angina.

The first recorded direct operation for cardiac ischemia was performed by Leriche and Fontaine in 1933, who in a dog excised part of an infarct produced by ligation of a coronary artery. The defect so created was repaired with a free graft of pectoralis muscle. Thereafter a large number of papers were published investigating the nature of the collateral circulation to the heart and establishing its potential role in surgical methods for revascularization of the myocardium. In the British Isles Laurence O'Shaughnessy, and in the United States, Claude Beck, pioneered simultaneously several studies of this nature. Beck was the first surgeon to perform (in 1936) an operation on the human heart for the relief of angina pectoris. In this early procedure a pedicle graft of the pectoralis major muscle was sutured to the myocardium.

A wide variety of tissues have been applied to the myocardium either in association with or independent of epicardial and pericardial abrasion. Beck's earliest studies were confirmed by Grassi who subsequent to acute ligation of one or more coronary arteries in the rabbit sutured pectoralis muscle and omentum to the myocardium. These authors indicated that vascular connections were formed between the grafts and the heart and that no degenerative changes occurred. O'Shaughnessy achieved the same goal using greater omentum. This procedure was modified by Key and Baronofsky who utilized small intestine.

Augmentation of the coronary blood supply by cardopneumopexy was introduced by Lezius who suggested that anastomoses could be demonstrated by X-ray studies following the injection of thorotrast into the pulmonary artery. Carter investigated the procedure and modified it slightly. Smith observed subjective improvement in patients following this operation. Garamella recommended modification of the procedure by segmental resection of the lung with apposition of the raw surface to the myocardium. It was suggested that greater protective benefits would ensure. Carver investigated the effect of basilar cardiac denervation in association with cardiopneumopexy and concluded that the fibrillation threshold following acute coronary artery ligation in dogs was lowered. Liebow and his group at Yale indicated increased bronchial collateral circulation to the coronary arteries following cardiopneumopexy with ligation of branches of both pulmonary arteries and pulmonary veins. It was thought that this technic could be used in the treatment of transposition of the great vessels. On this hypothetical basis Burford operated on several human cases of coronary artery insufficiency. His basic procedure consisted in ligating the lingular arteries hoping to force the bronchial arteries to expand. He was of the opinion that collateral vessels were provoked that linked up with adjacent coronary arteries.

In other approaches the use of skin grafts, spleen, and diaphragm appeared to warrant little real hope that these procedures could be usefully effective. Historically, because of the relative simplicity of the procedure, the applied physiological aspects of induced adhesions was widely investigated. Irritants such as aleuronat, asbestos, talc, phenol, powdered bone, sodium morrhuate, and abrasion with a dental burr were utilised to incite adhesions within the pericardial sac and with surrounding structures. Aleuronat was used as early as 1907 by Coenen to foster pleural adhesions. O'Shaughnessy used it for a similar purpose. Thomson in prolonged studies investigated the use of talc (magnesium silicate powder). Claude Beck, after investigating numerous irritating substances, chose asbestos on the thesis that it gave superior results. The observation of Burchell that "it is doubtful whether any experiment could be devised in which the heart could be made to beat as a result of perfusion of pericardial adhesions" is perceptive and important. It would be useful however to bear in mind the state of the art and the nature of the illness as it could only be evaluated in those days.

Implantation of a systemic artery into the myocardium was vigorously investigated by Arthur Vineberg and extensively examined in the experimental

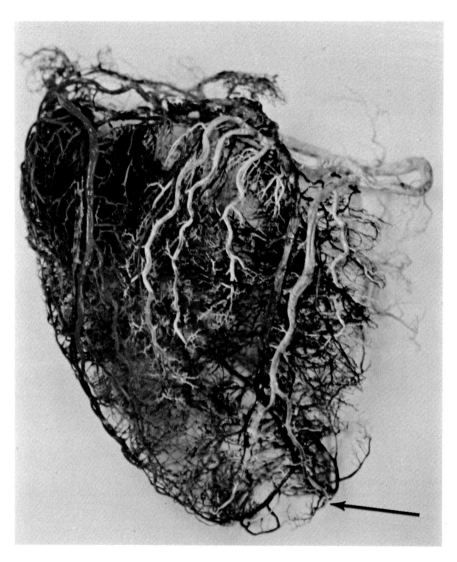

FIG. 4. Corrosion cast of Human Heart to demonstrate coronary arteries. Right coronary artery pattern. (Study of S. B. Day).

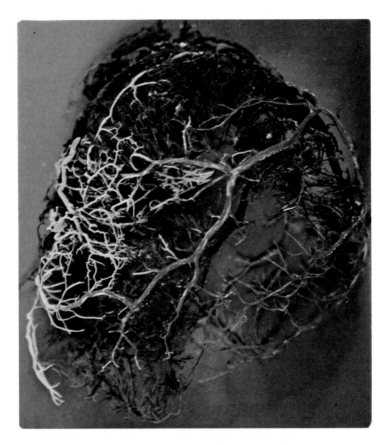

FIG. 5. Corrosion Cast of the coronary arterial system of a heart twelve weeks following implantation of the *Internal Mammary Artery* directly into the myocardium. When the yellow plastic material was injected into the systemic artery it flowed as shown into the coronary arterial vasculature, demonstrating Internal Mammary–Coronary Artery Anastomosis of the Vineberg operation. (Study of S. B. Day).

surgical literature. My own results in 1958 confirmed that the anatomical basis of the operation as proposed by Vineberg could undoubtedly be objectively demonstrated in some, (not all), experimental cases in the laboratory. These evaluations preceded the sophisticated cinecoronary angiographic technics that were to be developed in the early sixties. Coronary-mammary anastomosis appeared to be related to the heart's demand for arterial blood, as Vineberg had insisted. Vineberg had suggested, and my own results tended to concur, that the chance of an internal mammary implant branching would depend upon its being placed in an area where the pressure was lower than within itself. The greater the differential pressure between that within the implant and the pressure within the heart, the greater the likelihood of coronary-mammary union. This same rationale as proposed in the fifties serves no less equally as an explanation in the eighties.

The interesting experiments by Pratt in 1898 first suggested that the heart could be effectively nourished by means of flow of blood through the auricle into the veins and coronary sinus. The Langendorf heart preparation at the turn of the century made quite clear that the heart could be conceived as a biological pump, perfused with Ringer's lactate in lieu of blood, and in this sense viability could be maintained for the surgical exercise of transplantation.

Arterialization of the coronary sinus was advocated by Beck. In this procedure a vein graft was interposed between the aorta and the coronary sinus. At a subsequent operation the coronary sinus was partially or completely occluded. Blalock cautioned that such anastomoses in dogs frequently resulted in a back flow of the new source of arterial inflow into the right atrium. He further observed that complete occlusion of the coronary sinus resulted in an engorgement of the vessels of the myocardium, the net result of which was to provoke deterioration of the situation.

Fauteux pioneered pericoronary neurectomy in association with ligation of the great cardiac vein. A study of Siderys was relevant. Evaluating coronary ligation of the left anterior descending artery and retrograde coronary flow from the same artery in three group's of animals (a) control, (b) simple pericardiotomy and (c) ligation of the great cardiac vein, it was noted that coronary vein ligation and simple pericardiotomy equally protected the animals as compared to the control group. Clearly the conclusion that a simple pericardiotomy affords the same protection as "operation" threw into question the results of many experimental procedures.

The feasibility of end to end anastomosis between a coronary artery and a systemic artery was investigated by Gordon Murray in 1954. By an ingenious maneuver he was able to perform anastomosis at leisure by perfusing the distal segment of the coronary artery at a rate of not less than 20 ccs. of blood per minute through a polyethylene tube which had been inserted into a systemic

TABLE 1. Surgical Operations

A. Development of Collateral Circulation by Vascular Grafts
   1. Pectoralis muscle
   2. Omentum
   3. Lung
   4. Skin
   5. Intestine
   6. Spleen
B. Development of Collateral Circulation by Cardiopericardial Adhesions
   1. Aleuronat
   2. Asbestos
   3. Talc
   4. Phenol
C. Implantation of a Systemic Artery into the Myocardium
   1. Internal mammary artery
   2. Subclavian artery
   3. Splenic artery
D. Surgical Procedures Involving the Coronary Veins
   1. Ligation of the coronary sinus
   2. Ligation of the great cardiac vein and pericoronary neurectomy
   3. Arterialization of the coronary sinus
   4. Creation of artificial thebesian vessels
E. Neurosurgical Intervention
   1. Paravertebral injections of alcohol
   2. Ganglionectomy
   3. Posterior rhizotomy
F. Metabolic Intervention
   1. Ablation of the thyroid gland
G. Creation of a Low Pressure Fistula
      (Pulmonary Artery–Left Atrium Anastomosis)
H. Direct Surgery Upon the Coronary Arteries
   1. Resection and anastomosis of coronary artery
   2. Anastomosis of systemic artery to coronary artery
   3. Coronary thrombo-endarterectomy
   4. Aortocoronary Bypass

artery. Absolon and co-workers established the possibility of performing endarterectomy on the coronary vessels of man by practice observations performed on the human heart at autopsy dissection. Bailey reported an early clinical trial of direct surgery on the coronary vessels by doing a thromboendarterectomy of the anterior descending branch of the left coronary artery in man. Longmire reported experience of this operation in man. Studies using a low pressure fistula between the pulmonary artery and the left atrium to provoke intercoronary anastomosis were explored by Day. A small number of advanced clinical cases of ischemic heart disease, confirmed by preoperative coronary angiography, were operated on by Lillehei, utilizing this shunt technic. The more direct aorto-coronary by-pass procedure has been noted, and has been the most superior surgical procedure devised to date. It is the only relevant operation for today's consideration. All other procedures, (historical now), as described in this review are of great interest in illustrating the slow and involved course leading to the present position, which in fact has largely been made possible *by virtue of parallel developing technology in the cardiac surgery field.*

IV. THE PRESENT SITUATION

Almost without exception all of these surgical innovations proved unsuccessful. The one useful operation is the *aortocoronary by-pass procedure.* However even this surgical intervention is under re-evaluation. Earlier beneficial results appear not to be holding up over long term examination, and there are additional disadvantages of high cost and occasional post surgical complications. With the present cost, that may range as high as $10,000 per operative procedure, has come heated debate, for in view of some 70,000 such operations being performed yearly (in the U.S.), it is apparent that annual cost for this surgical treatment is rapidly passing into figures which cannot be contained. As things now stand, there is good evidence that drug therapy may be as valuable as surgical intervention, less traumatic, and as effective. Prolonged life has not been proven with surgical intervention, and scrupulous analysis again suggests that this disease is simply not "a surgical dilemma". Seventeen years ago in the Moynihan Dissertation I wrote, and I see now no reason to change, this viewpoint:

> It is the view of the author that no satisfactory solution of the problem of ischemic heart disease will be achieved unless there is an appreciation of the broad general, epidemiological, nutritional and biochemical background to the problem. Exact understanding of the interlocking multidisciplinary facets of the genesis of coronary artery disease will serve to indicate that surgical technique alone is likely to offer very little promise of affording other than palliation of the disease. It is to be understood, moreover, that no revascularising procedure can restore a myocardium that is no longer viable. A suitable surgical operation, while it may not be curative, should endeavor to add blood to the vascular bed of the heart before there are irreversible destructive changes in the cardiac muscle fibres.

CONCLUSION

As things stand, it is doubtful whether, save perhaps in very carefully selected cases, surgical intervention can fulfill the demand for *cure* of ischemic heart disease. Apart from the fact that aortocoronary by-pass operation is becoming almost prohibitively expensive, this sort of surgery, is now, *as it always was*, a *palliative* procedure at best. Surgery cannot stop the disease. There is probably no surgical "best" for coronary artery disease, save cardiac transplantation, and that, as it was in 1958, in the experimental surgical laboratory, remains still a research investigative operation, until such time as associated immunological problems can be absolved, and rejection of the transplanted heart certainly controlled. With respect to methods to redistribute available blood to the myocardium, these have no role in todays surgical approach. Thirty years experience by innumerable perceptive surgical researchers has taught perhaps what was always known—that coronary artery disease is better *prevented* than palliated. Surgical intervention, at a time when the heart muscle is worn out, and like any machine, in need of rest and "new parts," can afford only limited therapy. It is more than likely that the patient can never come to the surgeon "early enough." Almost inevitably, coronary artery disease will prove once again to be not a surgical disease!

References

1. Day, Stacey B. The Surgical Treatment of Ischaemic Heart Disease: An Experimental and Clinical Study with An Account of the Coronary and Intercoronary Circulation in Man and Animals, The Moynihan Prize Essay of the Association of Surgeons of Great Britain and Ireland, London, 1960.
2. Day, Stacey B. See pages 58–224 (551 References) in *The Idle Thoughts of a Surgical Fellow*, CEP, Canada 1968.
3. Preston, Thomas A. *Coronary Artery Surgery: A Critical Review*, Raven Press, 1977.

Peptic Ulcer Disease

STEPHEN B. VOGEL, M.D.
EDWARD W. HUMPHREY, M.D.
Department of Surgery
University of Minnesota Hospitals
Minneapolis Veterans Administration Hospital
Minneapolis, Minnesota

Peptic ulcers are mucosal ulcerations of the esophagus, stomach or duodenum. The term peptic ulcer disease (PUD) commonly refers to gastroduodenal ulceration and is thought to be related to the digestive action of hydrochloric acid and pepsin contained in gastric juice. In 1965, the mortality from this disease was estimated at over 12,500 deaths with an economic impact on society (including loss of income, disability, drugs and hospitalization) of over one billion dollars. Clearly, PUD represents a major health problem. This review will present several aspects concerning epidemiology, pathogenesis, and recent advances in the medical and surgical management of acid peptic disease.

At the turn of the century, the ratio of duodenal ulcer (DU) to gastric ulcer (GU) was approximately 1:20 but presently this ratio appears to be 10:1 in favor of DU. Simultaneously the peak incidence of DU appears to have passed and a decline in the incidence of both PUD and its complications may be occurring. Numerous factors such as race, climate, nutrition, sex, occupation, and genetic predisposition appear to affect the incidence of peptic ulcer disease. For instance, duodenal ulcers are most common in males, in patients with blood type O, and in relatives of patients with peptic ulcer disease. There are data to suggest a relationship between DU and occupations such as foreman, executives and other decision makers. In contrast, gastric ulcers are more prevalent among unskilled workers of the lower socioeconomic class and occur in an older population with the ratio of male to female of only 2:1 compared to 7:1 for duodenal ulcer. The familial incidence of both types of ulcers are similar and most patients with gastric ulcer are associated with blood group A.

In general, DU patients are hypersecretors of gastric acid. Such ulcerations are extremely rare in any patient with a low rate of acid production. This contrasts sharply with the gastric ulcer where, although acid must be present, ("no acid, no ulcer") these patients are usually hyposecretors of acid. Additionally, GU and gastritis commonly occur following chronic ingestion of aspirin and other antiinflammatory drugs.

The specific cause of peptic ulceration remains unknown. It is likely, however, that they result from the corrosive action of acid and pepsin which may overwhelm the natural defense mechanisms of gastric mucus, gastric inhibitory hormones, and gastroduodenal mucosal resistance.

The physiology of gastric secretion involves a complex neuro-hormonal inter-relationship between the parasympathetic innervation of the stomach (vagus nerve), secretion of the hormone gastrin (from the gastric antrum), and secretion of gastric inhibitory hormones from the intestine. This system controls both stimulation and inhibition of gastric acid secretion during the initial phases of digestion.

Significant advances and changes have occurred in the diagnosis and treatment of PUD involving newer techniques, advances in drug therapy and a more physio-logical approach to surgical management.

Radiologic evaluation of the stomach and duodenum by barium X-ray examination has been and remains a most important diagnostic technique. Recent advances, however, in the use of flexible fiberoptic endoscopes have enabled the gastroenterologist to adequately visualize the entire esophagus, stomach and duodenum with excellent diagnostic accuracy. The decision as to whether a gastric ulcer seen on X-ray is benign or malignant can now be made with upwards of 95% accuracy following endoscopic visualization and biopsy. Perhaps the most valuable use of fiberoptic endoscopy is in the diagnosis of upper gastrointestinal (UGI) hemorrhage. The physician can now detect the specific site of bleeding in over 90% of the cases and can alter and individualize the treatment with accurate knowledge as to the source of hemorrhage. Thus, intensive medical therapy is appropriate initially for the bleeding of diffuse gastritis but a massive hemorrhage from a gastric ulcer will be treated surgically. Finally, intensive investigation is underway using numerous transendoscopic techniques such as lasar photocoagulation and the application of tissue adhesives to control massive UGI bleeding (see section on Gastrointestinal Hemorrhage). In addition to accurate diagnoses, fiberoptic endoscopy will undoubtably play a major role in the treatment of the complications of peptic ulcer disease.

Current medical management of PUD is aimed at neutralizing gastric acid. In most cases this causes prompt relief of pain and promotes healing of the ulcer. Antacid therapy using aluminum and/or magnesium hydroxide compounds ef-fectively buffers acid but there are no data to prove that this therapy can prevent either complications or ulcer recurrence. Hospitalization and bed rest remain the most effective methods to heal an acute and nonresponding peptic ulcer. The role of dietary measures remains controversial, but many clinicians will attest to the value (perhaps placebo effect) of a bland diet and avoidance of foods and other substances such as tobacco, alcohol and caffeine that may cause dyspepsia.

The most promising and exciting development in medical therapy is the recent discovery of a new class of drugs, the histamine H-2 receptor antagonists. Treatment with these experimental drugs dramatically reduces gastric acid secretion by blocking the action of histamine on the gastric parietal cell. Cimetidine, the newest drug of this type, is being evaluated in controlled clinical studies and appears to be effective in relieving pain and promoting healing of the ulcer. Although somewhat premature, it is felt that these drugs may alter the pattern, incidence and complications of peptic ulcer disease during the next decade. Another class of therapeutic agents, the prostaglandins, have been shown to inhibit acid and pepsin secretion and stimulate mucous secretion from the gastric mucosa. In limited clinical trials, they have been shown to relieve pain in peptic ulcer patients, but more extensive and controlled investigation is required.

Surgery is usually performed for the complications of peptic ulcer disease including perforation, hemorrhage, obstruction, or intractability of the ulcer diathesis. Although a wide variety of surgical procedures including subtotal resection of the stomach have been performed in the past, the modern approach centers on altering the vagal and vagal-gastrin mechanisms of gastric acid secretion. Interruption of the parasympathetic pathway to the stomach (vagotomy) decreases maximal acid stimulation by about 60%. Resection of the antrum further reduces acid secretion up to 80-90% by removing the major source of serum gastrin. Vagotomy alone (with a drainage procedure to prevent gastric stasis) or vagotomy with antrectomy (resection of the distal one-third of the stomach) provides cure rates of approximately 90% and 98% respectively. These are the two most common elective operative procedures performed for DU. Certain surgical modifications are employed when confronted with the complications of hemorrhage (suture ligation of the bleeding site) or perforation (closure of the duodenal perforation). A newer and more selective approach to vagal denervation is currently being evaluated. Proximal gastric vagotomy, a procedure which denervates the entire parietal cell area of the stomach, is being evaluated. This new procedure obviates the need for gastric drainage following vagotomy, may be a more effective method for complete vagal denervation and may decrease the incidence and severity of postoperative symptoms. Prospective randomized clinical trials are now evaluating the efficacy of this new operation. When surgical intervention is indicated, gastric ulcer may be adequately treated by antrectomy (to include the ulcer) without vagotomy since these patients are hyposecretors of gastric acid.

Elective surgical procedures for peptic ulcer disease can now be performed with extremely low mortality and morbidity and represent the only chance for permanent cure of this disease. Medical therapy, although adequate in the majority of patients, is not curative. Our present hope for a nonsurgical treatment of PUD is the pharmacological reduction or suppression of gastric acid secretion.

References

1. *Current Concepts in Gastroenterology*, 1(3), October 1976.
   a. Walsh, J. Pathophysiology of gastric and duodenal ulcer, pp. 5–10.
   b. Morrissey, J. F. The use of fiberoptic endoscopy in the diagnosis and management of peptic ulcerative disease, pp. 11–14.
   c. Madsen, J. E. Jr. and Spiro, H. M. Medical therapy: New developments and what's good that's old, pp. 15–24.
   d. Nyhus, L. M. and Donahue, P. E. Surgical management of peptic ulcer disease: Indications and appropriate procedures, pp. 25–30.
2. Eisenberg, M. M. Physiological approach to the surgical management of duodenal ulcer. *Current Problems in Surgery*, Vol. 14, January 1977.
3. Sleisenger, M. H. and Fordtran, J. S. *Gastrointestinal Disease*. Philadelphia: WB Saunders Co., 1973, pp. 718–742.
4. *Principles of Surgery*, Schwartz, S. I., Editor-in-Chief. New York: McGraw-Hill Book Company, 1969.

Gastrointestinal Hemorrhage

STEPHEN B. VOGEL, M.D.
EDWARD W. HUMPHREY, M.D.
Department of Surgery
University of Minnesota Hospitals
Minneapolis Veterans Administration Hospital
Minneapolis, Minnesota

Gastrointestinal hemorrhage is still a common problem to physicians and remains a frustrating and catastrophic disease entity. Bleeding represents the initial symptom in approximately 30% of patients with gastrointestinal disease and often necessitates prompt diagnosis of the bleeding site, and immediate resolution of the problem. A multidisciplinary approach should include a physician, surgeon, and radiologist for adequate diagnosis and treatment. There are over 40 sources of gastrointestinal hemorrhage originating anywhere in the esophagus, stomach, intestine, colon or rectum. A careful history and physical examination of the patient enables the physician to predict the site and source of hemorrhage in many cases, but more elaborate diagnostic methods are necessary in others. A brief discription of the presenting signs and common sources of bleeding follow.

Gastrointestinal bleeding, either acute or chronic or mild to massive presents as *hematemesis* (vomiting of blood or blood-stained gastric juice), *melena* (passage of black, tarry stools), *hematochezia* (passage of bright red blood per rectum), or *occult bleeding* (chemical determination of hemoglobin in normal appearing stool).

Peptic ulcer of the duodenum accounts for $\frac{1}{2}$ to $\frac{2}{3}$ of all cases of upper gastrointestinal bleeding. A history of previous ulcers suggests this diagnosis but it is not unusual for bleeding to be the initial presenting sign. The next most common cause is a diverse group of lesions termed *gastritis*. Ingested agents such as aspirin, alcohol, or antiinflammatory drugs or the stress of trauma, burns, respiratory failure and sepsis strongly suggests gastritis as the source of hemorrhage. Another common source is *esophageal varices*, usually secondary to the portal hypertension of alcoholic cirrhosis. Upper gastrointestinal hemorrhage often presents as hematemesis but the patient having a slow, chronic bleed will present with anemia and occult blood in the stool. *Diverticulitis, cancer of the colon, and ulcerative colitis* represent the common sources of lower intestinal bleeding and usually present with hematochezia.

In treating the patient with massive hemorrhage, techniques of diagnosis and therapy proceed rapidly and simultaneously. During the acute phase, adequate volume replacement and blood transfusions are most important. Following this, and often before a diagnosis is made, other therapeutic measures including iced saline gastric lavage, antacids and intravenous and intragastric instillation of vasopressors are used to control hemorrhage. Each specific bleeding site has evolved its own form of nonoperative therapy but with the results that depend on the source and rate of hemorrhage.

During the past decade, newer diagnostic techniques have enabled clinicians to (1) accurately diagnose the bleeding site and (2) attempt to more selectively control the hemorrahage prior to, or in lieu of surgical intervention. Since 1963 the use of flexible fiberoptic instruments has enabled physicians to adequately visualize the esophagus, stomach, duodenum and colon. Fiberoptic endoscopy, can accurately detect the site of bleeding in over 90% of cases. Intensive research is underway using several transendoscopic modalities to nonoperatively control acute gastrointestinal hemorrhage. Various types of *tissue adhesives* (cyanoacrylates, polyurethane, and epoxy resins) that quickly harden from the liquid state are being evaluated. Their ease in application (via the endoscope) and their ability to coat various types of lesions make this research promising. *Thermal agents* such as electrosurgery, cryosurgery and transendoscopic lasar photocoagulation are being investigated. Additionally, numerous pharmacologic and sclerosing agents are being applied directly on, or injected into specific bleeding sites. It is hoped that the endoscopist will soon have available a safe, effective method of controlling gastrointestinal hemorrhage.

A technique often used following endoscopy is selective arteriography of the celiac, superior mesenteric, or inferior mesenteric arteries and their branches. Diagnostic accuracy of 90% has been reported when bleeding is at a rate of 3 ml/min. or greater. Additionally, the selective infusion of vasoconstrictive agents such as vasopressin and epinephrine has, in selected cases, either arrested bleeding or substantially slowed the rate of hemorrhage.

Finally, a new drug, *Cimetidine*, one of the histamine-H_2-receptor antagonists is currently being evaluated experimentally and clinically in patients with acute hemorrhage from gastritis and peptic ulcer disease. The drug blocks the action of histamine on the stomach and greatly reduces gastric acid secretion. It is hoped that controlled clinical studies will prove the efficacy of this new form of medical therapy for acute hemorrhage.

Surgical management has remained the ultimate method of control of moderate to massive gastrointestinal hemorrhage but with operative mortality rates of 5-25%. This is usually a manifestation of depressed hemodynamic, respiratory and cardiac function in these acutely ill and volume depleted patients.

In treating gastrointestinal bleeding the physician's therapeutic decision is a difficult one at best. He must weigh the risk of surgery against persistance of medical therapy in the face of a deteriorating clinical situation. There can be no greater tragedy than hemorrhage and death from a benign and often curable disease process whether in the operating room or on the medical ward.

Selected References

1. Sleisenger, M. H., Fordtran, J. S. *Gastrointestinal Disease.* Philadelphia: WB Saunders Co., 1973, pp. 718-742.
2. *Principles of Surgery.* (Schwartz, S. I. Ed.-in-Chief.) New York: McGraw-Hill Book Company, 1969.
3. Katon, R. M. Experimental control of gastrointestinal hemorrhage via the endoscope: a new era dawns. *Gastroenterology* 70: 272-277 (1976).

Identification of Increased Susceptibility to Neoplasia in the Gastrointestinal Tract of Man

MARTIN LIPKIN, M.D.
Memorial Sloan-Kettering Cancer Center
New York, New York

Both environmental and genetic factors contribute to the development of gastro-intestinal neoplasms in man. Studies with carcinogenic compounds have suggested that interactions between inherited and environmental elements may be involved in the evolution of the cellular changes leading to neoplastic transformation of gastrointestinal epithelial cells.

In colon and stomach of man, where gastrointestinal neoplasms appear with highest frequency, benign tubular and villous neoplasms are believed by some to be major sources of colorectal cancer in the general population. In adenomatous polyps and villous adenomas increased numbers of nuclei appear in the columnar epithelium lining the glands of these lesions, with atypias and excesses of mitotic figures. There is an increased probability of development of a malignant clone of cells as size increases. Abnormal karyotype with involvement of C and D group chromosomes, as well as heteroploidy have been reported.

A characteristic signaling a heightened susceptibility of some individuals to colorectal neoplasia is the development of multiple colonic neoplasms. In man, the incidence of carcinoma is higher in individuals who have survived a previous cancer of the colon than in the population at large. The subsequent lesions usually appear in the colon and rectum and occasionally in the stomach.

Relatives of patients with colorectal cancer appear to have an increased risk of three times that seen in the general population for developing colonic neoplasia. Individuals under age 40 developing colorectal carcinomas have been reported more likely to have a family history of colon cancer than those over age 40. Multiple primary colonic malignancies show a significantly earlier age of onset of the colonic cancer in relatives than in the population as a whole.

Cancer families also have been described because of the high frequency of primary malignancies in multiple anatomic sites (including colon), early age of onset, and apparent dominant inheritance. At present the inherited disease adenomatosis of the colon and rectum has been the most extensively studied, as this disease shows the best defined expressions of neoplastic transformation.

They lead most strongly to the early warning signal, the widespread formation of adenomatous polyps and to cancer. In this disease events occurring during neoplastic transformation of the colonic epithelial cells provide information related to the etiology and pathogenesis of colorectal cancer. Some of the changes that develop are associated with the appearance of atypical proliferative characteristics in the cells.

Individuals with increased susceptibility to gastrointestinal neoplasia, and also rodents that have been exposed to chemical carcinogens, both show similar changes in growth characteristics of gastrointestinal epithelial cells during neoplastic transformation. During progressive stages of abnormal development, cellular phenotypes appear in which epithelial cells have gained an increased ability to proliferate and to accumulate in the mucosa.

In human colon, in inherited adenomatosis of the colon and rectum (A.C.R.; familial polyposis), progressive phases of abnormal growth appear in colonic eipthelial cells; they develop an increased ability to proliferate and accumulate in the mucosa. An early phenotypic expression identifying a genotypic propensity for cell transformation is a failure of colonic epithelial cells to repress DNA synthesis during migration to the surface of the mucosa; cells then develop properties enabling them to accumulate in the mucosa initiating the formation of neoplastic lesions. Recent studies *in vitro*, also suggest that cutaneous fibroblast and epithelial cells of individuals with A.C.R. develop characteristics of transformed cells with high frequency.

In the diseases of the stomach in man, atrophic gastritis and intestinal metaplasia, increased proliferative indices also develop in gastric mucosa. Epithelial cells fail to repress DNA synthesis during migration through the gastric pits and undergo impaired maturation. Microautoradiographic patterns of DNA, RNA and protein synthesis become abnormal in association with increased pathology.

In intestinal cells of rodents, similar changes have been observed in distal colon of mice following administration of the carcinogen 1,2-dimethylhydrazine. Increased proliferative activity with decreased cell cycle duration also appear as the cells show atypia, fail to repress DNA synthesis and develop an ability to accumulate in colonic mucosa.

Individuals with increased susceptibility to gastrointestinal neoplasia, and those in lower risk categories, are now classified by cell phenotype based on analyses of gastrointestinal and cutaneous cells. These classifications are leading to new predictive indices to identify heightened degrees of susceptibility of individuals at increased risk for colon cancer, the state of development of their disease, and the contribution of environmental elements that modify or accelerate the progression of disease in man. Future programs designed to identify high risk population groups will require systematic analyses of mode and time of appearance of abnormal phenotypic expressions in these individuals and their

families. Analysis of the risk factors together with corresponding study of interactions that may take place with relevant carcinogens, should lead to new means
to identify individuals and population groups at increased risk of gastrointestinal
cancer. Preventive measures can be expected to improve with a consideration of
these factors.

References

1. Anderson, D. E. Genetic varieties of neoplasia. In 23rd Annual Symposium of Fundamental Cancer Research: Genetics Concepts and Neoplasia, University of Texas, M. D.
Anderson Hospital and Tumor Institute at Houston. Baltimore: Williams and Wilkins,
1970, pp. 85–109.
2. Berg, J. S. Geographic pathology of colon cancer. Presentation at the American Cancer
Society's Second National Conference on Cancer of the Colon and Rectum, Bar Harbor,
FL, 27–29, September 1973.
3. Lipkin, Martin. Susceptibility of human population groups to colon cancer. *Advances
in Cancer Research*, Volume 27, 1978, in press.

Relationship Between HLA Type and Frequency of Disease

IRA GREEN

Laboratory of Immunology
National Institute of Allergy and Infectious Diseases
National Institutes of Health
Bethesda, Maryland

It has recently been demonstrated that the frequency of a number of important
human diseases is associated with particular HLA type. This observation is particularly interesting because it is recognized that the HLA gene complex located
on the VI chromosome and its congeners in experimental animals for example
the H-2 gene complex in mice, located on the XVII chromosome, contain a
series of genes which regulate several aspects of the immune response.

HLA antigens of humans and H-2 antigens of mice are cell surface glycoproteins that act as tissue transplantation antigens. They also regulate a variety of
immunological interactions including cell to cell interaction that lead to anti-

body formation and certain aspects of cellular immunity. More specifically the H-2 gene complex of the mouse contains immune response genes that determine the ability of mice to make an immune response to synthetic polyamino acid antigens and low doses of ordinary protein antigens. Furthermore, the susceptibility of mice to the leukemogenic effects of Gross and Friend leukemia viruses was also found to be genetically linked to the H-2 type of the mouse. Partially because of these observations, similar associations were sought and then found between HLA type and the frequency of particular human diseases. There are now at least 40 human diseases that are associated with particular HLA types and the list continues to grow. In some cases the association is so high that the presence of a particular HLA type in an individual patient can be used diagnostically, as for example in the association of HLA B27 with ankylosing spondylitis and Reiter's Syndrome. In other diseases the association is real but not as striking—for example, in juvenile diabetes and psoriasis. Finally there are associations between HLA type and diseases that are of only borderline significance—as for example in acute lymphocytic leukemia and Hodgkins Disease. Examples of other diseases in which there is an association between HLA type and frequency of disease include uveitis, dermatitis herpetiformis, gluten sensitive enteropathy, systemic lupus erythematosus, rheumatoid arthritis, multiple sclerosis, optic neuritis, thyroiditis, and hay fever allergy.

A number of theories have been developed to explain these associations; it is likely that no one theory will suffice. In one particular association one theory will prove correct and in another association another theory will prove correct. Some of the currently fashionable theories include the possibility that the HLA antigens act as receptor sites for viruses, or that particular HLA antigens may closely resemble an etiological agent so that no immune response is generated. Perhaps the most popular hypothesis is that HLA type is closely related to Ir genes or more recently discovered genes in the same complex that suppress immune responses, and that the presence or absence of these genes are related to the presence or absence of immune responses which are involved in a disease process. Despite a great deal of effort none of these hypothesis has yet been established to be correct. The situation is obviously complex and the associations between HLA type and some diseases may depend upon as yet unthought of mechanisms. Also although the associations between HLA type and disease may be high, they are not absolute. The reasons for the latter are several; first the heterogeneity of individual diseases, second, the inability to accurately type for all HLA type; third the influence of other genes segregating independently of HLA; and fourth that the association between disease and HLA is not really to HLA antigens but rather to other genes that only sometimes themselves are linked to HLA.

The unexpected new fact that HLA types are associated with disease susceptibility promises to be one of the most active areas of research. The elucidation of the reasons for such associations will undoubtedly lead to many new insights into the etiology of human disease.

References

1. Green, I. Genetic control of immune responses. *Immunogenetics* **1**: 4 (1974).
2. Ellman, L., Green, I., and Martin, W. J. Histocompatibility genes, immune responsiveness, and leukemia. *Lancet* **1**: 1104 (1970).
3. Vladutiu, A. O. and Rose, N. R. HLA antigens: Association with disease. *Immunogenetics* **1**: 305 (1974).
4. Walford, R. L., Smith, G. S. and Waters, H. Histocompatibility systems and disease states with particular reference to cancer. *Transpl. Reviews* **7**: 78, 1971.
5. McDevitt, H. O. and Benacerraf, B. Genetic control of specific immune responses. *Adv. Immunol.* **11**: 31 (1969).
6. McDevitt, H. O. and Bodmer, W. F. Histocompatibility antigens, immune responsiveness, and susceptibility to disease. *Am. J. Med.* **52**: 1 (1972).
7. Bach, F. H. and VanRood, J. J. The major histocompatibility complex—Genetics and biology. *New Engl. J. Med.* **295**: 806 (1976).

Hormone Receptors

JOHN R. SHEPPARD, Ph.D.
University of Minnesota Medical School
Minneapolis, Minnesota

Biological signals such as hormones elicit physiological responses upon their interaction with highly specific cellular components called hormone receptors. There is a similarity between hormone receptors and enzymes in the respect that the receptor binds the hormone just as an enzyme binds its substrate. However, while the enzyme biochemically modifies the bound substrate, the hormone receptor, after binding the hormone, modifies a second molecular component thereby serving as a communications device.

The language of biological communication at the cellular level is chemical in nature and involves a multicomponent system which includes: a primary signal (e.g., hormone), a discriminator or receptor, a transducer, an effector component (e.g., cyclase) and a secondary signal (e.g., cyclic nucleotide). All five components are present at each level of biological communication and can be imagined to relate to one another at different levels like Russian dolls. Thus the whole organism (no matter how complex) maintains a dynamic communications system among the tissues and cells that compose the organism.

As a component of this communications system at the cellular level, hormone receptors are found at both plasma membrane and cytoplasmic locations. Polypeptide (e.g., insulin) and catecholamine (e.g., adrenalin or epinepherine) hormone receptors are located at the membrane while the steroid hormone (e.g., estrogen) receptors are cytoplasmic. There has been much recent progress in the identification and chemical characterization of these receptors using radiolabeled hormones or related ligands that bind to specific hormone receptors. The biochemical properties of insulin receptors, for example, have been studied using radiolabeled insulin. Isolation and purification of discrete hormone receptor molecules have been accomplished using affinity methods, e.g., the hormone is covalently attached to small inert beads and by adding these beads to a solution containing the receptors, the specific receptor will be preferentially bound to the beads. Prior to these advances, information concerning the hormone receptor's biochemical characteristics (e.g., association constants or kinetics) could only be studied indirectly using an easily measured physiological response to the hormone, e.g., muscle contraction.

A puzzling hormone receptor problem involves the *supersensitivity* of cells to specific hormones after a period of minimal hormone exposure. On the other hand, subsensitivity or *refractoriness* is induced in cells after prolonged or intense exposure to the specific hormone. It appears that the specific hormone receptors mediate these effects by changing qualitatively and quantitatively. The mechanism(s) by which the presence or absence of a biological signal can modify the hormone receptor remains to be determined and is being carefully studied. Furthermore, the mechanism by which the receptor (or discriminator component) is transduced or *coupled* to the effector component is another elusive problem that awaits future clarification. The adenylate and guanylate cyclase enzymes, which are activated via hormone receptors, are examples of effector components and are the subject of another essay in this volume on "Cyclic Nucleotides."

Delayed Type Hypersensitivity (DTH)

BERT ZBAR, Ph.D.
Bethesda, Maryland

DTH is a form of host response. The major characteristic of this response is implied in the name: a response takes at least 24 hours to develop after exposure of the host to foreign material (antigen). Such delayed responses are observed in animals previously exposed to (immunized) to the antigen(s). Response of human beings to intradermal applications of suspensions containing cowpox virus was one of the earliest examples of DTH.

With the identification of mycobacteria as the causative agent of human tuberculosis, a major avenue for study of DTH was opened. Koch injected liquids from cultures of tubercle bacilli into humans previously infected with tubercle bacilli and observed a delayed cutaneous inflammatory reaction (tuberculin reaction). No such reaction occurred in persons not infected with the tubercle bacillus.

Study of DTH reactions revealed characteristics in addition to delayed onset. DTH can be transferred from an immunized donor to an unimmunized recipient with cells obtained from the immune donor. Transfer cannot be achieved with serum from the immunized donor. This failure of serum to transfer DTH distinguishes DTH from antibody mediated immunological reactions. DTH reactions are characterized histologically by a prominent perivascular mononuclear cell infiltrate.

DTH reactions are observed frequently in clinical practice. The cutaneous inflammation in poison ivy, the inflammation associated with certain insect bites, the response to cutaneous application of vaccina virus, the rejection of a kidney hemograft, all are examples of DTH.

DTH reactions are useful to the clinician for detecting exposure of the host to infectious agents and are valuable as a measure of host immune competence.

Current concept of the mechanism for expression of DTH is as follows. The host is exposed to antigen (e.g. tuberculin, poison ivy). Lymphocytes with ability to react specifically with antigen come into contact with it. As a consequence of this interaction the lymphocyte produces a variety of soluble factors (mediators of DTH). These factors serve to amplify the inflammatory response by concentrating a population of macrophages at the antigen site. Vascular permeability is altered and tissue edema develops.

As indicated, the cellular events leading to expression of DTH by the host underlie a broad range of host responses. Investigators are seeking an understanding of the cellular and molecular basis of DTH to obtain control over this important immunological response.

The Immunoglobulins

RICHARD HONG, M.D.
Department of Pediatrics
University of Wisconsin
Madison, WI

The immune system of human defense prevents the establishment of infectious agents and other foreign substances in the body. Two kinds of lymphocytes play the key roles in this protection. They are designated T (thymus derived) or B (bone marrow derived) depending upon their lineage. The B lymphocytes synthesize and secrete proteins known as immunoglobulins (formerly termed gamma globulins).

The B lymphocytes are derived from pluripotential stem cells normally found in the bone marrow, where they arise in common with hematopoietic precursors. In order to respond to antigens, the B lymphocytes must have appropriate antigen recognizing surface receptors. In combination with the other major lymphocyte population, the T cells (helper cells) and the macrophages, antigen first causes a proliferation of the responding B cell population; at this point, the individual is "immunized." Further antigen exposure results in triggering the immune B cell population to synthesize and secrete the induced Ig. Specific antibodies can now be measured in the serum.

In order to fully protect the individual from an infectious agent, the secreted antibody must be able to activate the nine component complement system. As the various components are sequentially activated, leucocyte chemotaxis, increased vascular permeability, enhanced phagocytosis occur; finally particle lysis results in death of the agent. Without the complement system, antibody coating has minimal biological effect and cannot contain a rapidly replicating infectious organism.

The basic structure of the immunoglobulins has been known for some time and correlations of structure and function well defined. These features are described and illustrated in Table 1 and Fig. 1.

There are five major classes of immunoglobulins (Igs) in man, known as IgG, IgA, IgM, IgD and IgE. The general roles of Igs in human biology are listed in Table 2. The individual immunoglobulins subserve different functions.

IgM is a very large (1,000,000 M.W.) protein composed of 5 identical subunits. Because of its great dimensions, it is very efficient as an agglutinator, in enhancing phagocytosis, and in complement fixation. It appears first after antigen exposure and is mostly distributed in the intravascular compartment. These characteristics make it an ideal first line of defense ready for rapid and immediate immobilization of intravascular infections. Patients with selective IgM deficiency characteristically demonstrate a propensity to succumb to rapidly developing

TABLE 1. Nomenclature and Size of Ig's.

MONOMERS

POLYPEPTIDE CHAINS

Heavy Chain	Light Chain	Formula for Whole Molecule	Name	MW
γ	κ	$\gamma_2 \kappa_2$	IgG, γG globulin	140,000
	λ	$\gamma_2 \lambda_2$		
α	κ	$\alpha_2 \kappa_2$	IgA, γA globulin	160,000
	λ	$\alpha_2 \lambda_2$		
δ	κ	$\delta_2 \kappa_2$	IgD, γD globulin	16,000
	λ	$\delta_2 \lambda_2$		
ϵ	κ	$\epsilon_2 \kappa_2$	IgE, γE globulin	197,000
	λ	$\epsilon_2 \lambda_2$		

POLYMERS

POLYPEPTIDE CHAINS

Heavy Chain	Light Chain	Other	Formula for Whole Molecule	Name	MW
α	κ	SC, J	$(\alpha_2 \kappa_2)_2$ SC.J	Secretory IgA	370,000
	λ	SC, J	$(\alpha_2 \lambda_2)_2$ SC.J		
μ	κ	J	$(\mu_2 \kappa_2)_5$ .J	IgM or γM globulin	900,000
	λ	J	$(\mu_2 \lambda_2)_5$ .J		

From: Reed, C. E. ed., *Allergy: Principles and Practice*, St. Louis: C. V. Mosby Company, 1978. With permission of the publishers.

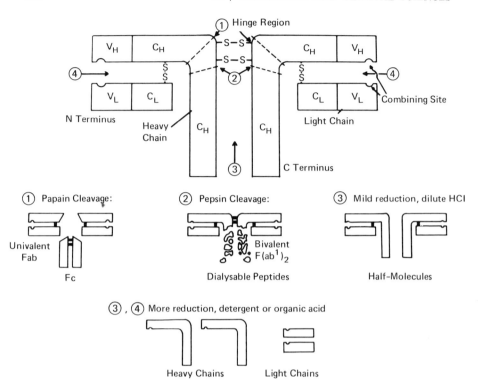

FIG. 1. Basic structure of Ig's. About $\frac{3}{4}$ of the heavy chain and $\frac{1}{2}$ of the light chain is very similar from molecule to molecule and is referred to as a constant (C) region. The remainder of each chain is very different and constitutes the variable (V) region. The antibody combining site is found in the variable region. Special characteristics of Ig classes (e.g., half-life, placental passage) are properties of the constant regions of the heavy chains (CH regions). A number of biologically active fragments can be prepared by treatment of the intact Ig molecule with various chemical reagents. The treatment and resultant products are shown above. The numbers in the upper figure show the site of cleavage by the various treatments and can be correlated with the numbers in the lower portion of the figure to see the resultant product. In general, the Fc portion (composed of portions of the CH regions) relates to nonantibody binding activities which are related to enhancement of the immune reaction (e.g., complement activation). The Fc portion also provides the structure for differentiation of the major classes—IgG from IgA, etc. The Fab portions of all Ig classes are essentially the same, except for the variable region. (Figure reprinted from, Reed, C. E. ed., *Allergy: Principles and Practice*, St. Louis: C. V. Mosby Company, 1978. With permission of the publishers.)

TABLE 2. Role of Igs in Human Biology.

- Prevention of viral infection (Rubeola, varicella, infectious hepatitis)
- Prevention of bacterial infection (staph, strep, pneumo, H. flu)
- Antibody-dependent cell mediated cytotoxicity
- Allergic (IgE) reactions
- Antigen-antibody complex disease
- Direct cytolysis
- Blocking factors in tumor immunology
- Blocking antibodies in allergic desensitization

From: Horowitz, S. and Hong, R., *The Pathogenesis and Treatment of Immunodeficiency,* Basel, S. Karger (1977, p. 5). With permission of the publishers.

overwhelming sepsis. The early response of IgM to infection has been utilized in certain diagnostic situations. For example, neonatal or intrauterine infection is inferred from elevated IgM levels in cord blood. Also specific IgM antibody directed against an infectious agent (e.g., rubella) is taken as evidence for newly acquired infection.

IgA is located primarily in the lamina propria underlying the epithelium of secretory surfaces (gastrointestinal tract, nasopharynx). It is also a polymer, composed of two identical subunits. When found in the secretions, IgA is attached to another protein molecule, a product of epithelial cells, termed secretory component (SC), forming secretory IgA or S-IgA. SC may help IgA to function adequately in the special areas where it must operate. The IgA molecule apparently coats the secretory epithelial surface on the luminal side. Here it prevents bacterial invasion by acting as a mechanical barrier to the attachment of the organism. In order to function in areas such as the lumen of the gastrointestinal tract, IgA must resist proteolytic digestion and firmly attach to the luminal coat. SC may provide the structural adaptation to permit these specialized needs. SC may also act as a receptor molecule which actually moves the IgA from the area of the lamina propria through the epithelial cell and onto the luminal surface (transport function). In situations of IgA deficiency, IgM or IgG can sometimes compensate for the lack. IgA is also found in the serum but only as a minor Ig constituent (10% of total Igs). In the serum, it is usually a monomer and does not contain SC. S-IgA is found only in trace amounts in the serum; its level may be increased when there is undue stimulation in secretory areas. In this latter event, S-IgA in the serum may simply represent spillover from the original site of synthesis. Ordinarily, serum IgA response is independent of secretory IgA response.

IgG is localized equally to extravascular as well as intravascular compartments. In the serum, it represents the major Ig (85% of total Ig). It is the only Ig which crosses the placenta to provide passive immunity to the fetus. Its long half life makes it appropriate for this purpose. IgG is the major response in a recall or second exposure to an antigen.

IgE has the capability of combining with surface receptors on most cells causing discharge of the pharamacologically active substances within, giving rise to symptoms of hay fever or asthma. In other situations, release of platelet aggregation factor (PAF) and histamine secondary to IgE surface triggering can prepare tissues for subsequent antigen antibody complex assault.

IgD may have little or no role as a circulating immunoglobulin. Its major function seems to be expressed as a surface receptor of the lymphocyte. Here it may play an important role in modulating the initial immunization process leading to responsiveness (immunity) or nonresponsiveness (tolerance). The biological characteristics of the Ig's are summarized in Table 3.

Igs are assessed qualitatively and quantitatively. Most deficiencies of the Ig's are easily detected as a marked diminution of the serum levels. The most serious forms of deficiency involve all three major classes-panhypogammaglobulinemia. Selective deficiencies are known of which the most common involves IgA. The proteins are readily measured by a technique known as single radial diffusion. As mentioned previously, IgA exists in secretions as well as serum. Direct measurement of secretions must be performed to be absolutely certain that selective deficiency of secretory IgA in the face of normal serum IgA does not exist. This type of deficiency is extremely rare however and serum measurements will detect nearly all IgA deficiency states.

Qualitative deficiency, i.e., a situation of completely normal levels of Ig without any known antigen combining function has not been unequivocally demonstrated in man. A form of qualitative defect is seen in B cell malignancy. Since a given B cell produces only one antibody and since the malignancy involves only one cell and its descendants, in multiple myeloma or chronic lymphatic leukemia a very homogeneous Ig is produced. As is common in cancer, the malignant expansion occurs at the expense of the noninvolved cell populations. Thus, with time all other antibody producing B cells are shut off and a de facto qualitative deficiency exists. The myeloma protein, although a specific antibody, has only a single specificity and the heterogeneous array of binding capability characteristic of normal Ig's has been suppressed. The laboratory characteristics of these disorders are: (1) homogeneous electrophoretic behavior of the protein resulting in "spikes"; (2) reactivity with only one of the major class of light chain antibody reagents (See Fig. 1); (3) marked diminution to absence of the noninvolved Ig's; and (4) unusually high levels of the involved Ig, suggesting loss of appropriate control mechanisms.

Another situation of "qualitative" deficiency in the face of normal to elevated serum levels of Ig is found with subgroup deficiency of IgG. There are four

TABLE 3. Biological Properties of Human Immunoglobulins.

Property	IgG 1	IgG 2	IgG 3	IgG 4	IgA 1	IgA 2	IgM	IgD	IgE
Serum level (mg/ml)	5-12	2-6	0.5-1	0.2-1	0.5-2	0.2	0.5-1.5	0-0.4	0-0.002
Localization in secretions	−	−	−	−	+	++	±	−	±
Half-life (days)	23	23	16	23	6	6	5	3	2
Placental transfer	+	+	+	+	−	−	−	−	−
Activation of C by classic pathway	++	+	++	−	−	−	+	−	−
Activation of C by alternate pathway	−	−	−	−	+	+	−	±	±
Intravascular/extravascular ratio		<1			<1		>1	>1	

++ Strong activity
+ Moderate activity
− No activity

From: Reed, C. E. ed., Allergy: *Principles and Practice*, St. Louis; C. V. Mosby Company, 1978. With permission of the publishers.

groups of IgG, 1-4, which although grossly very similar, show slight antigenic differences. These minor structural variations are associated with differences in biological capability. For example, IgG 1 and 3 activate complement while IgG 2 and 4 do not. Deficiencies of only one of the subgroups can be associated with infectious susceptibility although total IgG levels are normal.

Qualitative assessment can be performed by measuring for specific antibodies. One can utilize antigens to which the individual has been naturally exposed, e.g., blood group substances, bacterial antigens (streptolysin), antigens received during routine immunizations, or the patient can be actively immunized and his antibody response measured.

The body is constantly assaulted by organisms. Our survival as a species has required the evolution of a series of defense systems of amazing capability and complexity. The immunoglobulin system is a fascinating prototype.

References

1. *Basic and Clinical Immunology* (Fudenberg, H. H. et al., eds.) Los Altos, California: Lange Medical Publications, 1976.
2. *Clinical Immunobiology* (Bach, F. H. and Good, R. A. eds.), Vol. 1, New York, Academic Press, 1972, pp. 29–46.
3. Roitt, J, *Essential Immunology*, London, Blackwell, 1974.

The Immune System in Senescence

DIEGO SEGRE
College of Veterinary Medicine
University of Illinois
Urbana, Illinois

Age-related changes in immune function have been described both in man and in experimental animals. The association between immune disfunction and senescence has been observed with such frequency that a cause-effect relation between the two has often been assumed. Whether aging causes a decline in immune function or is itself the result of a malfunctioning immune system, however, has not been resolved. Both views have authoritative supporters.

Vigorous experimental research in recent years has brought about an increased understanding of the immune system. Thus, a reexamination of the relations of immune function to age is desirable. While it has long been known that, in general terms, immune activity is depressed in aged individuals, only recently has it become possible to study the effects of aging on the individual cellular components of the immune system.

Before describing the effects of aging on immunity, it is useful to attempt to define, if not aging itself, at least the aged animal that may properly be used for immunologic investigations. Aging is a continuous process, which proceeds at different rates in different individuals. For each individual the duration of life depends on the interaction of two variables: the effects of the environment and the intrinsic changes that occur during aging within a living system. Ideally, the effects of aging on immunity should be studied on a population of genetically identical individuals kept in a uniform environment. Under these conditions all individuals would presumably reach the same maximum age. In practice these

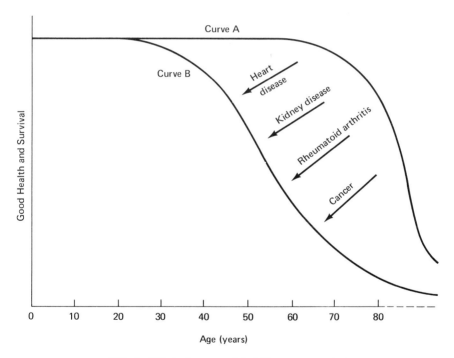

FIG. 1. Effect of age-associated diseases on survival.

Curve A: Persons who maintain health and vigor well into old age (14% of the population)
Curve B: Persons who succumb to infirmity and early death through age-associated diseases (86% of the population)

conditions are never attained; while mice of inbred strains approach genetic uniformity, environmental variations cannot be eliminated. Thus, chronological age cannot be equated with physiological age. It would be most useful if a simple indicator of physiological age were found, capable of measuring the portion of an individual's lifespan that has elapsed at any given time. In the absence of such an index of individual age, immunologic studies should be conducted on populations of genetically homogeneous individuals, such as inbred mice, which have reached an arbitrarily defined old age. The mean lifespan, that is the age at which 50% of the individuals in a population are expected to be dead, is an acceptable, if arbitrary, age at which the experimental animals may be considered old. In choosing a strain of mice for gerontological research, one must keep in mind that mice of long-lived strains are more apt to provide information relevant to the aging process itself; mice of short-lived strains, in contrast, are likely to develop pathological processes which may influence the variable being studied (e.g., immune function). For example, NZB mice, which develop autoimmune disease and succumb to it at a relatively early age, are not appropriate subjects for an investigation of the effects of aging on the immune function.

Age-related changes in humoral immunity have been studied more thoroughly than changes in cell-mediated immunity, in part because humoral immunity can be more easily quantitated. In general, the quantity of antibody produced in response to antigenic stimulation reaches a peak in young-adult individuals and declines gradually thereafter. For example, the primary response of mice of varying ages to heterologous erythrocytes peaks at 16 to 20 weeks of age, then declines gradually throughout the life of the mice. At 120 weeks of age, corresponding to the mean lifespan for the mice used in this investigation, the response was only 5% of the peak response. Transfer experiments, in which lymphoid cells from young mice were transferred to irradiated old mice, revealed that the age-related decline of immune potential was largely an intrinsic property of the cells, although the environment (young or old recipient mice) in which the cells were functioning played a minor role in determining the extent of the antibody response.

Three major cell types are involved in the humoral immune response: the B lymphocytes, which include the precursors of antibody-forming cells; the T lymphocytes, which must interact with B lymphocytes in order for the latter to respond to most antigens; and the macrophages or accessory cells, which play a role in the processing of antigen as well as in the B lymphocyte–T lymphocyte interaction. Any one of these cell types may, in principle, become the limiting factor that determines the level of immune responsiveness of the whole animal. Thus, it is profitable to inquire how the function of each cell type varies with advancing age.

Not much is known about possible age-related changes in macrophage function. Some studies indicate an increase in phagocytic activity of macrophages from old mice, as compared to macrophages from young mice. It is not clear, however, whether the phagocytic activity of the macrophages is related to their role in the initiation of the immune response. No information is available as to the integrity of the latter function in macrophages of old animals.

The frequency of B lymphocytes specific for a T-independent lipopolysaccharide antigen was found to be 2.5 times lesser in the spleen of old mice than in the spleen of young mice. No age-related changes were found in the frequency of B lymphocytes specific for foreign erythrocytes, a T dependent antigen, in the bone marrow of young and old mice. In another study, the ability of spleen cells from young and old mice to generate antibody forming cells in culture was investigated. In the presence of an excess of specifically activated T lymphocytes from young mice, the old spleen cells generated 2.5 times fewer antibody-forming cells than did spleen cells from young mice. Thus, although an age-related decrease in number and/or function of B lymphocytes has at times been found, it is not profound enough to account, of itself, for the marked reduction of immune potential exhibited by old mice.

The performance of B lymphocytes as a function of age has also been studied by measuring other parameters. In one investigation no difference was found between spleen cells from young and old mice in the kinetics of the secondary response to a hapten in a transfer system, nor in the antigen dose-response curve; yet, the number of antibody-forming cells in the spleen of old mice was profoundly depressed. A significant age-related reduction in the average avidity of antibody for a hapten has also been reported. Perhaps related to this latter finding is a report indicating that 10 times more tolerogen is required to induce B cell immunologic tolerance in old mice than in young mice. A decrease in the avidity of the immunoglobulin receptors of B lymphocytes would result in a decrease in the ability of these cells to bind the tolerogen, which would be compensated by increasing the tolerogen concentration.

The frequency of T-lymphocytes capable of interacting with B lymphocytes in the response to foreign erythrocytes (helper T cells) was found to be twice as great in the spleen of old mice than in the spleen of young mice. In another investigation it was found that the bone marrow of old mice contained increased numbers of T lymphocytes. Another report indicates that the same amount of tolerogen will induce T lymphocyte tolerance in both young and old mice. Thus, no obvious age-related T lymphocyte deficiency emerged from these studies that could account for the decline of humoral immune responsiveness with advancing age.

A recent report, however, indicated that immunization of old mice results in the development of a subpopulation of T lymphocytes capable of specifically suppressing the humoral immune response (suppressor T cells). In these experiments, spleen cells from immunized aged mice were transferred to irradiated recipient mice together with spleen cells from young mice immunized with the same antigen. Under these conditions, the secondary response of the young spleen cells was markedly depressed. Removal of T lymphocytes from the old mouse spleen cells abrogated the immune suppression. There was a negative correlation between the degree of immune suppression exerted by the old cells and the magnitude of the secondary response of the old cells themselves. Thus, there is evidence that excessive development of suppressor T lymphocytes following immunization is responsible, at least in part, for the decline of humoral immune potential observed in aged mice.

Further studies with this system revealed that it was often possible to reconstitute the immune potential of cells from old mice by removing the T lymphocytes from a spleen cell suspension and supplementing it with helper T lymphocytes from young mice. The restoration of the immune response under these conditions indicated that the B cells from the old mice were still functional and capable of mounting a vigorous immune response but were prevented from doing so by the suppressor T cells. An attempt was then made to analyze the lymphocyte function of aged mice. The results suggested that an increase in the number or activity of suppressor T cells is the first immunologic lesion of old age. This is followed by a deficiency in the helper T cell function and finally by loss of B cell function.

While the B lymphocytes, with the collaboration of helper T lymphocytes, are responsible for humoral immunity, the cell-mediated immune response is a T lymphocyte function. T lymphocyte activity may be measured in several ways; some of the tests provide a general measure of T lymphocyte function or activity, others are more directly related to T lymphocyte immune potential. Proliferation of T lymphocytes in response to mitogens has been found to decline with advancing age in parallel with the decline of immune potential. A more directly immunological measure, the mixed lymphocyte reaction, which corresponds to the recognition phase of cell-mediated immunity, has been found to decline with age by some investigators. Others found no age-related changes in mixed lymphocyte reactivity, although the mice used in this investigation should probably be classified as mature adults rather than old. In the latter investigation an age-related decline in graft-vs-host reactivity was also observed, but there was no decrease in the ability of "old" mice to become sensitized to dinitrofluorobenzene, a contact sensitivity-inducing chemical.

More experimental work will be necessary to clarify the relations between aging and lymphocyte function. The general picture that emerges from the experiments conducted with long-lived inbred mice is one of gradual decline of both humoral and cell-mediated immunity with advancing age. The T lymphocyte, which is responsible for cell-mediated immune reactions as well as for the regulation of both cell-mediated and humoral immunity, appears to be primarily involved in the age-related decline of immune potential.

Biochemical Structure of Laetrile

Laetrile* is a name that is applicable to at least three, and probably more, compounds belonging to a class of naturally occurring chemicals known as cyanogenic glycosides. As their name implies, these compounds can be hydrolyzed enzymatically or with hot mineral acids to yield a mole of hydrogen cyanide in addition to a saccharide characteristic of the glycoside in question.

The current controversy surrounds three particular cyanogenic glycosides, the structural formulae of which are given in Fig. 1 (A–C). Compound A is amygdalin (L-mandelonitrile β-D-gentiobioside) and is the material being currently identified as laetrile. The molecule consists of two glucoses joined by a β 1-6 linkage (gentiobiose) plus L-mandelonitrile. Amygdalin is found, along with emulsin, in bitter almonds (*Prunus amygdalus*) and in the seeds of the *Rosaceae* (plum, cherry, apricot, apple, etc.).

Compound B in Fig. 1 is prunasin (L-mandelontrile β-D-glucoside). It is found in the bark of wild cherry (*Prunus serotina*) in small quantities and is generally prepared synthetically by the oxidative removal of the first glucose moiety of amygdalin. Compound C, L-mandelonitrile β-D-glucuronoside, is not known to occur in nature but Krebs claimed to have synthesized it from amygdalin.[1]

In addition to amygdalin and prunasin, at least twelve other cyanogenic glycosides occur in nature and these are widely distributed in over 800 plants, usually in association with a glycolytic enzyme capable of completely hydrolyzing the particular glycoside to its constituent chemical parts.[2,3] The natural function of these widespread chemicals is uncertain; it may be that they protect

---

*Laetrile is a contraction of Laevorotatory mandelonitrile.

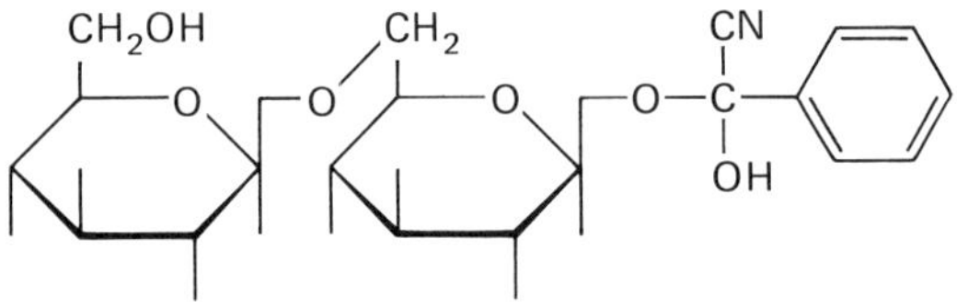

A. Amygdalin (L-Mandelonitrile β-D-gentiobioside)

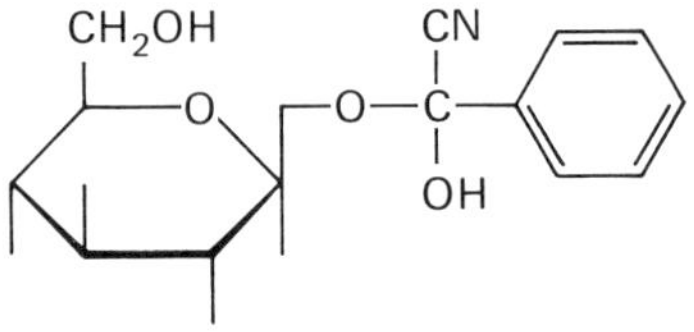

B. Prunasin (L-Mandelonitrile β D-glucose)

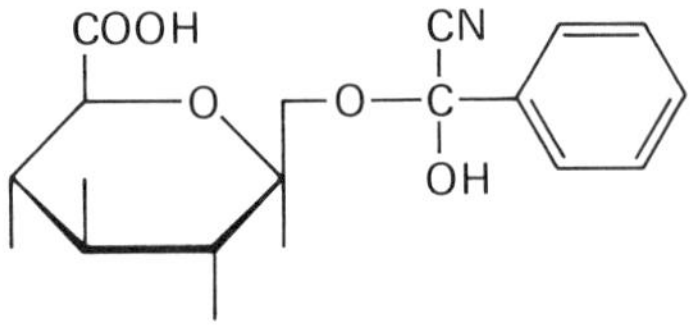

C. L-Mandelonitrile β-D-glucuronoside

Fig. 1. A. Amygdalin (L-Mandelonitrile β-D-gentiobioside)
        B. Prunasin (L-Mandelonitrile β-D-glucose)
        C. L-Mandelonitrile β-D-glucuronoside

the plant or germinating seedling from microorganisms and insects through the release of cyanide.

In the late 1940's, Krebs advanced the hypothesis that laetrile which was then identified as Compound C was selectively active against cancer cells by virtue of the fact that such cells had elevated levels of the primarily lysosomal enzyme β-glucuronidase which he claimed hydrolyzed this glucuronoside to produce cytotoxic cyanide. Further, he claimed that unlike neoplastic cells, normal uninvolved tissues were doubly protected from such cytotoxicity by having relatively little β-glucuronidase while having sufficient amounts of rhodanese, an enzyme which can detoxify any cyanide that is produced by conjugating it with sulfur to yield thiocyanate. Elevated levels of β-glucuronidase had in fact been observed in malignant tissues in comparison to normal or uninvolved adjacent tissues by

Fishman and Anlyan[4] and rhodanese was known to be present in a variety of normal tissues, the liver having the most abundant quantities[5]. The demonstration that all known samples of "laetrile" were actually amygdalin instead of mandelonitrile glucuronoside cast the whole proposal into doubt since *β-glucosidase* rather than *β-glucuronidase* would be required to cleave amygdalin for cyanide release at the tumor site.

References

1. Krebs, E. T., Jr. Brit. Pat. 788,855 and U.S. Pat. 2,985,664.
2. Armstrong, E. F. and Armstrong, K. F. *The Glycosides*. London: Longmans, Green and Co., 1931, pp. 67–72.
3. McIlroy, R. J. *The Plant Glycosides*. London: Edward Arnold & Co., 1950, pp. 20–24.
4. Fishman, W. H. and Anlyan, A. J. *Science* 105: 646, 1947; *Cancer Res.* 7: 807, 1947.
5. Rosenthal, O. The distribution of rhodanese. *Fed. Proc.* 7: 181, 1948.

Current Status of Disorders of the Lung

JOHN E. MAYER, JR., M.D.
EDWARD W. HUMPHREY, M.D., Ph.D.
Department of Surgery
University of Minnesota Hospitals
Minneapolis Veterans Administration Hospital
Minneapolis, Minnesota

The lung is vital to the survival of all complex terrestrial organisms including man. The understanding of the normal and abnormal function of this organ has been significantly advanced in recent years. The purpose of this discussion is to review pertinent aspects of its function as well as major disease processes which affect this function.

Blood flows within 0.1 micron of the alveolar gas through capillaries, and gas exchange occurs by diffusion across the alveolar-capillary membrane.[6] Blood flow ($\dot{Q}$) and ventilation ($\dot{V}_A$) must be closely matched to maintain the normal concentrations of oxygen and carbon dioxide in the arterial blood. The mechanisms by which this matching occurs are incompletely understood at the microscopic level but the effects of gravity, the concentrations of oxygen and carbon dioxide in blood and alveolar gas, and the pH of the blood are important.[25] Several recent studies have suggested that "local hormones" such as prostaglandins, histamine, or catecholamines may act as mediators of this local control.[2] Techniques using intravenous and inhaled radioactive gases such as X_e^{133} with multiple external gamma counters have allowed studies of ventilation and perfusion on intact humans at the macroscopic level.[25]

The major pulmonary disease confronting man in civilized industrial societies is a complex of disorders termed chronic obstructive pulmonary disease. This complex is actually three separate disorders: chronic bronchitis, emphysema, and asthma, all of which may coexist in a given patient.[15] These disorders share a common functional feature of increased resistance to air flow, especially during exhalation. There is an increased work of breathing from the increase in airway resistance and there are major disorders of $\dot{V}_A/\dot{Q}$ matching. The $\dot{V}_A/\dot{Q}$ mismatching results from uneven regional ventilation and possibly increased diffusion distances as lung substance is destroyed. Because of the large functional reserve of the lung, significant anatomic changes may occur without functional impairment to the organism. However, sensitive tests of pulmonary physiology allow the detection of these abnormalities at stages before major functional impairment has occurred and offer hope for the identification of individuals at high risk from this disorder. Continuous measurement of the nitrogen tension in expired gas gives an index on unevenness of regional ventilation since poorly ventilated alveoli have higher nitrogen concentrations and overventilated (underperfused) alveoli have low nitrogen concentrations. The change in compliance at fast rates of breathing which occurs in these disorders can also be measured.[7]

Chronic obstructive lung disease has been clearly shown to depend partially on environmental factors.[15] The mortality rate of heavy cigarette smokers from emphysema is 20 times greater than in nonsmokers and chronic bronchitis has only been found in cigarette smokers. Why cigarette smoking should cause these changes has not been clearly defined, but studies have shown decreased motion of the cilia, tiny hairlike projections in the respiratory tract which normally clear particles from the respiratory tree. Smoking has also been shown to cause overgrowth of glands in the lining of the airways (as in chronic bronchitis), and emphysema has been produced in cigarette smoking dogs. Because a higher incidence of chronic lung disease is found in urban areas with air pollution, this factor is probably also significant. Increased rates of this disorder are noted in coal, steel and flax workers.[3] These changes are difficult to separate, however, from the effects of cigarette smoking.

Certain patients have an early onset of a severe form of emphysema that is associated with a genetic deficiency of the enzyme α-1-antitrypsin. This substance, which is normally present in blood, inhibits the action of many proteolytic enzymes also present in the blood and tissues. It is postulated that the lack of this inhibitor results in destruction of lung tissue. Studies to date have shown this disorder to occur only in patients homozygous for the deficiency. Studies are in progress to determine if heterozygotes are also at greater risk of developing emphysema.[13] A second congenital disorder affecting the lung is cystic fibrosis in which the mucous produced in the lung has a markedly increased viscosity leading to increased susceptibility to infection. Survival rates in this disorder have improved through the use of antibiotics and physiotherapy.[22]

The treatment of these chronic obstructive disorders is limited at the present time to mechanical respirators, supplemental oxygen, and antibiotics to help combat infection.[15] All efforts are hampered by the underlying irreversible anatomical derangements in the lung. A possible area for future advance lies in lung transplantation.

Two acute types of respiratory failure are the respiratory distress syndrome of the newborn, and post-traumatic pulmonary insufficiency in adults. In the newborn syndrome, there is a failure of production of surfactant by alveolar type 2 cells. The surfactant normally offsets the increased surface tension predicted by the Law of LaPlace when radius decreases. Without surfactant those alveoli with small radii tend to collapse, while those of larger radius tend to over-expand, causing decreased compliance of the lung and major disorders of $\dot{V}_A/\dot{Q}$ matching. Several advances have been made in the prenatal diagnosis of this disorder and its treatment postnatally. Sample of the fluid surrounding the infant in utero have demonstrated that a low ratio of lecithin to sphingomyelin are associated with a high risk of its development.[8,9] The administration of certain steroids has been shown in experimental preparations to increase the production of surfactant and attempts are in progress to apply these experimental findings clinically. In the infant with developed respiratory distress syndrome, major treatment advances have been made by the use of continuous positive pressure ventilation to keep the airway pressure positive and so prevent the collapse of alveoli.[8]

A similar syndrome occurs in adults after severe trauma, infection or shock from other causes, although the exact mechanisms are still unclear. An excess accumulation of fluid in the lung occurs, leading to failure of $\dot{V}_A/\dot{Q}$ matching. Experimental evidence has suggested multiple blood clot emboli to the lung, over-activity of the sympathetic nervous system, or a circulating toxin as possible sources of the deranged lung function.[10,11] The significance of each of these factors remains under investigation. Current therapy employs the use of mechanical respirators and continuous positive pressure ventilation as in the newborns.[5,11] There have been some therapeutic trials with an artificial lung in attempts to tide the patient over the acute injury and allow reparative processes to operate. These trials have been characterized by problems of long term anti-coagulation and those caused by the interface between blood and nonbiological surfaces.

Infectious processes are responsible for a large percentage of all lung disease and are caused by a variety of organisms. The prevention and treatment of viral infections are hampered by the difficulties in growing the viruses outside the host. No drugs currently in use have been clearly shown to affect the course of viral disease although research efforts continue. Viruses are thought to play a role in the etiology of bronchiolitis in infants. They have assumed increased importance in patients with renal or cardiac transplants who have a decreased ability to deal with infection because of the necessary suppression of the immune system. The usually mild fungal infections of the lung have also assumed increasing importance in these patients.[4]

Tuberculosis, although decreased in frequency, still represents a major infectious disease of the lung. Specific drug therapy is available for most cases of tuberculosis and is responsible for a marked lowering in the mortality rate and the incidence of this disease as man is the only host for *mycobacteria tuberculosis*. Certain strains of tuberculous organisms such as the Battey strain are very resistant to all chemotherapeutic agents, and in these patients surgical removal of diseased areas is utilized.[14]

Antibiotics have had a similar impact on bacterial infection of the lung. The incidence of severe complications of bacterial infections such as chronic pockets of infection within the lung or chest cavity and destruction and dilatation of the airways by infection have been markedly reduced.[18] However, the massive use of antibiotics, especially in hospitals, has led to the evolution and emergence of strains of bacteria which are highly resistant to all antibiotics. Interestingly, epidemics of these resistant bacteria have been controlled by reducing the use of antibiotics which affect the normal bacterial flora. In this way the selective advantage favoring the highly resistant bacteria is removed and the epidemic can be controlled.[20]

A variety of environmental agents have been cited as damaging to the lung. As noted earlier, cigarette smoking has been linked to the development of emphysema. Air pollution also has been thought to be damaging to the lungs although the exact etiologic agents have not been identified. In both children and adults, however, there is an increased incidence of respiratory disease in urban areas with high levels of air pollution. Certain industrial exposures have also been associated with respiratory disease. Ammonia, nitrous oxide, and halogen fumes, and aerosolized heavy metals have all been cited as damaging. In coal workers a syndrome known as "black lung" occurs, probably resulting from chronic exposure to graphite and silica.[3] Agricultural workers may develop "farmer's lung" from an allergic response to inhaled *Micropholyspora faeni*, a mold which forms in silage.[19] A similar picture is seen with the use of certain drugs. Chronic exposure to asbestos has been associated with the development of cancer of the pleura, diffuse scarring in the lung, and may act synergistically with cigarettes to increase the risk of lung cancer.[3] Boeck's sarcoid is a pulmonary disease which remains an enigma. The etiology is unknown although infection with nontuberculous *Mycoplasma*, immune abnormalities, or infection with an unknown agent have been suggested.[16]

Cancer of the lung is found in increasing numbers in all civilized societies and is the most common cause of death from cancer in males. The association with cigarette smoking is clear-cut as the incidence in nonsmokers is 3.4 per 100,000 while in those smoking more than 40 cigarettes a day, it is 217 per 100,000.[1] In untreated cases, the average survival is less than one year after the time of discovery. The overall survival with all forms is approximately 6% at five years,[1,12] and the chance of survival has not been significantly improved over the past 20 years. The only means of therapy which offers much hope for long-term cure is surgical removal of the cancerous tissue, but less than half of the patients are diagnosed prior to distant spread of the disease and can be offered surgical therapy.[12] The diagnosis of lung cancer has been aided by the recent development of fiberoptic instruments which allow more peripheral areas of the lung to be visualized directly.[21] Of those patients operated upon for lung cancer only $\frac{2}{3}$ can have their cancers removed and of those who do have tumors removed, only 25% survive for five years.[12] The importance of the stage of the disease is clear as the survival in those who have evidence of spread to lymph nodes near the tumor is worse than in those who do not. In spite of these data, attempts to find cancers at earlier stages in high risk patients by the use of frequent chest X-rays has not resulted in improved survival.[23] Radiation therapy has produced a few long term survivors but currently is used principally for palliation of certain local complications of cancer of the lung. Various drugs have been used in attempts to palliate those with advanced disease or to attempt to improve the cure rates in those undergoing the surgical removal of cancer tissue. To date, no improvement in the survival has been seen in either of these groups with the exception of a slight prolongation of survival in those with small cell carcinoma. New directions in therapy include attempts to enhance the patient's immune response to his tumor through the use of nonspecific stimulating agents such as BCG.[24]

With the high incidence of respiratory failure secondary to chronic lung disease, there has been interest in lung transplantation. Success with transplantation of kidneys has added impetus to these attempts. As of January 1, 1975, 36 patients had undergone transplantation of part or all of one lung, but the longest survivor lived for only 10 months.[10] Several major problems have restricted the use of transplantation of the lung. The rejection of foreign tissue is common to all organ transplantation. Methods for artificial support of lung function are being developed, but cannot currently maintain patients for long periods while they are awaiting a suitable donor to become available. Infection is a particular problem because the lung is continuously exposed to infectious agents in the inspired air. The use of transplantation remains an intuitively attractive treatment method but many problems remain to be solved.

References

1. Adkins, P. C. Neoplasms of the lung. In *Surgery of the Chest*. (D. G. Sabiston, ed.) Philadelphia: W. B. Saunders, 1976, p. 443.
2. Bergofsky, E. H. Mechanisms underlying vasomoter regulation of regulation of regional pulmonary blood flow in normal and disease states. *Am. J. Med.* 57: 378 (1974).
3. Bouhys, A. and Peters, J. M. Control of environmental lung disease. *New Engl. J. Med.* 283: 573 (1970).
4. Brode, F. R. Pulmonary diseases in the compromised host: A review of the clinical and roentgenographic manifestations in patients with impaired host defense mechanisms. *Medicine* 53: 255 (1974).
5. Campbell, G. S. Respiratory failure in surgical patients. *Current Probl. in Surg.* 12 (1976).
6. Comroe, J. H. *Physiology of Respiration*. Chicago: Year Book Medical Publishers, 1974.
7. Cotes, J. E. *Lung Function*. Oxford: Blackwell Scientific Publications, 1975.
8. Farrell, P. M. and Avery, M. E. Hyaline membrane disease. In *Lung Disease: State of the Art*, Murray, J. F., ed., New York: American Lung Assoc., 1976, p. 1.
9. Greenfield, L. R. Surfactant in surgery. *Surg. Clin. N.A.* 54: 979 (1974).
10. Hardy, J. D. Lung transplantation. In *Surgery of the Chest*. Sabiston, D. G., ed., Philadelphia: W. B. Saunders, 1976.
11. Humphrey, E. W., et al. The adult respiratory distress syndrome. In *Trauma, Clinical and Biological Aspects*. Stacey B. Day, ed., New York: Plenum Publishing Corp., 1975.
12. Humphrey, E. W. Operative therapy for carcinoma of the lung. In *Advances in Cancer Surgery*, Miami: Symposia Specialists, 1976.
13. Hutchinson, D. C. S. Alpha-1-antitrypsin deficiency and pulmonary emphysema: The role of proteolytic enzymes and their inhibitors. *Brit. J. Dis. Chest.* 67: 171 (1973).
14. Johnston, R. F. and Wildrick, K. H. The impact of chemotherapy of the care of patients with tuberculosis. In *Lung Disease: State of the Art*. Murray, J. F., ed., New York: American Lung Assoc., 1976, p. 1.
15. Jones, N. L. and Campbell, E. J. M. Chronic airway obstruction due to asthma, bronchitis, and emphysema. In *Principles of Internal Medicine*, 7th ed. New York: McGraw-Hill, 1974, p. 1275.
16. Mitchell, D. N. and Scadding, J. G. Sarcoidosis. In *Lung Disease: State of the Art*, Murray, J. F., ed., New York: American Lung Assoc., 1976, p. 117.
17. Morgan T. E. Pulmonary surfactant, *New Engl. J. Med.* 284: 1185 (1971).
18. Murray, J. F. Bronchiectasis, lung abscess and broncholithiasis. In *Principles of Internal Medicine*, 7th ed., New York: McGraw-Hill, 1974, p. 1285.
19. Norman, P. S. Hypersensitivity reactions of the lung. In *Principles of Internal Medicine*, 7th ed., New York: McGraw-Hill, 1974, p. 1307.

20. Petersdorf, R. G. Nosocomial infections. In *Principles of Internal Medicine*, 7th ed., New York: McGraw-Hill, 1974.
21. Sackner, M. A. Bronchofiberoscopy. In *Lung Disease: State of the Art*, Murry, J. F., ed., New York: American Lung Assoc., 1976, p. 261.
22. Warwick, W. J., et al. Survival patterns in cystic fibrosis. *J. Chronic Dis.* 28: 609–622 (1974).
23. Weiss, W. Survivorship among men with bronchogenic carcinoma: three studies in populations screened every six months. *Arch. Environ. Health.* 22: 168 (1971).
24. Wessl, S. A. Immunological aspects of pulmonary neoplasms. In *Surgery of the Chest.* Sabiston, D. G., ed., Philadelphia: W. B. Saunders, 1976, p. 404.
25. West, J. B. *Ventilation/Blood Flow and Gas Exchange.* Philadelphia: F. A. Davis Co., 1970.

Mass Spectrometry in Biology and Medicine

ALBERTO FRIGERIO, Ph.D.
Head, Laboratory of Mass Spectrometry
Istituto di Ricerche Farmacologiche Mario Negri
Milan, Italy

In the last few years mass spectrometry has become an analytical method of prime importance in the identification and structural analysis of organic compounds. For some problems, such as the determination of components separated by gas chromatography, it is one of the few techniques able to give specific data from a sample at the nanogram (10^{-9} g) level. One of its major limitations is that it requires a vaporized sample.

Mass spectrometry has become an indispensible tool in organic chemistry, pharmacology, biochemistry, medicine, toxicology, food chemistry, forensic science, petrology, geochemistry, pollution studies, and many other sectors of research. Until 1940 mass spectrometry was only used for the analysis of gases and for the determination of the stable isotopes of chemical elements. It was later used to carry out quick and accurate analyses of complex mixtures of hydrocarbons from petroleum fractions; then, when it was demonstrated that a complex molecule could give rise to a well-defined and reproducible mass spectrum, interest in its application to the determination of organic structures was established. At present, the mass spectrometer is an indispensible tool in biology and medicine.

The coupling of the gas chromatograph with mass spectrometer has extended the applications of both techniques. These two methods are, in fact, highly complementary: The gas chromatograph is efficient in the separation of the constituents of a mixture but does not always give a full identification, whereas the mass spectrometer can identify a single compound but is less efficient in the study of complex mixtures. A very important factor is the comparable sensitivity of the two techniques. One of the advantages of the mass spectrometer is that it can record spectra from small quantities of substance. In fact, the technique gives more information for a nanogram of sample than any other method at our disposal.

In biological experiments, for example, very small quantities of very precious substances are often isolated from natural sources after accurate and patient purification by thin layer chromatography. The only possible method of analysis in such cases is mass spectrometry. With a careful study of the spectrum and with the help of a standard, it is possible to arrive at the structure of the product.

The mass spectrometer is an instrument that produces ions from a molecule, separates the ions as a function of their mass-to-electric charge ratio, and records and displays the relative abundance of these ions. In principle, the function of the mass spectrometer is relatively simple. The molecules of the substance, in a gaseous phase, become ionized, the ions produced are accelerated in an electric field of high potential, they become deviated in a magnetic field and then arrive at a collector, generating a signal the intensity of which is proportional to the number of ions arriving. The whole apparatus operates under high vacuum. The record of the signals constitutes the mass spectrum.

To illustrate the function of a mass spectrometer, imagine a stone being projected from a catapult toward a delicate vase. On impact the vase is shattered. If the pieces are carefully collected, the vase can be reconstructed from the fragments. In this example the vase represents the molecule, the catapult the filament, and the stone the bombarding electron.

Mass spectrometry has gained wide acceptance in biology and medicine: Of particular interest are the studies concerning the identification of metabolites formed from the ingestion of nutrients, pollutants, or drugs. Other investigations concern the identification of endogenous metabolites in living organisms, either animal or vegetable. Finally, there are medical applications of this technique for diagnostic purposes, which will, in time, make mass spectrometry a necessary tool in every major hospital.

Melanins

VACLAV HORAK, D.Sc.
Professor of Chemistry
Georgetown University
Washington, D.C.

Melanins are a family of macromolecular pigments frequent in man, animals, and plants. The name is derived from a Greek term for *black* and this color is characteristic for eumelanins ("good" melanins). However, chemically, biologically, and cytologically, closely related pheomelanins ("dusky" melanins) colored from brown to yellow also exist. Additional color variations result from physical phenomena and from the effect of accessory pigments acting as screens. Melanins are found in hair and fur (e.g., black, brown, red, blond); skin (e.g., man, reptiles); melanoma tissue, choroid tissue, substantia nigra in the brain, sclerotins of insect cuticula, feathers of birds, mushrooms, bacteria, sepia ink, soil, (humic acids), plant seeds and fruits, and wood (e.g., ebony). Many melanins have been prepared synthetically and derive names after the starting material e.g., d, l-dopa melanin, catechol melanin, etc. Melanins are isolated in the form of small particles or granules (e.g., 0.2 to 0.3 inch diameter for sepia ink melanin). Natural melanin is linked to protein (e.g., 10% protein in sepia melanin), which can be removed by acid hydrolysis (e.g., 30% HCl) but is resistant to proteolytic enzymes. The most abundant acids identified after hydrolysis of sepia melanin protein are aspartic acid, glutamic acid, glycine and valine. The detailed chemical structure of melanins is unknown, but it is certain that natural melanins are more complex than poly-5,6 indolequinone. Eumelanins contain nitrogen atoms which are part of indole and pyrrole nuclei.

While eumelanins contain only traces of sulfur (e.g., 0.2 to 0.3% in sepia ink melanin), pheomelanins from chicken feathers contain large proportions of sulfur (19% and more). The sulfur comes from cystine which is incorporated into melanin during melanogenesis. For the structural determination of melanins, degradation processes, particularly by alkali fusion at 200° to 250°, and oxidation by hydrogen peroxide and potassium permanganate have been extremely useful. The main degradation products from sepia ink melanin have been identified as 5,6-dihydroxyindole and its carboxylic acids and pyrrole carboxylic acids. Synthetic melanins are formed by oxidative processes (nonenzymatic or enzymatic) from catechols e.g., (d, l-dopa, catechol) via corresponding o-quinones which polymerize spontaneously. The starting molecules can be monophenols (e.g., tyrosine), or simply substituted benzenederivatives which first undergo

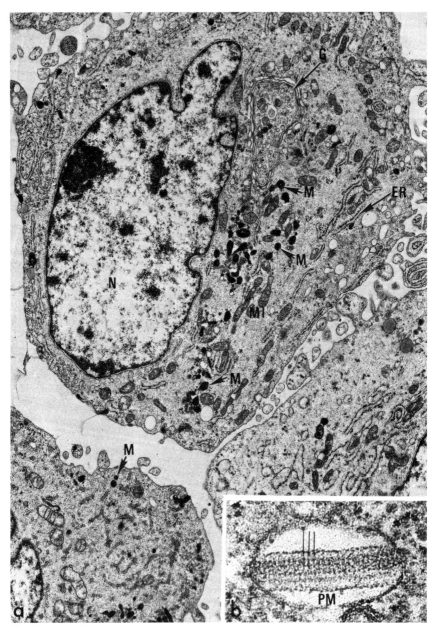

Ultrastructure of Melanoma. (Study of Dr. Nurul H. Sarkar)

enzymatic hydroxylation to the appropriate catechol. Natural melanins are formed by melanogenesis in melanocytes. This process, in which air oxygen is consumed, is catalyzed by polyphenoloxidases, e.g., tyroginase. Enzymatic melanogenesis involves formation of carbon dioxide which results from decarboxylation of aminoacids. Portion of the carbon dioxide comes from ring degradation processes caused by hydrogen peroxide. This finding is consistent with detection of pyrrol carboxylic acids in melanin degradation studies. The following is the simplified scheme of melanogenesis based on tyrosine as a starting molecule:

Tyrosine $\longrightarrow$ Dopa $\longrightarrow$ Dopaquinone $\longrightarrow$

5,6-Dihydroxyindole $\longrightarrow$ 5,6-Indolequinone $\longrightarrow$ Melanin.

Melanin can exist in both an oxydized form (polyquinone) and a reduced form (polycatechol). In aerobic medium, melanin exists predominatly as polyquinone. Melanin can easily be reduced. In partially reduced form it can assume the function of an "electron-exchange" buffer (oxidation-reduction buffer). The semiconductive properties of melanin were predicted theoretically. Spectroscopies in general are not useful for characterization of melanins. A strong ESR signal indicates presence of free radicals in the macromolecule. ESR g-values for different melanins have been determined as: dopa melanins 2.0036, sepia ink melanin 2.0030, melanoma melanin 2.0031, hair melanin 2.0043. Ability of melanins to adsorb certain types of organic molecules have been frequently documented. High affinity compounds toward melanins have polynuclear aromatic and heteroaromatic character. This phenomenon is responsible for toxicity of certain drugs (e.g., chloroquine, chlorpromazine) which accumulate in the eye to high levels and create toxic effects.

Malignant Melanoma

Nurul H. Sarkar, Ph.D.
Sloan-Kettering Institute for Cancer Research
New York, New York

Malignant Melanoma is a tumor which arises in the melanin producing cells of the epidermis, retina or epithelium of the alimentary canal. Melanoma cancer cell can invade the epithelium and can metastasize via the lymphatic system or the blood stream to virtually every organ of the body. Melanoma tumor cells from metastatic deposits can be grown in tissue culture with a reasonably high degree of success. Many such melanoma cultures continue to retain their differentiative functions of melanin production even after many *in vitro* passages. Continued melanin production provides a visual maker identifying the cultured cells as the progeny of the original tumor cells. Melanin can be seen directly under the light microscope in the cytoplasm of cells which produce large quantities of melanin. The cell pellets resulting from centrifugation of cell lines which produce too little melanin to be seen by light microscopy are often tan or brown indicating the presence of the pigment. However, using electron microscopy the specialized melanin containing organelles called premelanosomes (PM, Fig. 1b) and melanosomes (M, Fig. 1a) can be visualized. Figure 1a shows many pigment granules containing melanosomes which appear as very electron dense (black) bodies in the cytoplasm. An enlargement of a premelanosome showing striated structures arranged in a regular way is presented in Fig. 1b. M, melanosome; PM, premelanosome; N, cell nucleus; G, Golgi; ER, endoplasmic reticulum; MI, mitochondria. Fig. 1a, ×7500; Fig. 1b, ×57,000.

Monoamines as Neurotransmitters in the Central Nervous System

R. Samanin, Ph.D.
Istituto di Ricerche Farmacologiche Mario Negri
Milan, Italy

Among the substances assumed to play important roles in brain functions, noradrenaline, dopamine, and serotonin (5HT) have been extensively studied in recent years. Very sensitive enzymatic-isotopic methods have been developed by which these substances can be measured in minute brain areas of laboratory animals. These studies have confirmed the regional heterogeneous distribution of the amines, supporting the hypothesis that they are concerned with specialized functions in the brain.

    Important progress was made in this field as better knowledge was acquired of the properties and factors regulating the activity of tyrosine and tryptophan hydroxylases, considered to be the rate limiting enzymes in the biosynthesis of catecholamines and serotinin respectively. These studies have contributed to the development of kinetic models by which it is now possible to estimate the synthesis and turnover rate of these substances in different experimental situations. The turnover rate of an amine is considered a more reliable index of the functional activity of a monoaminergic neuron than the steady-state levels.

Particularly significant was the finding that physiologically released amines are mainly taken up into the presynaptic terminals by an active transport mechanism at the neuronal membrane. Various drugs, particularly the tryciclic antidepressants, are potent blockers of this mechanism and are therefore believed to increase the amount of transmitter available at the postsynaptic receptors. These findings further support the hypothesis that certain affective disorders may be due to disfunction in these brain monoaminergic systems.

More direct evidence that these monamines may serve as neurotransmitters is provided by histochemical fluorescence studies, by which neuron systems specifically containing these substances can be viewed in the brain. It thus becomes possible to act directly on the systems with lesioning or stimulating electrodes and to correlate the resulting changes of amine functions with changes of animal behavior and drug activity.

The discovery of neurotoxic compounds such as 6-hydroxydopamine and 5,6-dihydroxytryptamine, which cause selective degeneration of monoaminergic neurons, has also greatly contributed in this field. Research using all these means of selective manipulations of monoaminergic systems in the brain has confirmed the importance of dopamine's role in the regulation of certain aspects of motor behavior and, in general, the involvement of brain monoamines in central activities such as sleep, temperature regulation, sexuality, feeding, etc.

The following findings, which point to brain serotonin as instrumental in pain perception and narcotic actions, illustrate the evidence commonly obtained in support of monamines having a role in physiological situations and drug action. Parachlorophenylalanine (PCPA), a specific blocker of 5HT synthesis, and raphe lesions, which produce a selective degeneration of 5HT-containing neurons in the brain, both increase sensitivity to painful stimuli in rats and cats. On the other hand, stimulation of central 5HT neurons produces powerful analgesia which decrease and increase functional 5HT in the brain, respectively antagonize and potentiate morphine analgesia. A reduction of morphine analgesia is also found in animals treated with 5,6-dihydroxytryptamine, which causes a selective degeneration of central 5HT neurons. Moreover, intraventricular injection of 5HT potentiates the analgesic effect of morphine, and analgesic doses of morphine increase the turnover rate of brain 5HT.

An important aspect to be considered in interpreting studies with monoamines is that there are interactions between the neuronal pathways containing these substances in the brain. Although it has long been proposed that monoamines are reciprocally active in various situations such as the sleep-waking cycle, Parkinson's disease, temperature regulation, etc., only recently has biochemical and pharmacological evidence of interaction between these substances in the brain been obtained. This is a crucial area for future investigation since it may add greatly to the understanding of many physiological and pathological situations, and clarify some pharmacological and pathological aspects of psychotropic drugs.

Much work has been aimed at elucidating the intimate characteristics of how monoamines interact with postsynaptic receptors in the brain. Although these studies suffer from various drawbacks and methodological limitations, some progress has been made in this field. Combined cytochemical and electrophysiological techniques have shown there are inhibitory influences on neurons in various areas with known monoaminergic inputs. Thus a noradrenergic inhibitory input to cerebellar Purkinje cells and hippocampal pyramidal cells has been demonstrated. A dopaminergic inhibitory influence on striatal cholinergic neurons also appears to be well documented by electrophysiological, biochemical, and pharmacological studies. A tonic inhibitory serotoninergic influence on some portions of the limbic system has been also proposed on the basis of microiontophoretic studies. However, it is important to underline that, with these few exception, the exact sites and functions of monoaminergic-mediated synapses in the central nervous system are largely unknown.

Monoamine-sensitive adenylate cyclases have been found in brain areas rich in monoaminergic terminals, and various pharmacological and biochemical studies suggest that they might be associated with postsynaptic monoamine receptors. Although it remains to be determined whether such enzyme-active receptors are typical of all central monoamine-mediated synapses, these findings hold promise for extending our knowledge of the intimate biochemical mechanisms by which response to monoamines can be mediated inside the postsynaptic cells.

There is obviously no space in such a short overview to describe difficulties and methodological limitations, or to do more than underline the important discoveries being made in experimental and clinical research. Nevertheless, despite the fact that further work must obviously be done and new methods and strategies must be established, it is worth attempting, in conclusion, to summarize the present position. No concrete evidence is yet available of a neuron innervated directly by monoaminergic fibers where the release of monoamine causes ionic variations resulting in depolarization or hyperpolarization; however, the following reasons exist for currently considering monoamines neurotransmitters in the central nervous system (CNS):

1. They are unevenly distributed in the mammalian brain;
2. All the enzymes needed for their synthesis and degradation are present in brain

3. Monamines are preferentially localized in synaptic vescicles which are considered the main sites for storage and release of putative neurotransmitters in the CNS
4. Histochemical fluorescence studies have clearly demonstrated the presence of monoamine containing neurons in the CNS, whose stimulation releases these substances from nerve endings
5. Microelectrophoretic application of these substance can influence the electrical activity of various brain areas
6. Well known mechanisms of the peripheral noradrenenergic neurons, such as release, uptake and storage, appear to operate for monamines in the brain as well
7. Selective manipulations of these brain substances cause specific changes of behavior and drug activity in various animal species
8. Some psychotropic drugs markedly affect those monaminergic systems in the brain.

Feedback Mechanism in Physiological Systems

Okan Gurel
IBM Corporation
Westchester, New York

The name feedback is a term borrowed from the control engineering. A system may be controlled automatically by various ways. If a system described by its elements (variables in its mathematical equations) performs a function, it possesses a dynamics. It would be necessary to control this dynamics by either control variables or some additional elements (in mathematical terms called parameters). However, in a dynamic system there would usually be a set of input as well as a set of output variables with a dynamic activity producing the required output from the necessary input. In a variety of such systems the behavior of the system may be controlled by an interaction of the output set with either the input set or a set of any number of intermediate elements. This interaction forms another type of control which is properly called a feedback mechanism. Feedback could be of the type of either enhancing the dynamic activities, *positive feedback*, or deterring the activities, *negative feedback*.

In many biochemical systems, such as metabolic pathways, as well as gross physiological functions, examples of feedback mechanisms can be detected. It should be clear that not only a system may be controlled by either parameters or a feedback variable, but also both control mechanisms may enter into a particular systems dynamics. The importance of feedback concept in physiological systems come from the fact that without needing further elements (parameters), a system can control itself (autocontrol) by a proper feedback mechanism designed for this purpose. The proper feedback mechanism coupled with parameter variations can create not only stable equilibrium points (threshold values) for the variables (elements) of the dynamic system, i.e., steady state of a biochemical reaction, but also self oscillatory regimes, i.e., biological rhythms can be realized.

Some of the known phsyiological feedback systems can be mentioned as illustrations:

- Feedback of gonadal steroids on gonadotropin secretion; negative feedback of gonadal steroids on the release of gonadotropins FSH, the follicle-stimulating hormone and LH, luteinizing hormone. Under some circumstances, however, a positive feedback of estrogen and progesterone on LH secretion. Both estrogen and progesterone appear to have mainly a stimulatory or positive feedback action on the third gonadotropin, LTH, luteotropin, secretion.

- Feedback mechanism operating at the pituitary level regulates (stimulates or depresses) TSH (thyrotropic hormone, the thyroid stimulating hormone) secretion by the level of thyroid hormone in the blood.

- In glycolysis, fructose-1, 6-diphosphate (FDP) is not only an intermediate but also acts as a feedback to activate phosphofructokinase, an enzyme, entering the second step in the reaction starting with glucose (GLU) which yields fructose-6-phosphate (F6P), (FDP) as intermediate and glyceraldehyde phosphate (GAP) as the end product.

Psychopharmacology

FRANK M. BERGER, M.D., D.SC.
New York, New York

Psychopharmacology is the branch of science that utilizes drugs to increase knowledge of the workings of the brain. It aims to discover medicinal substances that would help to restore and maintain mental health. Drugs affecting the mind, such as caffeine, cocaine, marijuana, opium, and peyote, have been used since prehistoric times to produce abnormal mental states or altered states of consciousness. These psychoactive drugs, however, have been used primarily by normal people. They must be differentiated from the modern psychotropic drugs that are primarily used to treat mentally disturbed patients.

According to their clinical usefulness, psychotropic drugs can be divided into three groups: the antipsychotic agents, the antidepressants, and the anxiolytics. Each of these groups comprise a number of different drugs that differ from each other in chemical structure or mode of action. The antipsychotic agents, also called ataractics, neuroleptics or major tranquilizers, are used in the treatment of psychosis. These drugs counteract hallucinations, control excitement and assaultiveness, and facilitate social adjustment of withdrawn patients. Clinically, these agents proved to be of greatest value in the treatment of schizophrenia and of manic excitement.

The antidepressant drugs, also called thymoleptics or psychic energizers, counteract mental depression and restore vigor and life interest in these patients. The anxiolytics, also called minor tranquilizers or antianxiety agents, are used for the temporary control of hyperexcitability, nervous tension and insomnia.

The psychotropic drugs, unlike the nonspecific depressants of the central nervous system, selectively counteract the symptoms of mental disease. They are not "chemical straight jackets" or sedatives, but specific agents that appear to exert a specific effect on the underlying disease process. The psychotropic drugs do not exert their specific effects in healthy people. Accordingly, the antidepressants have a mood-elevating action only in depressed patients and antipsychotic drugs activate only withdrawn schizophrenic patients.

While the psychotropic drugs have greatly improved the outlook for the mentally sick, they have not wiped out mental disease. There are still mental diseases that do not respond to any of the available remedies, and some patients suffering from diseases that are usually responsive to the existing drugs may be refractory to them because of their inability to absorb or metabolize the drugs in the usual way. Nevertheless, through the use of these drugs most patients become amenable to treatment without the need for hospitalization. However, the significance of the new psychotropic drugs does not rest so much on their therapeutic efficacy as it does on the research that was stimulated by their introduction. Thus psychopharmacology has made it possible to study the role of neurotransmitters in the brain, and as a result of this, some understanding of the biochemical basis of mental disease has become possible at both the molecular and clinical level of affect.

Overview: Prenatal Diagnosis of Genetic Disorders

JAROSLAV CERVENKA, M.D., C.Sc.
Professor of Human Genetics
Division of Oral Pathology and Oral and Human Genetics
School of Dentistry

Professor, Dept. of Medicine
School of Medicine
University of Minnesota
Minneapolis, Minnesota

Studying the history of basic genetic research gives us a most interesting view on development of a scientific field which deals with the very fundamental aspects of life, evolution, speciation and, at the medical level, with the primary cause of disease.

However, reviewing the history of applied genetics, practical eugenics, we are confronted with controversial and disturbing examples of applied science. In one way or the other the eugenic efforts have been plagued by the interference of politicians, racist groups, and churches. But the primary cause for the ineffectiveness and sometimes naiveté of applied genetics in the pre-war era has been insufficient knowledge, i.e., the primitive state of genetic science itself.

During the last two decades the field of genetics and consequently medical genetics has been expanded and deepened considerably by development and application of new biochemical and cytogenetic methods. A large number of new genetic disorders has been discovered—autosomal dominant, recessive and

X-linked disorders exceeded 2000, not counting the polygenic traits. It has been estimated that one quarter of admissions to children's hospitals are due to these diseases and over two billion dollars are committed by society to care for patients with chromosomal disorders alone in the United States.

Thus considerable excitement arose in the medical community when a method for enabling prenatal detection of many genetic diseases was discovered. This technique consists of withdrawal of sample of amniotic fluid from the amniotic sac at about the 16th week of pregnancy and subsequent analysis of cells suspended in amniotic fluid or analysis of amniotic fluid itself.

Amniocentesis is a technique of withdrawal of amniotic fluid by penetration of the amniotic sac. Practically transabdominal amniocentesis is used almost exclusively in prenatal diagnosis while transvaginal amniocentesis is rarely employed due to the increased risk of infection of the uterine contents or induction of permanent fluid leakage. Transabdominal aminocentesis can be performed as an ambulatory procedure in the obstetrician's office. Following localization of the placenta and determining the position of the fetus by ultrasonic scanning or by palpation, aseptic conditions are strictly maintained. Under local anesthesia a 20 or 21 gauge (not larger) spinal needle with stylet is introduced through the abdominal wall into the middle of the uterus. The stylet is removed and amniotic fluid aspirated. Usually 10 to 20 cc of fluid are sufficient for most types of analysis. There is no evidence that withdrawal of this volume effects the development of the fetus if amniocentesis is performed following the 15th week of pregnancy. At sixteen weeks the total volume of amniotic fluid ranges between 150 and 280 ml.

Risk and Pitfalls. In a Canadian study of 1223 amniocenteses amniotic fluid leakage per vaginam occurred in 1.3% and uterine contraction, vaginal bleeding and spontaneous abortion occurred in 1.4%. In 634 cases studied in Massachusetts amniocentesis was implicated in 1.6% of major complications such as spontaneous abortion, fetal death and stillbirth. One should note, however, that in large studies of the "normal" general population spontaneous abortion, fetal death and stillbirth comprise about 3% of pregnancies. The risk to the fetus by amniocentesis performed by an experienced obstetrician at the 16th week of pregnancy, monitored by ultrasound with needles of the gauge 20-21 under aseptic conditions, is thus minimal and does not exceed the expectation of failure of random pregnancy without amniocentesis.

The risk to the mother is difficult to evaluate since no maternal mortality has yet been reported following amniocentesis in more than 5000-6000 pregnancies performed in the United States, Canada and Europe. In rare instances maternal problems have included hemorrhage, abdominal pain, uterine contractions and peritonitis due to perforation of viscera or urinary bladder and Rh isoimmunization due to fetomaternal hemorrhage in an Rh negative mother with an Rh positive fetus.

The optimal time for amniocentesis when tissue culture of amniotic cells is desired is around the 16th week of pregnancy.

Indications. Amniocentesis was first used for management of Rh incompatibility and this indication remains the most frequent. However, prenatal diagnosis of genetic disorders is becoming equally important. The main categories of genetic diseases prenatally diagnosed are (a) chromosome disorders (b) biochemical diseases—inborn errors of metabolism (c) X-linked disorders (d) neural tube defects.

Chromosomal disorders. About 75% of all amniocenteses for prenatal diagnosis of genetic disorders are performed in cases where the risk of chromosome anomaly in the fetus is increased. The majority are performed in older mothers. It has been calculated that the risk for chromosome meiotic nondisjunction resulting in a fetus with chromosome trisomy increases with the age of both parents. The data available for mothers show that at the age of 35-39 years the risk for trisomy 21 alone increases from $1:3000$ for younger mothers to approximately $1:100$ to $1:60$. The risk for mothers over 40 years of age increases to $1:30$ or $1:20$. The risk for other trisomies such as 18^+, 13^+ or polysomies of sex chromosomes has not been established but is considered to increase in analogy to trisomy 21.

In the family with a previous child with trisomy 21 (Down's syndrome), the risk of having another Down syndrome child increases to 2%, regardless of the age of parents.

Indication for amniocentesis is a pregnancy where one of the parents is a balanced translocation carrier. Most common are the carrier states for D/G translocation or G/G translocation resulting in translocation type Down's syndrome. These contribute about 3% of cases of Down's syndromes and about 7% of amniocenteses performed for chromosomal analysis.

Cytogenetic diagnosis from fetal cells requires tissue culture of cells for 14 days to four weeks. It is a reliable method assuming that laboratory procedures are performed by experienced and highly skilled staff capable of evaluation of possible complications such as a high degree of polyploidy, the emergence of an abnormal cell clone in tissue culture, contamination of fetal cell culture by the mother's cells, the possibility of genuine mosaicism, the effect of Mycoplasma contamination, etc. Due to low growth potential of amniotic fluid cells a second tap has been required in about 10-20% of cases.

Biochemical diseases. About $1:100$ live births are affected by "inborn error of metabolism." Many are associated with mental retardation and some have invariably fatal outcome. Over 40 disorders are presently detectable by enzyme activity assays of amniotic fluid cells grown in tissue culture. Most are inherited in an autosomal recessive manner, i.e., if both parents are heterozygous carriers the risk that a child will be homozygous affected is 25%.

Among the lipidoses prenatal diagnosis has been made in Fabry disease, Gaucher disease, gangliosidosis type 1, type 2, Tay-Sachs disease, Sandhoff disease, Krabbe disease, metachromatic leukodystrophy and Nieman-Pick disease A.

Of mucopolysaccharidoses Hurler syndrome, Hunter syndrome and Sanfilipo A syndrome have been diagnosed.

Of amnioacid disorders citrullinemia, cystinuria, argininosuccinic aciduria, maple syrup urine disease and methylmalonic aciduria have been diagnosed.

Among disorders of carbohydrate metabolism, prenatal diagnosis has been made in galactosemia and glycogen storage diseases type II and type IV.

Of other types of metabolic disorders diagnosed prenatally: adenosine deaminase deficiency, congenital nephrosis, cystinosis, hypophosphatasia, I-cell disease, Lesch-Nyhan syndrome, lysosomal acid phosphatase deficiency and xeroderma pigmentosum.

A number of other biochemical disorders possibly are detectable since the defective enzyme is known and assays for detecting the aberration have been well established.   Intensive research is being carried out on the detection of cystic fibrosis, Huntington chorea and a number of less frequent disorders.

X-linked diseases.   In three disorders in this category we can reliably distinguish the affected and nonaffected fetus: Fabry disease, Lesch-Nyhan syndrome, and Hunter syndrome.   In others the geneticist can only supply information concerning the sex of the fetus and present the risk estimate to the parents. In the case of a male fetus from a carrier mother the risk is 50% that the child will be affected.

The fetal sex may be detected by Barr body (X-heterochromatin) analysis or Y-body (fluorescing portion of the Y chromosome) analysis. However, general concensus is that the only reliable way of cellular sex detection is chromosome analysis (karyotype) and this method must be employed in all cases of prenatal sex determination. Variation in the size of the Y chromosome could confuse the Y-body test results, low values of Barr body counts due to technical factors and various aberrations of sex chromosome morphology and number could contribute to errors unless karyotypes are assembled.

Neural tube defects.   Human specific alphafetoprotein is synthesized by embryonal liver, yolk sac, and gastrointestinal tract, peaking between the 12 and 14 weeks of gestation followed by a steady fall toward term. Since fetal cerebrospinal fluid contains alphafetoprotein an open neural tube defects result in leakage of cerebrospinal fluid into the amniotic sac, elevated alphafetoprotein levels at the 16th week of gestation suggest an affected fetus. Since about 90% of neural tube defects are open, this phenomenon has been employed in prenatal diagnosis of anencephaly, meningocele, meningoencephalocele and major spina bifida.

The method is still being evaluated since a number of conditions may yield a positive diagnosis: twinning, congenital nephrosis, fetal death, imminent fetal death, severe Rh immunization and possibly other conditions. However, combined with an ultrasonogram and eventually roentgenography, alphafetoprotein assay is a widely used and useful tool in prenatal diagnosis of open neural tube defects.

Conclusion. At present, extensive research is being performed on fetoscopy. Using fiberoptic amnioscopy the fetus may be photographed and visually examined. Moreover, this technique may be modified to allow withdrawal of sample of fetal blood. Refinement of this methodology would immensely increase the impact and usefulness of prenatal diagnosis of sickle cell anemia, thalassemia, hemophilia, and various hemoglobinopathies.

Recommended Review Literature

1. *Antenatal Diagnosis*, A. Dorfman (ed). Chicago, London: The University of Chicago Press, 1972.
2. Milunsky, A. *The Prenatal Diagnosis of Hereditary Disorders.* Springfield, Ill.: Charles C Thomas, 1973.
3. *The Prevention of Genetic Disease and Mental Retardation*, A. Milunsky (ed). Philadelphia, London, Toronto: W. B. Saunders Co., 1975.

Protein Phosphorylation

JEANNE M. WEHNER
JOHN R. SHEPPARD
University of Minnesota Medical School
Minneapolis, Minnesota

Covalent modification of proteins by methylation, acetylation, hydroxylation, and phosphorylation are biochemical mechanisms which regulate the protein's structure and function. Protein phosphorylation is of intense current interest because some of these reactions are catalyzed by *cyclic nucleotide dependent* protein kinase enzymes. Thus this subject assumes a position of major importance when one recognizes that many hormones (via cyclic nucleotides) use this biochemical mechanism to elicit physiological responses. (See the articles entitled "Hormone Receptors" and "Cyclic Nucleotides" in this volume for more detail.)

The cyclic AMP-dependent protein kinase is composed of two distinct subunits. One called the *regulatory* (R) subunit binds cyclic AMP. The other subunit (C) *catalyzes* the transfer of the terminal phosphoryl group of ATP to serine or threonine residues in the protein substrate. The two subunits (regulatory and catalytic) are normally associated and form an inactive R-C complex. However, in the presence of sufficient concentrations of cyclic AMP, the regulatory subunit binds cyclic AMP and the R-C complex dissociates into free C (which is then able to phosphorylate protein) and an R-cyclic AMP complex. In the absence of cyclic AMP the R and C subunits reassociate to the inactive state.

The cyclic AMP dependent protein kinases are widely distributed throughout many phyla of the animal kingdom and are found in a variety of tissues. Another class of related enzymes, called phosphoprotein phosphatases, catalyzes the removal of phosphate from the protein substrate and are also widely distributed. Cyclic nucleotides may also influence these dephosphorylating enzymes. Thus physiological events involving cyclic nucleotides are controlled by the phosphorylation and dephosphorylation of specific proteins.

The protein substrates of the cyclic AMP-dependent kinases are numerous and vary in their cellular function. Some substrates are soluble enzymes involved in carbohydrate and lipid metabolism. Other phosphorylated proteins, such as contractile proteins, function structurally. The cellular location of these proteins is also variable. They may reside in the cytosol, nucleus, or membranes of the cell.

The degradation and synthesis of glycogen is regulated by hormones and mediated by phosphorylation and dephosphorylation of several metabolic enzymes. Glycogen phosphorylase, which catalyzes the first step in the breakdown of glycogen into glucose, is converted to its active form by phosphorylation of a specific serine residue. The enzyme responsible for this event is itself converted to its active form by phosphorylation. Another enzyme, glycogen synthetase, which catalyzes the synthesis of glycogen from glucose, is simultaneously converted to its inactive form by phosphorylation during physiological conditions favoring glycogen breakdown. In lipid metabolism, the hormone-regulated release of free fatty acids from storage as triglycerides is cyclic AMP dependent. The enzyme controlling fatty acid release, triglyceride lipase, is converted to its active form by phosphorylation.

In the nucleus of the cell, both histone and nonhistone proteins are phosphorylated; this may modulate the interaction between these proteins and DNA, thereby regulating gene activity. RNA polymerase is also phosphorylated in a cyclic AMP dependent fashion and may provide a mechanism for induction of enzyme synthesis by regulating RNA synthesis.

Proteins associated with plasma membranes are also substrates for protein kinases, e.g., in the mammalian nervous system, myelin basic protein found in the myelin sheath is phosphorylated. The actions of some neurotransmitters may also be mediated by the action of protein kinases on specific proteins in the synaptic membrane. Synaptic membranes contain all of the essential molecules to support such a theory: a neurotransmitter receptor coupled to adenylate cyclase, a cyclic AMP-dependent protein kinase, phosphoprotein phosphatases, and the protein substrates (one of which is the neurotransmitter receptor itself).

Other interesting roles for cyclic AMP-dependent protein phosphorylation include the postulated effect of cyclic AMP on muscle contraction mediated by the phosphorylation of the contractile protein troponin, and the phosphorylation of tubulin, the major microtubule protein. Physiological events involving cyclic AMP as a biological regulator are numerous and protein phosphorylation provides a biochemical mechanism by which simple molecules such as cyclic nucleotides can elicit diverse physiological responses.

Radioimmunoassay

JOHN LANGONE, Ph.D.
Bethesda, Maryland

The radioimmunoassay (RIA) technique, introduced by Yalow and Berson in 1960, is one of several competitive protein binding assays available to the clinician and laboratory investigator. RIA is based on the competition between a radiolabeled and unlabeled antigen for a limited number of binding sites on a specific antibody. In practice, a constant amount of radiolabeled antigen is incubated with an aliquot of test sample and antibody, then antibody-bound or free label is determined after separation by one of several available techniques. Antigen concentration is obtained by comparing the degree of competitive inhibition observed with that found in known standard solutions. By this procedure, antigen often can be detected at the picomole level in unprocessed physiological fluids. Prerequisites for any RIA are specific, high-affinity antibody, pure standard antigen, and radiolabeled antigen of high specific activity. Generally successful immunization methods are available and several techniques have been developed which allow the routine laboratory preparation of ^{125}I-labeled antigen derivatives.

Since certain diseases are characterized at least in part by abnormal levels of endogenous compounds, many RIA's of clinical value have been developed for peptide and protein hormones, plasma proteins, and many biologically important low molecular weight molecules including T_3, T_4, c-AMP, folic acid, vitamins, and prostaglandins. In particular, the discovery that certain tumor-associated antigens are similar to fetal or placental enzymes or proteins has inspired a widespread effort to determine if RIA can be used as a procedure to diagnose certain types of cancer. Initial enthusiasm based on results obtained with RIA's for carcinoembrionic antigen (CEA), human chorionic gonadotrophin (HCG) and α-fetoprotein (α-FP) has waned somewhat mainly due to the more recent findings that none of the antigens is a specific indicator of malignancy. Low levels of these antigens generally are found in the normal population, and abnormally high levels often are found in individuals suffering from noncancer related disease.

Assays for several of the endogenous steroid hormones appeared shortly after the first protein assays and over the past decade especially, increasing numbers of assays for low molecular weight compounds or haptens have been reported. Unlike the development of RIA's for immunogenic proteins, hapten assays depend on the initial covalent binding of the nonimmunogenic small molecule to an antigenic carrier. Thus, several general methods have been developed to introduce reactive functional groups into chemically unreactive compounds that can be used for this purpose. Since metabolism often plays an important role in the biological effects of a drug, RIA has found wide application in clinical pharmacology. Hapten assays are available for several classes of drugs including opiate alkaloids, barbiturates, amphetamines, and several antineoplastic agents (e.g., methotrexate and adriamycin). In some cases, specific RIA's for biologically active metabolites enable accurate pharmacokinetic data to be obtained without the extensive sample processing required by other techniques. Although most hapten assays have been developed for clinically important drugs and drugs subject to abuse, recent interest in the use of RIA to determine levels of environmental contaminants in physiological fluids indicates that epidemiology may be a new and promising field for RIA.

No doubt, future developments in RIA will continue to center around clinical needs. In this regard, completely automated systems capable of performing the several steps involved in carrying out an assay on large numbers of samples are beginning to appear commercially. However, basic research in metabolism, disposition and mechanism of action of pharmacologically active compounds and environmental agents also will be fruitful areas.

What appears to be the next generation of immunoassay is termed Enzyme Linked Immuno Sorbent Assay (ELISA). In this assay, active enzyme covalently bound to hapten is inactivated by steric inhibition when the hapten is in turn bound (noncovalently) to specific antibody. Enzyme activity, determined by a simple spectrophotometric measurement is the basis for quantification and does not rely on the radioactive tracers and accompanying facilities required for RIA. RIA and ELISA should continue to grow as important analytical techniques in both clinical and research laboratories.

References

1. S. A. Berson and R. Jalow, in *Immunobiology*, R. A. Good and A. W. Fisher (eds.). Stanford, Conn.; Sinover Associates, Inc., 1971.
2. Butler, V. P., Jr., and Beiser, S. M., *Adv. Immunol.*, **17**: 255 (1973).
3. *Principles of Competitive Protein Binding Assays*, W. E. O'Dell and W. H. Daughaday (eds.). Philadelphia: S. B. Lippincott Co., 1971.
4. Radioimmunoassay and saturation analysis, *Brit. Med. Bull.*, **30**: 111 (1974).

Screening for New Drugs

EUGENE N. GREENBLATT, Ph.D.
Group Leader, Neuropharmacology
Central Nervous System Disease Therapy Research Section
Lederle Laboratories
Pearl River, New York

Screening may be defined as a procedure or a specified set of procedures by which new chemical entities are tested in relatively large numbers for possible utility in disease states of man or other species. A more practical or working definition would perhaps be a method(s) of eliminating inactive or uninteresting compounds using the fewest test procedures or animals, the least amount of substance under investigation, and still maintain the highest scientific accuracy.

Historically, the earliest known large scale screening program was that carried out in the original search for a chemotherapeutic agent to cure syphilis. The 606th compound tested, arsphenamine or salvarsan (Ehrlich's "Magic Bullet") showed a satisfactory margin of safety between effectiveness and toxicity in test animals and subsequently became the treatment of choice before the advent of penicillin.

The program conducted in the United States during World War II to find an antimalarial substitute for quinine is an example of mass screening conducted by the government and by commercial laboratories. It is estimated that more than 15,000 compounds were screened during this effort. The discovery of chloroquine and plasmoquine was a direct result of this mass screening program.

There are, in general, two types of screens: *in vitro* and *in vivo*. Although penicillin was discovered by accident, all the newer antibiotics have primarily been screened by *in vitro* methods from soil samples collected from all over the world. New antibiotics, which may have been synthesized in the laboratory, are still mainly screened by *in vitro* methods.

In vivo testing presents greater problems and therefore greater challenges. First, the disease state found in man may not be found in animals, e.g., schizophrenia, headache and asthma. Secondly, is the inherent variability that exists among animals of any particular species or strain. The first problem leads to two general types of *in vivo* screens:

(a) empirical
(b) rational

The second problem leads to rigid statistical design and analysis.

Empirical testing occurs when there are no suitable animal models (see above). In this case screens are designed on the basis of the effects of known agents on a single response or in a battery of tests, and on the resulting profile. Rational testing is where the disease state is present, or at least can be simulated in the animal species tested, e.g., the study of antihypertensive drugs in spontaneously hypertensive rats, or in animals infected in the laboratory with a specific disease inducing organism, e.g., mycobacterium tuberculosum for the study of antitubercular drugs.

There are three kinds of test designs irrespective of whether a screen is carried on *in vitro* or *in vivo*:

(a) simple
(b) "blind" testing
(c) programmed

Simple screening refers to the case where a compound is synthesized for testing in a single procedure, in a single species for a single disease state. An example of a simple screen in addition to the antihypertensive screen (see above) would be a screen for an antipyretic effect. Procedures have been developed where the antipyretic actions of aspirin can be detected in a single rat whose temperature has been elevated by the subplantar injection of a brewers' yeast suspension.

Blind screening is designed to provide a pharmacological identity to a compound or group of compounds with no prior knowledge of activity. This type of screen should also be so designed to eliminate compounds which are inert. Blind screening was common practice during the last two decades when new drug development was flourishing. Methodology was developed to detect the presence of or observe various pharmacological activities, e.g., central, diuretic, analgesic, etc. effects in as few as two mice each at several dosage levels as well as determine some level of toxicity. The tests designed were highly efficient and accurate. For the purposes of screening, efficiency is defined as eliciting the most information from the fewest number of test animals; while accuracy refers to the ability of the tester to select actives in all cases and to reject inactives in a high percentage of the cases. The risk of missing a "winner" must be nil, while the risk of accepting a small number of "losers" can be greater, since subsequent testing would indicate the lack of desired activity. This would be called a "false positive."

In recent years, blind screening has evolved to *programmed screening*, i.e., compounds synthesized and tested in specific programs. Even in programmed screening, compounds would receive widespread testing after the designated testing has been completed, since new drugs are often discovered by serendipity. An example of programmed screening would be testing molecular modifications of Valium®, the drug with the largest sales volume in the United States. Although the testing is still empirical in nature, batteries of tests are designed to select the "therapeutic" effects and to detect as well, the "toxic" or side effects. These tests may be neuropharmacological, psychopharmacological or neurochemical or any combination of the three.

Regardless of which type of screening a compound is subjected to (single, blind, programmed), activity in one or more tests should be reported with some degree of rapidity to the synthetic chemists to guide further chemical modification of the compound. This leads to a study of structure–activity relationships where improvement, e.g., increased in desired activity and a decrease in less desired activity. It is important that communication lines are always open between chemists and biologists—today's results can be tommorow's synthetic direction.

Statistical analyses of collected data utilizing a reference standard (positive control) and a placebo play an important role in the design of most screening procedures. For example, if the screen employed is one which measures a quantitative response, e.g., change in locomotor activity, one must know precisely what is a significant decrease or increase from control (placebo) values. The level of significance depends upon the screen and on the accumulated data with reference agents and controls.

Since most of the compounds tested are inactive, sequential designs are often used to keep the number of animals and amount of compound used to a minimum. In a sequential procedure, unquestionable inactives are eliminated at an initial stage; the compound is then subjected to the necessary number of stages, each step having a more rigid criteria than the last until a desired level of activity for final acceptance or rejection is reached.

Whatever method is ultimately selected for screening, and this is wholly dependent upon the particular needs of an individual, research team, institution or company, it must be sufficiently selective to detect all clinically active compounds in that therapeutic family; and wherever possible, doses in the screen should have some correlation with doses in the species for which use is intended. The screen must also meet financial or budgetary needs. Screening can be a very costly operation, but the rewards for success are great. A private testing firm has estimated charges for a two-day antihypertensive screen (see above) at $70 to $100 per compound, depending upon the number screened. This figure does not include the cost of synthesis or purification of the compound which is always an unknown factor.

As the knowledge of pharmacology, physiology, biochemistry and behavior expands, screening should constantly be modified, altered, and revised to fit needs and to incorporate the new knowledge wherever appropriate. The ever constant problem of developing a screen, wherein no reference drug exists, is most difficult and demands from those involved in screening to continue to use ingenuity to conquer those disease states which still remain beyond our grasp—cancer, alcoholism and heart attacks to name but a few.

Sensory Physiology

HENRY TAMAR, Ph.D.
Indiana State University
Terre Haute, Indiana

Sensory pathways show much structural similarity. The afferent nerve impulses originating from specialized receptor cells or sensory neurons pass across synapses and through intermediate neurons whose cell bodies lie in relay stations (relay nuclei). The impulses then reach one or more specific areas of the sensory cerebral cortex, which for the most part have not yet been satisfactorily delimited.

Although photoreceptors can interact through direct coupling, the network of synapsing nerve fibers which lies beyond a sensory system's receptors makes up the primary structural basis for interaction between the receptors and between them and nearby cells. Lateral inhibition is the important form of receptor interaction, but lateral excitation may also take place. In certain systems receptors stimulate cells central to them and are then inhibited by these cells. Receptor self-inhibition by a circuitous route across a synapse is also known.

In the visual, auditory, and somatosensory relay stations of the thalamus, a three-neuron synaptic complex (the triadic coupling), may be the principal neuronal arrangement. In a triadic coupling a sensory afferent neuron not only synapses with a next-higher-order relay neuron but also with an interneuron which then inhibits the relay neuron through dendrodendritic synapses.

Another neuronal arrangement, the reciprocal dendrodendritic synapse, is also important in sensory structures, such as the olfactory bulb. In this case a neuron, by way of a dendrodendritic synapse, typically excites an interneuron and is inhibited by this interneuron through a similar connection.

Both triadic couplings and reciprocal dendrodendritic synapses appear to produce significant inhibitory effects. They also both probably function by the transmission of electrotonic potentials across synapses between dendrites, a mechanism which seems to be widespread and important in the brain.

In sensory systems the afferent tracts conveying nerve impulses toward the cerebral cortex are in a general sense paralleled by efferent tracts which typically consist of discrete fiber bundles lying near the afferent nerves. Such efferent tracts, which innervate sensory neuron nuclei and also extend to the periphery, produce inhibition or facilitation.

There exist secondary and accessory sensory pathways and systems, associative and collateral connections, and efferent tracts which are still poorly known or are being investigated.

The first process in a receptor's response is transduction, the conversion of the stimulus energy into electrical energy. It is believed that the first electrical potential, the receptor potential, generally arises due to the inward movement of sodium ions through the cell membrane. However, it is still far from certain by what means this ion movement may be initiated. In mechanoreceptors, mechanical deformation of the transducer membrane has been suspected to open sodium pores. In taste and olfactory receptors, receptor molecules in the membrane, proteins or perhaps phospholipids or glycolipids, have been thought to undergo conformational changes on complexing with stimulus particles (thus eventually opening pores). Bitter substances appear to stimulate by a unique mechanism, and may have to pass into the interior of taste cells to elicit responses.

Adaptation is a decrease in sensitivity to a maintained stimulus and could prevent fatigue. It reduces the input from static environmental features and thus helps to emphasize changes which are normally more significant to an organism. The synapses in relay stations show adaptation and cause perceptual adaptation to take longer than receptor adaptation. In invertebrate photoreceptors an increase in calcium ions is essential for light adaptation, and calcium may also be involved in vertebrate photoreceptor adaptation.

Inhibition, one of the most important sensory processes, enhances contrast or discrimination and is instrumental in stimulus localization. It eliminates irrelevant and less important sensory information and, when a certain stimulus has come to attention, it reduces impulse activity in the other sensory pathways. Inhibition further introduces feedback into sensory systems, allowing them to respond more delicately, and seems to control the interplay between sensory and motor neurons in the brainstem. If inhibitory interneurons in sensory relay stations are themselves inhibited through efferent fibers, facilitation should result. At higher stimulus intensities inhibition may have a vital protective function and may maintain sensitivity.

Accumulation of potassium ions at central nervous system neurons due to repeated activity may inhibit transmitter release from the presynaptic terminals contacting these neurons.

Invertebrate visual ommatidia and retinal receptors may be less susceptible to lateral inhibition at low levels of excitation, enabling them to function more effectively during weak illumination.

The functional significance of many efferent tracts is still only partly understood.

It is also because of selectively-responsive neurons that only the more useful fraction of the sensory input is perceived.

Convergence and divergence of impulses occur at all levels of sensory systems, resulting in considerable redundancy but also making transmission less susceptible to interference. At higher levels, convergence (spatial and temporal) produces facilitation or amplification. The last, linked with inhibition, may raise signal-to-noise ratios and create narrow neural routes which lead to localization and lower thresholds. One can conceptualize sensory potentials to be repeatedly transformed, by interaction and recombination at different pathway levels, due to convergence.

There are two concepts presently held as to how sensory information is encoded. According to the older "labelled-line" theory, each sensory neuron's activity informs the organism of the action of only a particular stimulus. Each sensory neuron is thus labelled for the organism. Sensations would be produced by combinations of such labelled responses, each neuron's impulse frequency indicating the intensity of the stimulus effective for it. The "across-fiber pattern" theory states that different response rates in a number of parallel nerve fibers, interpreted together, identify a stimulus. An increase in stimulus intensity causes a higher impulse frequency in each fiber, while the across-fiber pattern (now at a higher frequency) remains relatively unaltered. However, a change in stimulus quality results in a new pattern.

Responsiveness to more complex and significant stimuli appears in higher auditory, visual, and gustatory centers. Thus investigated presumed auditory decoding neurons in the cat's medial geniculate body respond to either a specific vowel, a certain consonant, the direction of frequency change or an environmentally-important noise.

Stress

Hans Selye
Institut de médecine et de chirurgie expérimentales
Université de Montréal, Montréal, Quebec, Canada

Stress is the nonspecific response of the body to any demand. "Nonspecific" is the key term in this definition and requires elucidation. In some respects, every demand made upon our body is unique, that is, specific. Heat, cold, joy, sorrow, muscular exertion, drugs and hormones each present a particular problem whose solution calls forth highly individualized responses. But all these agents have one thing in common: they bring an increased demand for readjustment, for performance of adaptive functions which reestablish normalcy. This heightened demand is nonspecific, that is to say, independent of the specific activity which caused the increased requirement. The essence of stress is the demand for activity as such, and the intensity of this demand is all that really matters—it is irrelevant whether the agent or situation provoking it is pleasant or unpleasant. While it is difficult to see how such basically differing conditions as sorrow or joy or heat or cold can produce identical biochemical reactions in the body, nevertheless, this is the case. The fact that certain reactions are totally nonspecific and common to all types of exposure can now be demonstrated by highly objective biochemical determinations.

When the body is exposed to continuous stress for long periods, it necessarily goes through the three phases of the "General Adaptation Syndrome" (GAS). The first stage is called the *alarm reaction* and is characterized by adrenocortical enlargement with histologic signs of hyperactivity, thymicolymphatic involution with certain concomitant changes in the blood count (eosinopenia, lymphopenia, polynucleosis), and gastrointestinal ulcers, often accompanied by other signs of damage or shock.

Of course, if the stressor (stress-producing agent) is so severe that continued exposure is incompatible with life, the organism will die within a few hours or days during the alarm reaction; otherwise, a *stage of adaptation* or *resistance* will ensue, since no organism can be maintained continuously in a state of alarm. The adaptive stage is characterized by the vanishing or diminishing of the initial symptoms—the body has achieved optimal adaptation. After still more prolonged exposure to the stressor, this acquired adaptation is lost and a third phase, the *stage of exhaustion*, is entered into, since the "adaptation energy" or adaptability of an organism is apparently finite.

These three phases are analogous to the three stages of man's life: childhood (with its characteristic low resistance and excessive responses to any kind of stimulus), adulthood (during which adaptation to most commonly encountered agents has occurred and resistance is increased), and finally senility (characterized by irreversible loss of adaptability and eventual exhaustion) ending in death.

The pathways through which the stress response is mediated are extremely complex. Apart from specific changes, the first effect of a stressor acting upon the body is to produce a nonspecific stimulus. This may be a nervous impulse, a chemical substance or lack of an indispensable metabolic factor, and is referred to simply as the *first mediator* because we know nothing about its nature. We are not even certain that it has to be an excess or deficiency of any particular substance; it is possible that various derangements of homeostasis can initiate the stress response. Undoubtedly, in man, with his highly developed central nervous system (CNS), emotional arousal is one of the most frequent initiators of somatic stress; yet it cannot be regarded as the only factor capable of causing stress, since stress reactions (that is, nonspecific responses common to all demands made upon a living organism) also occur in the most primitive animals in the absence of a nervous system. But even in man, conscious psychic disturbances are not indispensable, since typical stress reactions can occur in patients exposed to stressors (for example, trauma, hemorrhage) while under deep anesthesia. Indeed, anesthetics themselves (for example, ether) are commonly used in experimental medicine to produce stress, and the stress of anesthesia is a serious problem in clinical surgery.

Although we have still to identify the first mediator(s), we do know that eventually stress acts upon the hypothalamus and particularly the median eminence (ME). This action appears to be regulated largely by means of (mediated through or modified by) nervous stimuli coming from the cerebral cortex, the reticular formation and the limbic system (especially the hippocampus and amygdala).

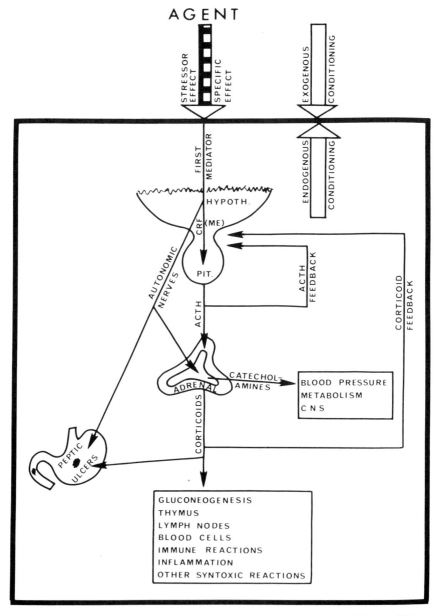

FIG. 1. *Principal pathways mediating the response to a stressor agent and the conditioning factors which modify its effect.* As soon as any agent acts upon the body (outer frame of the diagram) the resulting effect will depend upon three factors (broad vertical arrows point-

The incoming nervous stimuli reach certain neuroendocrine cells, most of which are located in the ME. These act as transducers, transforming nervous signals into a humoral messenger, the corticotrophic hormone-releasing factor (CRF) which can be demonstrated histochemically in the ME region and can also be extracted from it. Oddly enough, the posterior pituitary contains the highest concentration of CRF, and it has been isolated from this source in pure form, thus permitting the determination of its chemical formula as a polypeptide. Yet, we have no absolute proof that the CRF-active material extracted from the hypothalamus is identical with that obtained from the posterior lobe since only the structure of the latter has been definitely ascertained. Although vasopressin (antidiuretic hormone) possesses considerable CRF activity, it is not identical with CRF; this has been shown by the well-documented differences in their chemical structure and physiologic activity.

CRF reaches the anterior lobe through the hypothalamo-hypophyseal portal system that originates in the ME region within a network of capillaries into which CRF is discharged by the local neuroendocrine cells. It is then carried down through the larger veins of the pituitary stalk to a second capillary plexus in the pituitary.

The hypothalamus does not stimulate the adrenocorticotrophic hormone (ACTH) secretion of the anterior lobe through nervous pathways descending in the pituitary stalk but rather through blood-borne substances carried by way of the portal veins. That is why transection of the stalk inhibits ACTH secretion only before vascular connections between the hypothalamus and the gland are reestablished. If regeneration of these vessels is prevented by interposing a plate between the cut ends of the stalk, this pathway is permanently blocked.

Both *in vivo* and *in vitro* experiments have proven that CRF elicits a discharge of ACTH from the adenohypophysis into the general circulation. Upon reaching the adrenal cortex, it causes secretion of corticoids, mainly glucocorticoids, such as cortisol or corticosterone. These induce glyconeogenesis, thereby supplying a readily-available source of energy for the adaptive reactions necessary to meet the demands made by the stressors. In addition, they facilitate various other

---

ing to the upper horizontal border of the frame). All agents possess both nonspecific stressor effects (solid part of arrow) and specific properties (interrupted part of arrow). The latter are variable and characteristic of each individual agent. They are inseparably attached to the stressor effect and invariably modify it. The other two heavy vertical arrows, pointing toward the upper border of the frame, represent exogenous and endogenous conditioning factors which largely determine the reactivity of the body. It is clear that since all stressors have some specific effects, they cannot elicit exactly the same response in all organs; furthermore, even the same agent will act differently in different individuals, depending upon the internal and external conditioning factors which determine their reactivity. (From *Stress in Health and Disease*, Reading, Mass.: Butterworths, 1976)

enzymatically-regulated adaptive metabolic responses and suppress immune reactions as well as inflammation, assisting the body to coexist with potential
pathogens (syntoxic reactions). Furthermore, the glucocorticoids are responsible
for the thymicolymphatic involution, eosinopenia and lymphopenia characteristic of acute stress. Curiously, glucocorticoids are needed for the acquisition of
adaptation primarily during the alarm reaction, but not so much to maintain the
adjustment during the stage of resistance. ACTH plays a comparatively minor
role in the secretion of mineralocorticoids, such as aldosterone, which is regulated mainly by the renin-hypertensin system and the blood electrolytes whose
homeostasis is in turn influenced by them.

This chain of events is cybernetically controlled by several biofeedback
mechanisms. Whether an excess of CRF can inhibit its own endogenous secretion is still doubtful because its lifespan in the circulating blood is very short.
On the other hand, there is definite proof of an ACTH feedback (short-loop
feedback) by a surplus of the hormone, which returns to the hypothalamo-
pituitary system and inhibits further ACTH production. We have even more
evidence to substantiate the existence of a corticoid feedback mechanism
(long-loop feedback) in that a high blood corticoid level similarly inhibits ACTH
secretion. It is still not quite clear to what extent these feedbacks act upon the
neuroendocrine cells of the hypothalamus, the adenohypophysis or both.

A second important pathway that mediates the stress response is carried
through the catecholamines liberated under the influence of an acetylcholine
discharge, at autonomic nerve endings and in the adrenal medulla. The
chromaffin cells of the latter secrete mainly epinephrine (EP), which is of
considerable value in that it stimulates mechanisms of general utility to meet
various demands for adaptation. Thus, it provides readily available sources of
energy by forming glucose from glycogen depots and free fatty acids from the
triglyceride stores of adipose tissue; it also quickens the pulse, raises the blood
pressure to improve circulation into the musculature, and stimulates the CNS.
In addition, EP accelerates blood coagulation and thereby protects against excessive hemorrhage should wounds be sustained in conflicts. All of this is helpful
in meeting the demands of "fight or flight."

The Thyroid Gland

N. Vellani F.R.C.S.(I)., F.R.C.S.(C)
Sarnia, Ontario, Canada

The thyroid is an endocrine gland situated low in the anterior aspect of the neck and is concerned mainly with the maintenance of metabolic equilibrium, though it does have other functions.

DEVELOPMENT

Early in fetal life the gland commences development as an outgrowth of the pharynx involving the third, fourth and sixth branchial arches at the site marked by the foramen cecum in the posterior part of the tongue; thence, it grows downward in the midline to the thyroid cartilage where it spreads to occupy its final position. The importance of this development is that, on rare occasions, the only functional thyroid tissue in the body may be represented by a swelling at the back of the tongue and inadvertent removal of this "lump" would render the person permanently hypothyroid, necessitating life-long therapy with exogenous thyroid hormone; it is important also, in recognizing other thyroid congential abnormalities.

ANATOMY

The gland consists of two, roughly pyramidal, lateral lobes extending from the mid-thyroid cartilage to the sixth tracheal ring; they are connected across the midline by the isthmus which covers the second, third and fourth tracheal rings. It has two capsules, the true and false; superficially, it is related to two muscles, the infra-hyoid group and sterno-cleido-mastoid; medially are two tubes (the esophagus and trachea), two nerves (the recurrent and external laryngeal), and two muscles (the crico-thyroid and inferior pharyngeal constrictor). Posteriorly are two arteries (the common carotid and termination of the inferior thyroid).[1] Its blood supply is by means of two major arteries, the superior and inferior thyroid, and two minor ones, the thyroidea ima which is inconstant, and ac-. cessory arteries from vessels to the esophagus and trachea. All the main arteries may be ligated and the gland yet have reasonable blood supply from the accessory arteries.[2]

PHYSIOLOGY

The daily intake of iodine is approximately 150 μg of which two thirds is excreted in the urine and one third used to synthesize thyroid hormone. The iodine is coupled with tyrosine to form iodotyrosines and then tri-iodothyronine and thyroxine, T_3 and T_4 respectively; various factors, drugs, and chemicals may interfere with these processes and are, at times, utilized for therapeutic purposes, e.g., in hyperthyroidism to block the synthesis of thyroid hormone. The thyroid gland contains at least one month's supply of hormone which is stored as the protein thyroglobulin in the acini of the gland; this is in contradistinction to other endocrine glands which store hormones inside their cells. Release from the cell is effected by pinocytosis and the action of intracellular lysozomes which eventually results in the production of T_3 and T_4.[3] T_3, though far less than T_4 in concentration, is turned over much more rapidly, which in conjunction with its greater biological activity (3:1) makes it greatly more effective than its concentration would imply. The concentration of thyroxine in circulation is approximately 7 mg/100 ml and that of tri-iodothyronine 0.3 mg/100 ml; the remainder is bound to the plasma-binding proteins. There is evidence that a large proportion of T_4 is converted into T_3 peripherally.[4] In the circulation, T_3 and T_4 are carried bound to plasma proteins called thyroid binding globulin, albumin and pre-albumin, and any alteration in the concentrations of these binding proteins, as occurs with the birth control pill and liver disease, will inevitably lead to distorted values for T_3 and T_4 because the test results are dependant on the binding proteins, excepting the free thyoxine level. It is possible to have genetic disorders wherein the binding proteins may be elevated or depressed. T_3 and T_4 are metabolized by the kidney and liver and excreted in the urine and feces.

The hierarchy of control of hormone secretion commences with the hypothalamus which secretes the thyroid releasing factor (T.R.F.) a tripeptide. This reaches the pituitary via the portal system and results in the secretion of the thyroid stimulating hormone (T.S.H.) a glycoprotein which acts upon the thyroid gland to produce the thyroid hormone. Control of these processes is by a feedback mechanism. The thyroid hormone is necessary for the development of the central nervous system, skeleton and normal growth; its action primarily controls intracellular metabolism although the details are not yet known.

HYPERTHYROIDISM

Also known as thyrotoxicosis, this condition is due to an excess of hormone, either T_3 or T_4 though the former is rare and great circumspection is needed before the diagnosis is made, e.g., the T_3 level may be raised but this could be due to a response to equilibrate the overall $T_3 T_4$ effect should the T_4 level be depressed, but not necessarily below normal. Hyperthyroidism can be due to Grave's disease, a diffuse, symmetrical enlargement of the thyroid gland, an "hot" nodule or rarely, thyroid or other malignancy. Grave's disease is the commonest cause and evidence is accumulating that it may be due to an autoimmune process,[5] as other such conditions, e.g., pernicious anaemia, idiopathic thrombocytopaenic purpura and Hashimoto's disease have been found in relatives of patients with Grave's disease;[6] also, thyroid antibodies are found in the vast majority of patients. Classically, this disease results in a generalized, smooth, symmetrical goiter with the ocular sign whereby the sclera is visible all round the iris; this is in contradistinction to a toxic nodule, multinodular goiter or toxicity without any goiter; (the latter is more likely in the elderly where the picture may be clouded by such facts as that they may present in cardiac failure and with few, or no, other signs of thyrotoxicosis) finally, certain malignant tumors may cause hyperthyroidism.

The disease is most common in females in the third, fourth, and fifth decades; in a textbook case the person presents with a diffuse goiter, a staring gaze, shaky hands and excessive sweating; this, however, is not always the case; symptoms may be minimal and consist of a multiplicity of complaints such as altered menstruation, diarrhea, polyuria, dyspepsia, alopecia, chest pain and emotional disturbances; other symptoms include heat intolerance, irritability, lymphadenopathy, weakness, weight loss despite increased appetite, palpitations, eye complaints, fatigue and pressure effects such as dysphagia and dyspnoea. The commonly found signs are a goiter, tachycardia which is almost invariable and may be associated with an arrhythmia, a warm, moist skin, staring eyes with the sclera clearly visible around the iris, a fine tremor, weight loss, thinning of hair and skin, gynaecomastia, osteoporosis and pretibial myxoedema which is a thickening of the skin and subcutaneous tissues usually in the upper leg; the sudden onset of atrial fibrillation can be the initial sign. In the elderly, symptoms may manifest atypically as with cardiac failure, excessive fatigue and weakness; it should be noted that ocular signs do not accompany toxicity due to an adenoma.

DIAGNOSIS

The basal metabolic rate (B.M.R.) which measures oxygen consumption under basal conditions, is raised; however, this is a fairly crude test though some feel that serial B.M.R. measurements can be more indicative of control of the hyperthyroidism than serum hormone estimations. The protein bound iodine (P.B.I.) is also raised but this is not used much now as an index as it can be affected by some drugs (e.g., cough medicines with iodine), the contraceptive pill, hereditary defects, liver and renal disease. The serum T_4 level is increased as is the free thyroxine and T_3 resin uptake, which is an indirect calculation of T_3 by measuring the unbound thyroid binding globulin sites and T_4. The level of the thyroid binding proteins can affect other tests and therefore no one test, with the possible exception of the free thyroxine, is an accurate reflection of the thyroid status of a person; renal and liver disease, as well as other factors, can affect the serum T_4 level reading; therefore, the free thyroxine is a more reliable index and if the facilities for its estimation are not available, calculation of the free thyroxine index is an admirable substitute; as noted above, it is possible to have a normal T_4 but an elevated T_3, the so-called T_3 toxicosis. In Grave's disease, in the radio-iodine studies, there is a generalized increased uptake whereas with a toxic nodule only the nodule would show increased activity; also, the uptake cannot be suppressed by the administration of T_3 unlike other causes of increased uptake such as hereditary defects of hormone synthesis, iodine deficiency or the recovery phase of sub-acute thyroiditis.[3] The T_3 suppression test should not be performed on patients with a compromised cardiac status as it may precipitate cardiovascular complications. The long-acting thyroid stimulator (L.A.T.S.), a gamma globulin and possible thyroid autoantibody, is raised in Grave's disease but not in toxicosis due to an adenomatous hyperfunctioning nodule; also, the thyroid stimulating hormone (T.S.H.) is almost always undetectable "confirming the importance of abnormal stimulators or autonomous hyperfunction as the main factor in this disorder."[7] Finally, administration of the T.R.F. results in secretion of the T.S.H., but in hyperthyroidism there is naturally, a failure of this response; a normal response almost eliminates the possibility of hyperthyroidism, put simply, an elevated T.S.H. virtually excludes the diagnosis of hyperthyroidism.[8]

TREATMENT

Three forms of therapy are available: (1) drugs, (2) surgery, (3) radioactive iodine. Drug therapy may consist of the use of carbimazole, neomercazole methimazole or propyl thiouracil all of which act by preventing the conversion of inorganic to organic iodine or coupling, of the iodotyrosines, treatment by this modality is necessary for 12 to 18 months and has a 50% success rate. Periodic white blood cell counts need to be done as agranulocytosis is one of the complications encountered; other complications are various drug reactions, patient noncompliance because of the length of treatment and failure of cure; during therapy the symptoms abate, albeit slowly, and suppressibility may return; should this be the case there is a higher chance of remission; should this not occur within 6 to 9 months, surgery should be considered.[3] Symptoms of an hyperkinetic nature, e.g., sweating, tachycardia and tremor may be controlled by the β-blocker propranolol and/or tranquilizers.

Surgery may be performed only after the patient has been rendered euthyroid as otherwise the danger of a "thyroid storm" with it attendant morbidity and mortality is greatly increased; subtotal thyroidectomy or hemi-thyroidectomy is probably the treatment of choice. The drawbacks of surgery are:

1. a mortality of approximately 1% even in expert hands.
2. ensuing hypothyroidism in up to 40% if followed for up to 20 years.
3. persisting hyperthyroidism.
4. hypoparathyroidism.
5. recurrent laryngeal nerve damage with consequent voice changes.

Its advantages are in the rapid abatement of toxicosis, relief of pressure symptoms, and the possible exclusion of malignancy; surgery is the best treatment if there are pressure symptoms or if there is a suspicious nodule.

Treatment by radioactive iodine is usually restricted to adults although some centers treat teenagers by this method which is simple and safe; however, following therapy, the patients must be followed carefully as up to 50% will develop hypothyroidism in time; additionally, up to 50% may be hyperthyroid at the end of a year and hence require a second, and occasionally more, treatments with radioactive iodine; often the second treatment may be given after four to six months following the initial one. In some centers the thyroid gland is heavily irradiated to induce hypothyroidism and the patient subsequently placed on life-long replacement therapy. The possibility of inducing carcinoma by this form of treatment has not totally been excluded.

For the person with a nodule or multinodular goiter who is euthyroid, treatment tends to be medical though if there is any fear of malignancy surgery is indicated; a nodule which is "cold" on radioactive iodine studies (that is, does not take up radioactive iodine) is much more likely to be malignant than an "hot" one, as is also the case if it is enlarging rapidly. Following thyroidectomy exogeneous hormone is given in doses to suppress the T.S.H. secretion and thus, further thyroid activity; multinodular goiters respond well to suppression therapy though they should be followed closely for the first few months to ensure adequate response; should this not occur, surgery is indicated.

HYPOTHYROIDISM

The major cause of this condition is previous treatment for other thyroid disorders; in the treatment of hyperthyroidism by any modality a strong possibility exists for the eventual development of hypothyroidism or myxedema; it may also follow surgical treatment of a multinodular goiter or carcinoma and may ensue from thyroiditis; finally it may be congential, or spontaneous in origin; when the former, it could be due to defective hormone synthesis or absent thyroid and in the latter, of unknown cause. Congenital hypothyroidism is known as cretinism and in this condition early diagnosis is essential as mental development, as well as skeletal, is dependent on adequate levels of hormone; should this be lacking, mental retardation is sure to follow and lost ground cannot be retrieved. Clinically, the picture is one of delayed mental development, delayed skeletal development and thus short stature, coarse brittle hair, dry skin, a protruberant abdomen often with an umbilical hernia and a protruding tongue. X-rays show delayed bone age and epiphyseal stippling; the T_4 is low and the T.S.H. raised; this latter is the earliest sign of hypothyroidism. The outlook for full development is much better with the later onset of the condition but with the congenital type normal mental development is hardly, if ever, achieved. It is possible to have myxoedema with a goiter; this could occur in circumstances where there is iodine deficiency which results in increased T.S.H. secretion and in time, a goiter.

In adults, hypothyroidism is more common in females, and is often associated with circulating autoantibodies; as noted above, it may follow previous therapy for thyroid disease or thyroiditis; rarely it is due to insufficient T.S.H. secretion by the pituitary and is known as secondary hypothyroidism; a normal plasma cortisol would tend to mitigate against this diagnosis as usually other anterior pituitary hormone deficiencies occur; even more rarely the hypothalamus does not secrete T.R.F., resulting in tertiary hypothyroidism; a normal T.S.H. response to T.R.F. excludes secondary hypothyroidism and implicates the hypothalamus.

As noted above, the earliest indication of decreasing thyroid function is an elevated T.S.H.; in addition, the T_3 and T_4 are decreased as is the radioactive iodine uptake; increase in T_4 and ^{131}I uptake after administration of T.S.H. would favor secondary hypothyroidism. Many patients will have Hashimoto's Disease (thyroiditis) and therefore, the occurrence of other autoimmune diseases will not be unexpected.

The clinical picture is that of slowed metabolism; fatigue, bradycardia, weight gain, cold intolerance, dry skin, coarse hair, loss of the outer $\frac{1}{3}$ of the eyebrows, puffy facies, anemia which may be macrocytic though usually is not, E.K.G. changes, constipation, anorexia, atrophic gastritis and achlorhydria in up to 50% of patients[3] and mental deterioration in the shape of sleepiness, dullness and forgetfulness; often there is an obvious increase in the relaxation time of the deep tendon reflexes especially that of the Tendo Achillis; ptosis is not uncommon and neither are menstrual irregularities and infertility. It is extremely important to differentiate psychiatric states from that due to myxedema, as many are the patients with hypothyroidism who have been placed in psychiatric institutions on a mistaken diagnosis.

Treatment in the vast majority of cases is by means of synthetic T_4 (*l*-thyroxine) as its composition is constant and its effect therefore predictable; also, in view of the fact that T_4 is converted to T_3 peripherally combination therapy is not necessary; however much as dessicated thyroid preparations may be decried, there is the occasional patient who apparently does not respond to T_4 but does to dessicated thyroid. Some authorities use T_3 though it has been shown that this achieves a rapid peak and far less prolonged, and therefore less physiological, effect, none-the-less, some patients will respond only to T_3 treatment.

The dosage should increase in small increments over a period of weeks; in the elderly and in severely myxedematous patients, increases should occur over a longer period because of the hazard of precipitating cardiac failure, angina, arrhythmia or even myocardial infarct. Therapy is for life and is usually satisfactorily controlled by clinical monitoring of the weight, sleeping pulse, and general state of the patient, although some recommend periodic assessment of serum T_4. Treatment of secondary hypothyroidism without establishing the diagnosis may provoke a supra-renal crisis.

It is possible for a patient to go into, or present with, myxoedema coma; this is a serious condition but its high mortality has been reduced since the introduction of intravenous thyroxine; the mortality used to be as high as 75% and the condition is characterized by severe hypothermia, sometimes as low as 25°C, together with other features of myxedema.

THYROIDITIS

Infections: Bacterial and fungal infections are very rare.

de Quervain's Disease (sub-acute thyroiditis).

There are a multitude of synonyms for this condition; though its etiology is unproven, it is generally thought to be viral in origin; it has not been associated with autoimmune disease but has been associated with viral disease. It is preponderant in females and in the fourth, fifth, and sixth decades, and is self-limiting over a period of weeks to months; it presents with neck pain and a fever; often the thyroid gland is enlarged and symmetrically so, and may be quite hard to palpation as well as tender. Since the process is an inflammatory one involving the thyroid, transient elevations of serum thyroxine are not uncommon; this point should be borne in mind in the differential diagnosis of hyperthyroidism.[9]

In mild cases, treatment is with drugs like aspirin and in more severe cases with corticosteroids. Complete recovery is usual.

HASHIMOTO'S DISEASE

Now considered an autoimmune process, this condition is closely linked to Grave's disease and is characterized by a firm, symmetrical, thyroid enlargement which may or may not, be bosselated; if untreated, the majority will result in myxedema, supposedly as a consequence of the lymphocytic infiltration and fibrosis. This disease is probably the commonest cause of a goiter in the country. The most effective treatment is a suppressive dose of *l*-thyroxine up to 0.2 mg daily, the rationale being that suppression of T.S.H. secretion will modify the thyroid enlargement. There is said to be an increased incidence of lymphomata associated with this condition.

CARCINOMA

Carcinoma is four times commoner in females. In children, the majority give an history of having been radiated about ten years earlier for thymus, tonsil or acne conditions.

The tumors may be differentiated or undifferentiated; of the former, by far the greatest proportion are papillary in type and these, more often than not have elements of the follicular type interspersed; other forms are the Hurthle cell and medullary carcinoma; the undifferentiated tumors are small or large cell; multicentricity is characteristic of carcinoma of the thyroid.

Presentation is in the form of a painless nodule, one not responding to suppression, secondary deposits in the neck or pressure symptoms such as dysphagia, dyspnoea or dysphonia. Very rarely, it may present with hyperthyroidism or tetany due to a calcitonin secreting medullary carcinoma; this type may be associated with parathyroid adenoma, phaeochromocytoma, mucosal and neurofibromata and cafe au lait spots.[8]

The T_3 and T_4 are usually normal and the scan may show a "cold" nodule, which is known to be more frequently associated with malignancy. Papillary tumors, especially if detected early, are compatible with an almost normal life span; anaplastic tumors carry a grim prognosis.

Treatment is by total thyroidectomy followed soon after by a ^{131}I uptake to detect residual thyroid tissue and secondaries; should this be positive, treatment is supplemented by radioactive iodine therapy.

References

1. Garry, T. A. Personal communication.
2. Lee McGregor, A. *A Synopsis of Surgical Anatomy*. 1936, p. 20.
3. Ingbar, S. H. Editing "Thyroid today," pp. 13–16.
4. Braverman, L. E., Ingbar, S. H., Sterling, K. Conversion of thyroxine (T_4) to tri-iodothyronine (T_3) in athyreotic human subjects, *J. Clin. Invest.* **40**: 855 (1970).
5. Fialkow, P. J. *Progress in Medical Genetics,* 6: 117–167, New York: Grune & Stratton.
6. Bartuska, D. Evolving concepts of the genetics of thyroid disease, *Amer. J. Med. Sci.* **266**: 249–252 (1973).

7. Catt, K. J. ABC of endocrinology, VI. The Thyroid Gland, *Lancet* p. 1387, June 27, 1970.
8. Burrow, G. N., in Symposium on Current Concepts of Thyroid Disease. *Med. Clin. N.A.* Edited by Burrow, G. N., **59**: 5: 659: 1975.
9. Volpé, R., in Symposium on Current Concepts of Thyroid Disease. Edited by Burrow, G. N. See pages 1163-75, *Med. Clin. N.A.* **59**: 5: 1975.

Transfer Factor

RICHARD HONG, M.D.
Department of Pediatrics
University of Wisconsin Center for Health Sciences
Madison, Wisconsin

Transfer factor (TF) is a low molecular weight dialyzable material obtained from leucocytes. It is a water soluble factor obtained from leucocytes which have been disrupted by repeated freezing and thawing. It can be inactivated by heating at 56° for 30 minutes; however, it resists digestion by pancreatic ribonuclease and is extremely stable when stored in the frozen state. Properly prepared fractions have retained their potency for five years. The material does not resemble immunoglobulins nor any large fragments of immunoglobulin molecules. Although it can confer immunologic specificity, it does not contain either the relevant antigens nor does it contain the histocompatability antigens of the donor leucocytes. Transfer factor is essentially nonantigenic.

Despite its relatively small size, transfer factor is a very potent biological molecule. It appears to have the capability of transferring to individuals with negative skin tests positive cutaneous reactivity to antigens to which the donor is sensitive. It is usually necessary to use, as a donor, an individual with a very high degree of cutaneous reactivity to a given antigen. By and large, individuals who have negative skin tests cannot confer positivity to other individuals; thus, the transfer is specific. Some exceptions to this general statement have been observed, however. Transfer factor confers upon the recipient only a positive test of T cell reactivity. Effects on B cell activity, i.e., increases in specific antibody titers, have not been observed.

Supported by grants from NIH (HD-07778, AI-10404, AI-11576) and the National Foundation.

The mechanism of action of TF remains obscure. The specificity of the transfer requires that the active principle carry some structure which is unique to a given antigen. Studies a few years ago suggested that transfer factor was a polypeptide comprised of approximately 12 amino acids plus three or four RNA bases. It seems unlikely that such a small polynucleotide would have the capability of providing a specific genetic message for a particular antigen. More recently, the NIH group has further characterized the active moiety and RNA does not seem to play a role in the transfer. This information is more easy to reconcile with the existing data as it is difficult to envision how such a small amount of nucleotide could provide the specific message for a particular antigen. Kirkpatrick proposes that transfer factor consists of a mixture of polypeptides. The amino acid sequence in these small polypeptides dictates the antigenic specificity in the same way that a small stretch of ordered amino acids dictates the antibody specificity of the immunoglobulin molecules. It can be shown mathematically that with a polypeptide of six amino acids, even if five of the six amino acids must be changed in order to generate a new specificity, there can be over 500,000 different combinations. Thus, the polypeptide material appears to have the capability of generating sufficient variability to account for a tremendous number of antigens. What the polypeptide does is, of course, open to conjecture. Apparently, transfer factor acts mostly upon the T cell system which controls fungal, viral and acid fast infections as well as causes transplant rejection. Transfer factor seems to have no effect upon creating a new population of antibody producing (B cell) lymphocytes. From the foregoing, it is tempting to hypothesize that transfer factor may represent the elusive T cell receptor and that it becomes incorporated on the membrane of T cells lacking this specificity. In that way, it confers upon them the ability to react with an antigen to which they have not been previously exposed.

Transfer factor was largely a laboratory curiosity for many years until the mid-1960's when it was shown that children with chronic candida infections which could not be controlled by intravenous Amphotericin therapy could be maintained symptom-free for varying lengths of time if transfer factor was given in addition to the fungicide. This remarkable and dramatic control of a grossly disfiguring disorder led to a resurgence of interest in transfer factor as a form of immunotherapy. Since that time, transfer factor has been employed in a wide variety of diseases. The clinical experience with transfer factor in a number of disorders is summarized in Tables I and II. It can be seen that transfer factor may have some role as supportive treatment in many disorders. It does not seem to have curative powers on its own nor can it significantly restore immunity in cases of profound immunodeficiency. In evaluating the real value of transfer factor, it is necessary to look at the clinical benefit. In many cases where impressive laboratory changes have been observed, the patient has gone on to die, nevertheless.

TABLE 1. Immunodeficiency.

Disorder	Number Treated	Comment
Wiskott-Aldrich	18	Clinical improvement reported in 9/18
		5/11 improved eczema; 3/10 increased platelet count
		? associated with histiocytic malignancy
Nezelof Syndrome	7	No clear benefit
Hypogammaglobulinemia	1	Increase in IgG to levels obtained by injections (130 mg/dl)
		Decrease in IgM
Ataxia-telangiectasia	5	Skin conversion in 4/5
Severe Combined Immuno-deficiency (SCID)	3	0/3; may have produced a lymphopro-liferative disorder in one.

TABLE 2. Cancer*.

Worker	Number Treated	Type of Cancer	Comment
Vetto	35	Various	13: tumor regression; ↓ pain
			Arrest of metastases, but only of 2-12 months duration
Bukowski	6	Malignant melanoma	No benefit
	3	Renal carcinoma	
Bearden	6	Osteogenic sarcoma	Used with chemotherapy and/or radiotherapy
			3 are disease free 206-496 d after amputation
Silva	9	Melanoma	No effect in 8/1; 1 stable lesion
			↓ in disseminated vulvar neoplasm; death from other metastases
			No benefit in thymoma, ovarian carcinoma
Maini	3		
Silverman	1	Waldenstrom's macroglobulinemia	Decrease in IgM
		Multiple myeloma	No change

*From: Ascher, M. J., Gottlieb, A. A. and Kirkpatrick, C. H. (eds). *Transfer Factor: Basic Properties and Clinical Applications.* New York: Academic Press, 1976.

TABLE 3. Miscellaneous Disorders.

Disease	Number Treated	Comment
Chronic Aggressive Hepatitis	5	4 transaminase lowered
		4 histologic improvement 9 mo.–2.5 yrs. after last dose transfer factor
"Viral Diseases"	13	Includes SSPE, neonatal herpes, herpes zoster; 8/13 improved
Leprosy	9	6/9 improved
Coccidiomycosis	35	12 improved
		19 did not respond
		4 deteriorated
CMC	22	12/22 responded
		Usually requires pretherapy with amphotericin
		Two reports of fetal thymus transplant benefit after TF
Juvenile Rheumatoid Arthritis	36	No benefit differing from placebo group
Bechet's Disease	6	4 benefited; occasional transient exacerbation of lesions after TF
Multiple Sclerosis	12	2 improved

A note of caution should be sounded regarding the use of TF in immunodeficiency syndromes. Lymphoproliferative disorders and histiocytic disorders of the skin have been reported in a number of patients after transfer factor therapy. It is not fair to incriminate the transfer factor as a cause of the malignancy inasmuch as the patients involved do show a predisposition toward the development of cancer. However, the appearance of the malignancy was shortly after the institution of transfer factor therapy and one must carefully weigh the possibility of inducing this complication in a situation of predisposition.

In summary, transfer factor is a biologically important substance which has the capability of conferring antigen responsivity to immunologically naive individuals. It appears to have the capability of helping to eliminate certain infectious agents in conditions of poorly defined T cell deficiency. Its benefit in profound immunodeficiency states and malignancy is more equivocal and must await further study.

References

1. Ascher M. S., Gottlieb A. A., and Kirkpatrick, C. H. (eds). *Transfer Factor: Basic Properties and Clinical Applications.* New York: Academic Press, 1976.
2. Horowitz S. D., and Hong R. H. *Monographs in Allergy* **10:** 118 (1977).
3. Kirkpatrick C. H., and Smith T. K. *Proceedings of the Eleventh Leucocyte Culture Conference* (in press).
4. Lawrence H.S.: Advances In Immunology **11:** 195 (1969).

Human Warts

NURUL H. SARKAR, Ph.D.
Sloan-Kettering Institute for Cancer Research
New York, New York

There are many kinds of warts differing in their appearance, age groups in which they occur, their location on different parts of the body and other characteristics. However, they all have the same basic pathological features and consist of a localized hyperplasia of epithelial cells with sharply defined boundaries.

It has been shown by electron microscopy that skin warts located above the stratum of granulosum contain many cells with numerous intranuclear virus particles. At the lower keratin level there are only a few cells, but these cells contain large aggregates of papilloma virus.

Typically, thin sectioning of laryngeal papillomas and genital "wart" type lesions shows a highly cellular structure with no recognizable virus. Only rarely can micrographs of intranuclear papilloma be obtained.

The electron micrograph, shown in Fig. a, represents cells from a "wart" type genital lesion. The section represents typical epithelial cells, which are connected with each other by clearly defined trilamilar structures, described as desmosomes (arrows in Fig. a). Fig. b represents a higher magnification view of the desmosomes (D). N, nucleus; Fig. a, X1125; b, X45,000.

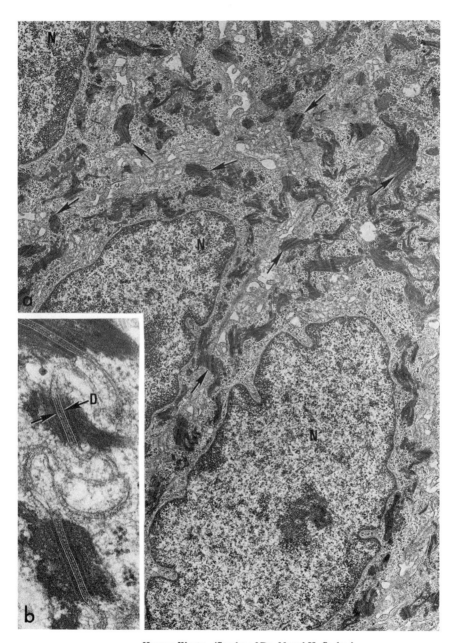

Human Warts. (Study of Dr. Nurul H. Sarkar)

Murine Breast Cancer Virus

NURUL H. SARKAR, Ph.D.
Sloan-Kettering Institute for Cancer Research
New York, New York

The only mammalian breast cancer known at the present time to be caused by a virus is murine mammary cancer. The occurrence of breast cancer varies widely among the various inbred laboratory strains of mice; in some strains 100% of the mice develop tumors between 8-12 months of age. In other strains only a low percentage of the mice get tumors even after two years. The occurrence of breast cancer among wild mice is thought to be 3-5% of the mice in the general mouse population. However, it is found that the genetic information of the endogenous virus is ubiquitous in most of the strains of mice that have been examined. The virus, described as murine mammary tumor virus (MuMTV) is transmitted mainly horizontally (infectiously) to immature mice via the mother's milk. Foster nursing of females from high tumor incidence strains by mothers from low tumor incidence results in a drastic reduction of tumor incidence among the nursing mice and their offspring. However, foster nursing cannot completely eliminate the development of mammary tumors of certain strains of mice since these strains also transmit virus vertically as an inherited autosomal trait. The virus is replicated mainly in the mammary epithelium. The milk of mice from strains with a high incidence of cancer is very rich in virus. Cells from mammary tumors cultured *in vitro*, also produce a small amount of virus. MuMTV is an enveloped virus. It is composed of single stranded RNA (about 1%), protein (70%), lipid (30%), and a small amount of an enzyme, unique for the enveloped RNA tumor viruses, the reverse transcriptase. The internal component of the virus, designated as the core (C, insert Figure) has four polypeptides. The core is assembled in the cell cytoplasm near or at a distance from the cell membrane. The cores from the interior position of the cell move to the cell membrane. These cores as well as those formed near the cell membrane bud off from the cell membrane. Although the virus acquires its membrane from the cell, the viral membrane (M) is distinct from the cell membrane biochemically, immunologically, and morphologically. The surface of the viral membrane is covered with projections (P).

The large micrograph in the figure represents a prelica of a cultured mammary tumor cell. The spherical particles (V), about 100 nm in diameter, are seen to bud from the cell surface. The border of the cell is designated B. The direction of the shadow cast by evaporated metals (during preparation of the cell replica) is shown by the arrow. X67,500; inset, X26,250.

Murine Breast Cancer Virus
Plate A

Plate B shows the surface architecture of MuMTV, purified from the mouse milk. The virions were deposited on specimen grids, freeze dried, and shadowed with platinum-carbon at an angle of 45°. The direction of shadow is shown by the thick arrow. The virons, although observed as shpere by other electron microscopic techniques, appears here as icosahedral, (six corners of a virus are pointed by six arrows) and the viral surface is found to contain evenly spaced indentations each of which are surrounded by six other indentations, approximately 21nm apart. ×120,000 (Nurul H. Sarkar).

Murine Breast Cancer Virus
Plate B

Herpes Virus—Nucleocapsid

Nurul H. Sarkar, Ph.D.
Michael I. Bernard, Ph.D.
Sloan-Kettering Institute for Cancer Research
New York, New York

Herpesviruses are large (200nm in diameter), spherical, DNA-containing, lipid-enveloped viruses. Within the viral envelope the major internal structure is the nucleocapsid (hexagonal in cross-section, approximately 100nm in diameter). The surface of the nucleocapsid (Figs. a and b) is composed of 162 capsomeres, 150 of which are hexagonal in cross-section and 12 of which are pentagonal. The capsomers are arranged in a $5:3:2$ symmetry. Within the nucleocapsid is a core structure containing the double stranded viral DNA ($\pm100 \times 10^6$ daltons). Current information indicates that the nucleocapsid contains six proteins.

Within the infected cell, viral proteins and glycoproteins are synthesized in the cytoplasm, to the exclusion of normal cellular functions. The viral proteins migrate into the nucleus where the nucleocapsids are assembled. The nucleocapsids become enveloped as they bud through the nuclear membrane. The final effect upon the host cell is lysis and cell death.

Herpesviruses are very common and their effects upon their natural hosts range from mild infections to severe disease. Cross-species infections may be especially severe, as in the case of the B virus of monkeys which is innocuous in monkeys, yet is fatal in man. Herpesviruses are also implicated as the probable etiologic agent of several naturally occurring cancers: the Lucke herpesvirus of frogs is associated with the Lucke renal adenocarcinoma, Marek's disease virus of chickens and the associated lymphoma. In man there is strong evidence linking the Epstein-Barr herpesvirus to Burkitt's lymphoma: there is statistical evidence indicating an association between herpes simplex virus type 2 and cervical carcinoma; and there is evidence indicating that other human herpesviruses may be able to induce cancers following suitable manipulation of the virus.

An unusual characteristic of the herpesviruses is their ability to cause latent infections in which the viral genome is present (generally in nervous tissue) but inactive. Numerous factors may cause reactiviation of the viral genome and recurrence of the associated disease (herpes labialis is the most common example).

There are five known herpesviruses of man: herpes simplex type 1 (herpes labialis, herpes encephalitis), herpes simplex type 2 (genital variant of type 1), varicella-zoster (chicken pox and shingles), cytomegalovirus (congenital cytomegalic inclusion disease and other cytomegalic inclusion diseases)

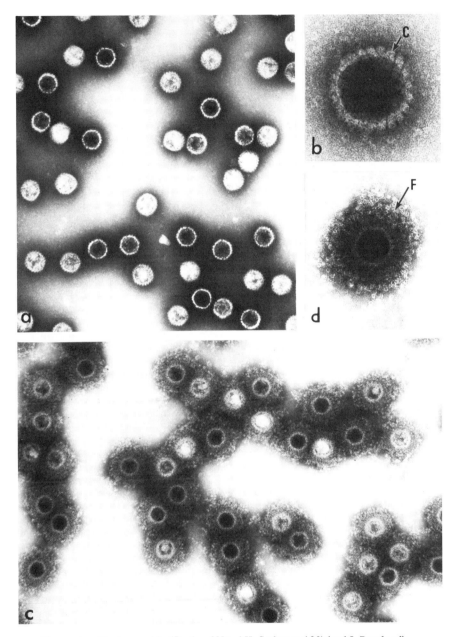

Herpesvirus-Nucleocapsid. (Study of Nural H. Sarkar and Michael I. Bernhard)

and the Epstein-Barr herpesvirus (Burkitt's lymphoma and infections mononucleosis).

Current techniques allow the purification of inact virions or nucleocapsids. The figure shows a preparation of purified nucleocapsids. The capsomeres are clearly visible in Fig. b. Using the technique of immunoelectron microscopy it is possible to localize viral antigens on the nucleocapsid surface, as is illustrated in Figs. c and d.

(a) Herpes simplex virus type 1 nucleocapsids (100nm in diameter) showing capsomeres on surface;

(b) Higher magnification of a nucleocapsid, arrow (c) pointing to capsomere on nucleocapsid surface.

(c) Herpes simplex type 1 nucleocapsids in positive immunoelectron microscopy reaction with rat antiserum prepared to purified equine herpesvirus type 1 and hybrid anti-rat/anti-ferritin serum. Ferritin granules are clearly visible surrounding the nucleocapsids. A higher magnification view (d) shows detail with arrow (f) indicating ferritin. Figs. a and c, ×60,000; Fig. b, 28,500; Fig. d, 10,500.

Murine Leukemia Virus

NURUL H. SARKAR, Ph.D.
Sloan-Kettering Institute for Cancer Research
New York, New York

Murine leukemia virus, (MuLV) often described as a type C particle, is the causative agent of leukemia in mice. Particles similar in ultrastructural details are widespread among vertebrates and some are known to cause a variety of disease. There are numerous biologically, biochemically, and immunologically well characterized mouse leukemia virus strains. The most widely studied strains, are the Gross virus, and the virus strains of Friend, Rauscher, and Moloney.

The size of the virus particles varies between 95–110nm in diameter. The virus has two easily recognizable structural elements, the membrane and the core. The viral core is composed of about four polypeptides, single stranded RNA and RNA-dependent DNA polymerase (reverse transcriptase), which plays a major role in the integration of the viral genetic information into the host cell genome.

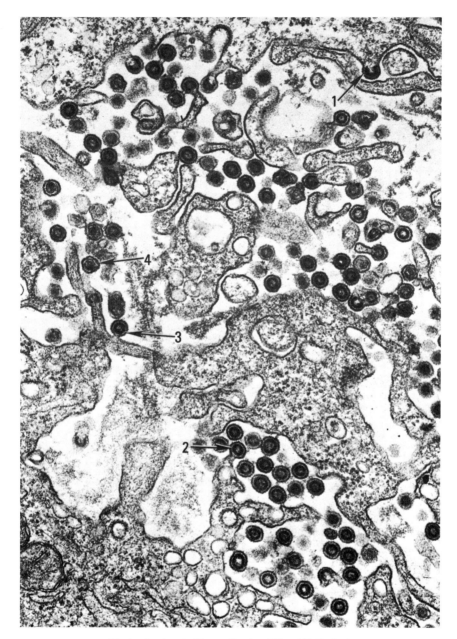

Murine Leukemia Virus. (Study of Nurul H. Sarkar)

Type C particles arise by the process of budding at the plasma membrane as is illustrated in the accompanying micrograph of a mouse cell section. The earliest stage in the development of this virus is characterized by a localized curved protrusion of the cell membrane associated with an underlying electron dense material. As the budding progresses the protrusion of the cell surface becomes greater and the electron dense material appears as a crescent like structure (marked in Figure). Next, the electron dense material underneath the cell membrane is completed to a sphere described as viral core, (which in section appears as a donut), and a definite amount of cell membrane is organized to enclose the core completely (Stage 2 and 3). Finally, the particles are pinched off from the cell membrane and the core component of the virus is rearranged to appear as collapsed structure, often angular in profile. This form of the particle is called mature leukemia virus. X45,000.

Slow Virus Infection

CARLOS LOPEZ, Ph.D.
Sloan-Kettering Cancer Research Institute
New York, New York

The concept of "Slow Virus Infections" was developed by Bjorn Sigurdsson in a series of lectures given in 1953. By definition, these are infections which develop insideously after a prolonged but rather predictable period of time. The clinical course is protracted and the disease is always fatal. Although virus-induced malignancies fit this definition, by common consent they are not included. Slow virus infections can be considered as two different groups: (1) slowly progressive infections caused by conventional viruses in which the immune response plays a significant role and (2) the subacute spongiform viral encephalopathies caused by unconventional agents. The study of animal models has provided much information on the agents which cause these infections and the pathogenesis of both types of slow virus infections.

The immune response appears to play two important roles in slowly progressive infections caused by conventional agents: A deficiency of the immune response is thought to allow the persistence of the virus infections and the immune response to virus antigens causes the immunopathology. Aleutian disease of mink is caused by a parvovirus which replicates in macrophages of inoculated mink. Although all color phases of mink are affected, those homozygous for the autosomal recessive Aleutian gene are the most susceptible to the disease. Initial viral replication is rapid and high concentrations of virus can be found in spleen and liver ten days after inoculation. Sixty days and more after infection, infectious virus can be found in the serum of mink as a nonneutralized virus-antibody complex. Tissue lesions are caused by the deposition of complexes in the kidneys and blood vessels. Most mink die of renal failure due to glomerulonephritis or due to rupture of blood vessels which demonstrate arteritis. Immunosuppression prevents tissue lesion formation indicating their immunopathologic nature. Infection of the appropriate hosts with other viruses can cause persistent infections and a slowly progressive clinical picture. Viruses such as lymphocytic choriomeningitis virus, lactic dehydrogenase virus, and equine infectious anemia virus have very different physical/chemical properties but can each cause a persistent infection. In each case, the humoral immune response to the virus causes the immunopathology. These viruses also have in common, the capacity to replicate in hosts' macrophages, a property which may allow the establishment of the persistent state.

Maedi and visna are slowly developing infections of sheep which are caused by a retrovirus. This virus replicates rapidly after inoculation but signs of disease are not usually apparent until two years later. Most infected sheep demonstrate the pulmonary disease, or maedi. Visna is a sporatic neurologic disease caused by infection with the same virus. Breeds of sheep have been shown to vary in their susceptibility to this viral infection. Histologically, maedi is characterized by infiltration of reticular and lymphoid cells into the interalveolar septa of the lungs. This eventually obliterates the alveoli and the sheep die of anoxia. The infiltration and proliferation of lymphoid cells in the brain (visna) is associated with widespread demyelination of the white matter which leads to paralysis. These are persistent virus infections in which a lymphoreticular response causes the immunopathology. Presumably, an immunologic defect allows the persistence of these infections.

In man, subacute sclerosing panencephalitis (SSPE) is a progressive neurological deterioration which follows years after a primary measles virus infection. Measles virus antigen can be detected in brain cells and can be isolated by special culture techniques. The means by which measles virus persists is not known but is thought to be because of an immunologic defect. The pathologic lesions appear to be caused primarily by the immune response to the persistently infected brain cells—perhaps complement mediated immune cytolysis. Similar infections with rubella virus have also been described.

Progressive multifocal leukoencephalapathy (PML) is a rare demyelinating disease of the brain of man caused by a papovavirus. Patients with PML always have some concurrent condition which is highly immunosuppressive and apparently allows expression of a persistent infection. In this infection, an immune response appears to play only a minor role in the pathology.

The subacute spongiform virus encephalopathies are slow virus infections caused by unconventional agents. Scrapie is a degenerative disease of the central nervous system of sheep. Clinical signs include changes in behavior, signs of cutaneous irritation, tremor, incoordination of gait, aimless wandering, visual impairment, and terminal prostration. Pathologic lesions consist of degeneration of neurons, hypertrophy and hyperplasia of astrocytes, and variable spongy alteration of gray matter. There is no evidence of an inflammatory response nor overt destruction of myelin. All animals demonstrating clinical signs of scrapie, die, usually six weeks to six months after insideous onset. The scrapic agent has been shown to be remarkably resistant to heat and formalin and relatively resistant to ether and low or high pH. Its resistance to gamma and ultraviolet irradiation indicate an unusually small inactivation target size. This has lead to speculation about the very nature of this agent. Although still not known, scrapie agent may be a viroid: a small piece of replicating nuclei acid tightly bound to, and protected by, cell membrane material. This agent replicates in lymphoid tissue (for months or years) without causing detectable lesions and eventually spreads to the brain. Mink encephalopathy is similar to scrapic and may be caused by the same or a similar agent.

Kuru is a disease endemic to the fore people of New Guinea. The histopathological lesions are similar to those described for scrapie and the properties of the agent which causes this infection are also similar. The host range of Kuru is different, however, requiring that it be studied in Chimpanzees and monkeys. Kuru is clinically characterized by cerebellar ataxia and shiveringlike tremor which progresses to complete motor incapacity and death about one year after onset. The practice of ritual cannibalism and subcutaneous self-inoculation with highly infectious brain material appears to have been the mode of spread of the Kuru agent and the reason for its being endemic in New Guinea. Cessation of this practice has reduced dramatically the incidence of this infection.

Creutzfeldt-Jakob (C-J) disease, one of the presenile dementias of man, also is caused by an unconventional agent similar in physical/chemical properties to scrapie. Cases of C-J disease are usually sporadic throughout the world although several families have been documented with multiple cases indicating a genetic component. The latter were also transmissable to chimpanzees. The origin and mode of spread of the agent of C-J disease are unknown. Since the presenile dementias were found to be caused by an unconventional agent, the possibility that other insideous neurological disorders are also due to such agents must be considered.

Immune Response to Viruses

CARLOS LOPEZ, Ph.D.
Sloan-Kettering Cancer Research Institute
New York, New York

The interactions between an invading virus and its host are complex and, depending on the method by which the virus spreads, either the humoral immune response or the cell-mediated immune response (CMI) will have the major responsibility of clearing that infection. The virus/host interaction can, on the other hand, result in immunologic injury which makes up some or most of the pathology associated with the infection.

Primary and secondary immunodeficiency disorders have been used as "experiments of nature" for defining the various aspects of the immune response required for controlling specific virus infections. Patients with agammaglobulinemia were found to have little or no difficulty with most virus infections. Since, however, they sometimes became reinfected with viruses, the lack of specific antibody was expressed as a lack of immunity to a second challenge with the same virus. There were, in addition, certain virus infections which gave agammaglobulinemic patients special trouble. Hepatitis virus infections in these patients are severe and often life-threatening indicating that a competent humoral immune system is necessary for resistance to hepatitis virus infections.

Patients with primary or secondary deficiencies of the cell-mediated arm of the immune response have great difficulty with many virus infections. Virus infections which, in normal individuals, are usually very mild or without clinical symptoms, are often severe and life-threatening in patients with a congenital impairment of the CMI response. Cancer patients, for example, whose CMI response is suppressed by their disease or the cytotoxic drugs used for its treatment, and bone marrow or renal allograft recipients whose CMI is suppressed by cytotoxic drugs, steroids, and antilymphocyte serum de.nonstrate a high incidence of virus infections, some of which are severe and life-threatening. Specific dysfunction of neutrophils has not been associated with severe virus infections.

The way a virus spreads is an important determinant of the immune mechanism responsible for controlling that infection. Virus infections which spread predominantly extracellularly, that is, virus replication in cells leads to rupture and release of virus which infects other cells, are usually controlled by neutralizing antibody. Virus infections which can spread from cell-to-contigious cell cannot be stopped by neutralizing antibody and are the responsibility of the CMI system. The CMI is complex and involves interactions between subpopulations of lymphoid cells and protein mediators, which make up the communication system between the cells. The CMI is thought to control the spread of virus infections in the following ways: (1) Lymphocytes might recognize viral antigens on the surface of virus infected cells and kill these cells prior to release of new infectious virus, (2) Lymphocytes might recognize specifically bound antiviral-antibody on the surface of infected cells and kill these cells prior to viral spread, (3) Lymphocytes might recognize viral antigens and produce protein mediators which induce macrophages to kill virus or virus infected cells, and (4) Lymphocytes might, in response to viral antigen, make and secrete interferon which limits the infection by reducing, at the site of infection, the number of infectable cells. The CMI response may also work in other ways.

Immunogenetic studies indicate that immune response genes within the major histocompatibility region of the mouse (H-2) play an important role in resistance to certain virus infections of mice. These genes are thought to determine the immune response required for control of the invading virus. Other genes, outside of the H-2, also have been implicated in resistance to virus infections of mice. Some are thought to function by restricting virus replication in macrophages or structural cells. Although the specific function of other genes is not known, they probably affect the strength of the immune response. Study of these mechanisms will, in the future, probably indicate the specific deficiencies in otherwise normal individuals who suffer from serious virus infections.

Recently, lymphocyte killing of virus infected cells was shown to be restricted to cells sharing histocompatibility antigens with the cytotoxic lymphocytes. These experiments indicate that lymphocytes probably recognize viral antigen in combination with host cell antigen. The virus infection of the cells either induces altered cell antigens or virus antigens are incorporated into cell membrane in close proximity to host cell antigens. Lymphocytes then only kill cells expressing the "altered self" antigen or combination of "self plus virus antigen."

In certain virus infections, the immune response to the virus infection, and not the lytic effect of the virus, causes the pathology associated with the infection. Lymphocytic choriomeningitis (LCM) virus infection of mice with a suppressed CMI response develop a carrier state with only minimal pathology. In immunocompetent mice, LCM infection leads to a rapidly fatal infection characterized by an acute inflammatory response in neurological tissues. The CMI response to infection, rather than the infection, caused the acute pathology and symptoms. The immunopathology associated with dengue shock syndrome appears to be caused by the humoral immune response. A second infection with another of the four serotypes of this virus leads to a massive secondary antibody response to cross-reacting antigens. Since the antibody is to *nonneutralizing* antigens, the second virus infection proceeds unabated. Antigen/antibody complexes are formed which activate complement and eventually cause histamine release, increased vascular permeability, and hypovolemic shock.

After primary infection, certain viruses have the capacity to establish a latent infection which can be reactivated at some later date. Herpesvirus infections appear to be inevitably followed by latent infections in the face of an immune response. These latent infections can be reactivated by certain stimuli including immunosuppression. The humoral immune response appears to help maintain latency while the CMI appears to be responsible for controlling reactivated infections.

Certain virus infections develop into persistent infections in the presence of a humoral immune response. The antibody response does not control the virus infection and often participates in the immunopathology caused by the infection. Presumably, persistent virus infections are established because of a deficiency in the host immune system required for controlling that virus infection.

Infection with a variety of viruses alters immune function. Virus infections have been shown to suppress CMI and humoral immune responses. This appears to be caused in some cases by a direct cytolytic effect of the virus on cells of the lymphoid system. Suppression, in other cases, appears to be due to the induction of suppressor lymphocytes which interfere with the function of other lymphocytes. The capacity to suppress immune responses is probably an important contributor to the severity of infections caused by those viruses. Virus infections have also been shown to adversely affect neutrophil function. The latter is probably responsible for bacterial infections accompanying certain virus infections.

Vasopressin

WEN-HSIEN WU, M.D., M.S.
Associate Professor and Director of Research
Department of Anesthesiology
New York University Medical Center
New York, New York

Chief, Anesthesiology Service
Veterans Administration Hospital
New York, New York

VLASTA ZBUZKOVA, Ph.D.
Research Scientist
Department of Anesthesia
New York University Medical Center
New York, New York

Vasopressin (VP), an octapeptide, is also known as antidiuretic hormone (ADH). It is synthetized within special hypothalamic neurosecretory cells in supraoptic nuclei along with neurophysin, a carrier protein. Neurosecretory granules containing VP bound to neurophysin migrate within the axons to their terminals in the posterior pituitary at a rate of 2-3 mm/hour. When an adequate stimulus depolarizes the membrane of neurosecretory axon terminals, calcium ions influx across the cell membrane, thereby activates the process of "exocytosis" resulting in the release of VP and neurophysin into portal venous blood.

PHYSIOLOGY

The primary action of vasopressin is to preserve fluid homeostasis by limiting renal excretion of solute-free water through the distal convoluted tubules and the collecting duct. Major mechanisms regulating VP release are:

(1) Osmoreceptors, located in anterior hypothalamus (Verney 1947), are stimulated by increase in plasma osmotic pressure (i.e., dehydration).
(2) Volume or stretch receptors, located in the left atrium and intrapericardial portion of pulmonary vein, are stimulated by hypovolemia such as hemorrhage.
(3) Aortic and carotid baroreceptors are stimulated by hypotension.

VP is initially bound to specific receptor sites on the contraluminal side of the renal cells. Coupling of the receptor and hormone activates the adenylate cyclase and thereby increases production of cyclic adenine monophosphate (AMP) from adenine triphosphate (ATP). Cyclic AMP is then translocated to the luminal cell membrane and activates the membrane-bound protein kinase. This kinase in turn causes phosphorylation of membrane proteins. This reaction increases membrane permeability to water.

Plasma VP was 22 μU/ml after 72 hours of dehydration in man. The VP dose required to induce maximal antidiuresis varies with the species. It is currently accepted that 40 μU/Kg of body weight produces antidiuresis, 4,000 μU/Kg or higher produces vasoconstriction in man. However, actual plasma VP levels produced by these doses is not yet clear.

PHARMACOLOGY

VP is a potent vasoconstrictor which reduces blood flow to various vascular beds including coronary, splanchnic, and renal. It also decreases cardiac output and heart rate, and increases blood pressure. Myocardial infarction has been observed after VP injection. High plasma VP level (260 μU/ml) in man during cardiopulmonary resuscitation has been observed by us.

Plasma VP levels can also be elevated by: (1) pain, (2) anxiety, (3) visceral traction, (4) strenuous exercise, (5) heat exposure, (6) endogenous substances, such as epinephrine, dopamine, renin or angiotensin, and (7) drugs such as nicotine, barbiturates, narcotics, and inhalation anesthetic agents. Plasma VP levels can be depressed by: (1) water load, (2) cold exposure, (3) drugs such as ethanol, reserpine, chlorpromazine and diphenylhydantoin, and (4) narcotic antagonists such as oxilorphan.

PATHOLOGY OF VASOPRESSIN

A. *Diabetes insipidus* is the disorder of water conservation which results either from inadequacy of VP release in response to physiological stimuli (central diabetes insipidus) or from renal unresponsiveness to the action of VP (nephrogenic diabetes insipidus). Defects may exist in: (1) the mechanism responsible for sensing plasma osmolality (the osmoreceptors), (2) the neurons of supraoptic nuclei where VP is formed, or their axons in which the hormone traverses to the neurohypophysis, (3) the storage site of VP in the neurohypophysis, and (4) the action of VP on the renal tubule.

B. *Syndrome of inappropriate ADH secretion (SIADH)*. The syndrome is a disorder in which there is a continual release of VP (ADH) unrelated to plasma osmolality in the absence of adrenal, pituitary and renal disorders. Patients with this syndrome are unable to excrete a dilute urine, develop fluid retention and consequently dilutional hyponatremia and cerebral dysfunction. SIADH may occur in association with a large number of clinical disorders such as: carcinoma of the lung and larynx, lymphoma, cerebral lesions, or drug-induced pulmonary infection. It may occur in the postoperative period especially in the elderly patients without known causes. Therapy of choice at present is fluid restriction with close observation.

METHODS FOR VP MEASUREMENT

Three methods are commonly used to measure VP concentration in biologic fluid.

(1) Bioassays
   a. Rat antidiuretic assay: urine flow is recorded in hydrated, ethanol-anesthetized rat and antidiuresis is used as a response to intravenous VP injection. This assay is capable of measuring 0.25 μU VP.
   b. Rat pressor assay: blood pressure is recorded in urethane anesthetized rat and its hypertensive response to intravenous VP is the assay parameter. This assay is capable of measuring a minimum of 5 mU VP.

(2) Renal function

Glomerular filtration rate (GFR), renal plasma flow, serum and urinary osmolality are measured. Free water clearance (C_{H_2O}) is calculated. Decrease in C_{H_2O} without changes in osmolar clearance and GFR is used to indicate increased VP activity.

(3) Radioimmunoassay (RIA)

It is the most advantageous method. The smallest detectable amount is reported to be 0.1 pg (0.04 μU) VP. This RIA is not widely available and is not used as a routine assay due to the difficulty in producing high affinity antibodies.

Normal values of plasma VP level in man is accepted to be 1-3 μU/ml. But they vary with: (1) posture—minimum in supine position and higher in standing position; (2) diurnal and nocturnal rhythm—higher during the night than during the day; (3) seasonal changes—high during the summer and low during the winter seasons.

The half life of VP in circulation varies with species. In man, it varies between 1 and 20 minutes, and depends on actual plasma VP levels: low levels have longer half lifes and high levels shorter ones.

Index

435

Poison ivy, 353
Poisons, 220
Policy issues for rural areas, 186
Pollution, environment, 85
Polyanions, 312
Polycistronic messenger, 327
Polydrug abuse, 79
Polypeptide, 352
Polyps, 347
Polyurethane, 344
Poor housing, 91
Population
  child, 69
  in the U.S.A., 98
  prison, 218
  trends, 184
Portal venous blood, 431
Positive health, 176
Positive self image, 116
Post-industrial society, 20
Posterior pituitary, 401
Postsynaptic receptors, 379
Potassium ions, accumulation, 397
Practice of nursing, 103
Practice setting, 169
Preceptors office, 170
Pregnancy, 133
  control, 57
Pregnant diabetics, 324
Pregnant woman, 120
Premature delivery, 123
Premelanosomes, 378
Preparation for nursing, 111
Prevention of cancer, 270
Prevention of pregnancy, 124
Preventive dental care, 82
Pre-written library functions, 25
Prilocaine, 254
Primary care MD's, 160
Primary health care, 126
Primary immunodeficiencies, 300
Primary physician, 168
Principals of nursing, 103
Print, 25
Printed materials, 13
Print outs, 10
Prison health, 217
Privacy, 9
Procaine, 253
Procaine anesthesia, 242
Process evaluation, 95
Process of communication, 3
Profession of social dynamics, 24
Professionals to rural areas, 187
Profile of child, 68

Program equations, 24
Programming language, 22, 25
Programming theory, 29
Program, nursing, 111
Programs, health in U.S.A., 80
Prokaryotes, 285
Promote health, 11
Promotion and fame, 12
Promotion of health, 85
Properdin, 311
Prostoglandins, 343
Protein binding, 390
Protein bound iodine, 405
Protein kinase, 321
Protein kinase enzymes, 388
Protein phosphorylation, 388
Protein synthesis, 327
Provider-consumer, health, 176
Proxemics, 47
Prunasin, 365
PSRO's, 41, 173
Psychiatric Mental Health, 112
Psychic energisers, 77, 383
Psychobiology, 128
Psychological needs, 19
Psychology, 43
Psychopharmacology, 383
Psychotropic drugs, 381
Public health Nursing, 112
Public Law 93-641, 102
Public opinion, 129
Public speaking, 7
Publication, 12
Publish or perish, 12
PUD, 341
Pulmonary disease, 310, 368
Punched cards, 32
Pyran copolymer, 231

Quality of environment, 19
Quality of life, 61, 134
Quality of survival, 178
Quality primary care, 71
Quest for kicks, 77
Queuing theory, 20

Racial bias, 117
Radiation, 264
Radical neck dissection, 282
Radiographic images, 45
Radiological health, 91
Radio, 7
Radio announcing, 7
Radioimmunoassay V.P., 390, 434
Radiolabeled insulin, 352

Stress factors, 101
Stress–induced diseases, 100
Stroke, 316
Structure of clerkship, 170
Structure of family, 98
Student clerkship, 170
Student's life experiences, 86
Stuttering, 51
Substance P, 238
Subtotal resection, 343
Suffering, 104
Suicides, 91
Super–efficient systems, 14
Super industrialized society, 80
Supernumerary coronary arteries, 334
Supersensitivity of cells, 352
Suppressor cells, 234
Suppressor T-cells, 234
Surface receptor, 358
Surface tension (anesthetics), 246
Surfactant, 369
Surgery,
    experimental, 329
    technology of, 242
Surgical anesthesia, 242
Surgical practice, physiological
        principles, 329
Surgical treatment of ischemic heart
        disease, 330
Survival, 99
    of a cell, 294
    rate, cancer, 281
Swamps, 89
Sweat duct unit, 284
Symbol system, 6
Symbology, 24
Symbols, 3
    internalizing of, 3
Sympathetic blockade, 243
Synapses, 396
Synthesis, DNA, 325
Synthesis of disciplines, 103
Synthesis of proteins, 286
Synthetic cannabinoids, 239
Synthetic peptide chemistry, 237
System, 21
System response to trauma, 61
Systems, 20
    biochemical, 382
    rural health care, 210
Systems intervention, 23

T3, 404
T4, 404
Taboo, death and sex, 151
Tabulating Machine Co., 32

Team building, 100
Team, child abuse, 68
Technical writers, 13
Technological innovations, 192
Technologies for heart surgery, 330
Technology, computer, 10
Technology innovations, 180
Technology of surgery, 242
Teen-age pregnancy, 121, 124
Teenagers,
    abortions, 57
    in U.S.A., 124
    sexually active, 125
Teeth, 82
Tel-AUD, 52
Tel-C, 51
Tel-Communicology, 48
Telecommunication, 10, 37
Teleconsultation, 45
Telediagnosis, 45
Telemedicine, 45
Telemetry, 45
Telephone encounters, 45
Teletype, 42
Television, 6, 7
Tel-Plus, 52
Template, 327
Tenseness, 66
Terminal, drug allergies, 41
Terminals, 36
Tetracaine, 253
Tetrachlorodibenzodioxin, 271
Thalamus, 396
Theory of probability, 20
Thermal agents, 344
Thermal sealing, fumes, 220
Third coronary artery, 334
Thirst with water drinking, 322
Thomas' calculator, 32
Thymicolymphatic involution, 402
Thymidylic acid, 325
Thymine, 325
Thymoleptics, 383
Thymus and aging, 233
Thymus gland, 91, 233
Thymus hormones, 236
Thymus transplantation, 236
Thyroid antibodies, 405
Thyroidea ima, 403
Thyroid gland, 403
Thyroid hormone, 404
Thyroiditis, 409
Thyroid releasing factor, 404
Thyroid, treatment, 406
Time resources, 100
Time-sharing system, 22